T0280469

Drought and hunger in Africa:
denying famine a future

Dedicated to John V. Granata
of Providence, Rhode Island
Physicist,
factory worker,
philosopher,
friend

Drought and hunger in Africa:

denying famine a future

Edited by
MICHAEL H. GLANTZ

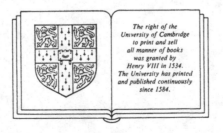

The right of the
University of Cambridge
to print and sell
all manner of books
was granted by
Henry VIII in 1534.
The University has printed
and published continuously
since 1584.

CAMBRIDGE UNIVERSITY PRESS
Cambridge
New York   New Rochelle
Melbourne   Sydney

CAMBRIDGE UNIVERSITY PRESS
Cambridge, New York, Melbourne, Madrid, Cape Town, Singapore, São Paulo, Delhi

Cambridge University Press
The Edinburgh Building, Cambridge CB2 8RU, UK

Published in the United States of America by Cambridge University Press, New York

www.cambridge.org
Information on this title: www.cambridge.org/9780521368391

First published 1987
First paperback edition 1988
Re-issued in this digitally printed version 2009

A catalogue record for this publication is available from the British Library

Library of Congress Cataloguing in Publication data
Drought and hunger in Africa.
Includes Index.
1. Famines – Africa.   2. Droughts – Africa
3. Food supply – Africa.   4. Agriculture and state – Africa.
I. Glantz, Michael H.
HC800.Z9F33   1987   363.8′096   86-9736

ISBN 978-0-521-32679-7 hardback
ISBN 978-0-521-36839-1 paperback

CONTENTS

PREFACE

This volume is based on presentations at a colloquium held in Boulder, Colorado, in August 1985. Papers presented have been revised in light of the discussions at the colloquium and two papers have been added. The colloquium, entitled 'Drought and Hunger in Africa: Denying Famine a Future', was part of the 25th anniversary commemoration of the founding by the US National Science Foundation of the National Center for Atmospheric Research (NCAR). NCAR is managed by the University Corporation for Atmospheric Research (UCAR), which is a consortium of 55 North American universities.

Many research activities at NCAR relate directly or indirectly to droughts in Africa. Basic research on atmospheric processes ranges, with regard to levels of analysis, from the local level (how land-use practices can affect the local patterns of precipitation; can the atmospheric processes be modified to produce rain artificially) to the national level (forecasts of seasonal climate anomalies for given human activities) to the global level (how carbon dioxide and other trace gases might affect atmospheric temperatures and regional rainfall patterns; how El Niño-Southern Oscillation events relate to climate anomalies elsewhere on the earth).

The volume is divided into four parts. Part I includes a general introduction to the problems and prospects for development in Africa presented by Bradford Morse, the recent Director of the largest operational organization in the UN family, the United Nations Development Program (UNDP). This Part also includes the geophysical overview of drought in Africa presented by meteorologist Eugene Rasmusson, who is with the Climate Analysis Center (National Oceanic and Atmospheric Administration, Washington, DC) and a case study of drought in Ethiopia, presented by Workineh Degefu (head of the Ethiopian Meteorological Service Agency).

At the colloquium, Degefu also represented Dr G.O.P. Obasi, Secretary General of the World Meteorological Organization.

Also included in this Part are two chapters that provide an overview of two issues not directly addressed in the remaining chapters. The chapter by Michael Glantz, Head of NCAR's Environmental and Societal Impacts Group, was inspired by the concern of many observers that, despite the influx of large quantities of relief grains and the high visibility in the media concerning the plight of Ethiopian peasants, drought continues to be seen only as a transient phenomenon and not a recurrent, ever-present constraint on the development process. The chapter by Michael Horowitz and Peter Little of the Institute for Development Anthropology serves as an excellent summary of the plight and role of pastoralists trapped in the midst of the agrarian crisis. The authors examine the relationship between pastoral poverty and ecology and challenge the view that the pastoralists with their herds are responsible for the degradation of rangelands.

Part II contains chapters that focus on internal and external factors which foster or hinder processes that can turn food production shortfalls into famine. Michael Lofchie, Director of the African Studies Center at UCLA, lays out the framework for the internalist–externalist schools of thought about the causes of the current crises in African agriculture, noting that the internalists 'tend to believe that the basic cause of these crises can be found in the economic policies pursued by African governments since independence', while the externalists 'place their principal emphasis on adverse features of the international economic environment'. Lofchie challenges the polarization between these views and asserts that both perspectives shed light on some of the complex causes of the present crises in Africa. He then attempts to disentangle these factors by focusing attention on the internal constraints to agricultural production.

Robert Cummings, Director of African Studies at Howard University, discusses primarily internalist perspectives on the current food production crisis, focusing on the Lagos Plan of Action (LPA), a document drawn up by African leaders for Africans. He notes that the LPA does not sufficiently address environmental issues. The LPA recognizes the problems that exist within African countries, some of which are self-generated while others have been imposed from the external environment.

The chapters by Timothy Shaw (Centre for African Studies, Dalhousie University) and Randall Baker (International Program, University of Indiana) approach the issue of African hunger and famine primarily from externalist perspectives. Shaw states in no uncertain terms that the roots of the current Africa-wide crisis 'lie in Africa's inheritance of impoverished, extroverted and underdeveloped economies compounded by adverse post-independence changes in the global division of labor'. This situation, he contends, makes Africa extremely vulnerable 'to drought as well as to dependence and dominance'.

Baker focuses on the international debt situation and how it adversely affects Africa's chances to cope with its agrarian crisis. He calls on the International Monetary Fund and other assistance agencies to increase the overseas development assistance for African states and to grant them a lengthy period of recovery and rehabilitation. Baker suggests changes needed to be carried out by donors, lenders and recipients.

The third Part contains case studies from various parts of Africa; Nigeria, Botswana, Zimbabwe, Ethiopia, Kenya and Burkina Faso. Michael Watts (University of California at Berkeley) discusses the role and effectiveness of households in combating constraints on food production processes, including prolonged drought episodes. He emphasizes the dynamic character of peasant and pastoral populations, as well as the knowledge of local conditions and adaptive flexibility to deal with them that resides with the small farmers.

Michael Bratton (Department of Political Science, Michigan State University) and James McCann (African Studies Center, Boston University) focus attention on the effectiveness of households and smallholder organizations in Zimbabwe and in Ethiopia, respectively. These case studies present the impacts on society of prolonged droughts in two countries that could be argued to represent two ends of a continuum of national well-being in Africa: Zimbabwe, on the one end, is in relatively good financial condition; Ethiopia, at the other, is viewed by most economic indicators as one of the poorest nations in Africa. Both provide recommendations to improve responses to drought in their respective countries.

Della McMillan (Center for African Studies, University of Florida), discusses planned settlements in Burkina Faso that were created in response to the 1968–73 drought in the West African Sahel. She analyzes the development of one settlement and the methods to assess the effectiveness of such settlements in alleviating the problems of shortfalls in agricultural production. John Cohen (Harvard Institute for International Development) and David Lewis (Cornell University) focus on the national reponse to drought. They analyze the effectiveness of the Kenyan government's policies and actions in averting a potential famine during the 1984 drought. They consider the Kenyan government's response to have been an exemplary one.

There are two additional chapters in Part III. William Torry (Department of Sociology and Anthropology, West Virginia University) investigates the effectiveness of food rationing in stateless societies versus state-organized food-rationing systems in averting famine during prolonged drought episodes. He discusses situations in Botswana and in drought-prone parts of India. Michael Scott (Oxfam America) presents an overview of the activities and role of non-governmental (or private voluntary) organizations in African emergency relief operations and in development activities. He compares the advantages and limitations of the voluntary organizations and of agencies of donor governments in dealing with African famine relief and development issues.

In the final Part, three additional case studies are presented, which suggest that there may be lessons for African leaders in how other leaders have dealt with drought, famine and food crises in the past. J. Gus Liebenow (Department of Political Science, Indiana University) discusses how Malawi managed to avoid an agrarian crisis of the magnitude that exists in many African countries today. Liebenow's chapter, along with other examples cited in the previous chapters, such as Bratton's on voluntary peasant organizations in Zimbabwe, suggests that there are lessons for combatting African drought and hunger within Africa itself.

Michelle McAlpin (Department of Economics, Tufts University) documents the response of the Government of India to recent Indian droughts and identifies lessons for African governments confronted by recurrent food production problems, including prolonged droughts. Just a few decades ago, India was a country racked by drought-related food shortages and famines. Yet, India is viewed in the mid-1980s as a success story with respect to coping with food-production problems and avoiding famine.

The chapter by Lillian M. Li (Department of History, Swarthmore College) discusses how the international community dealt with chronic famines in China in the 1920s and 1930s. Li points out that, although China was referred to as 'the land of famine' 50 years ago, it has successfully averted famine in the past decades. She identifies lessons that might be applicable to those organizations involved in international relief activities in sub-Saharan Africa today.

In a final chapter Payne, Rummel and Glantz (Environmental and Societal Impacts Group, National Center for Atmospheric Research) provide a brief overview of the discussions presented at the colloquium and in the preceding chapters. They highlight some of the debates that exist among physical and social scientists concerned with the plight of Africans in the face of recurring droughts that have plagued development efforts in several parts of Africa since about 1968 (about two-thirds of their post-independence period).

Interest in issues related to drought and hunger in Africa has been high in the past few years. It is important that these issues remain in front of the policy-makers, the media and the public in the years to come. Only with sustained concern and attention might the long-standing problem of hunger be successfully resolved. We must guard against the erosion of interest and concern once the rains have returned, referred to elsewhere as 'the paradox of good news'. Authors in this volume have proposed solutions for dealing with drought, famine and the agrarian crisis in Africa. Discussion of these and other solutions proposed in the development literature must continue if there is to be any hope of denying famine a future in Africa.

Acknowledgments

I would like to acknowledge the support of Michael Lofchie (UCLA, African Studies Center) and James McCann (Boston University, African Studies Center) in helping me organize the Drought and Hunger Symposium that was held at the National Center for Atmospheric Research in Boulder, Colorado, in mid-August 1985. Without their interest in and support for this effort, the symposium could not have been held. There are several other 'unsung heroes' that brought this meeting to fruition. The first is Bernard O'Lear of the Scientific Computing Division at NCAR. As a member of the UCAR-NCAR 25th Anniversary Committee, he alone carried the banner to commemorate its creation by focusing its resources and attention on a key issue of our times, the African famines as exemplified by the recent situation in Ethiopia and other parts of sub-Saharan Africa. He was the spark that brought the issue to light and his tenacity helped to attain the necessary resources to hold the symposium. The second person is Maria Krenz, who had been part of the organization of this symposium from the earliest moment. Not only was she involved in the planning but was responsible for carrying out the logistics of the meeting itself. In addition, she has been integrally involved in the editing of the manuscripts contained in this volume.

Clearly, many have supported this effort: Ann Garrelts, Assistant to the NCAR Director; Jan Stewart and Beverley Chavez, who assisted with the symposium logistics and with the preparation of the manuscripts; and Dr Walter Orr Roberts, former Director of NCAR, former director of the Aspen Institute's Food and Climate Forum, and chairman of the UCAR 25th Anniversary Committee. Finally, I would like to thank Dr B. Morse, Administrator of the UN Development Program and Director of the UN Office for Emergency Operations in Africa, for his involvement with our

colloquium; as well as Dr G.O.P. Obasi, Secretary General of the World
Meteorological Organization, and Dr M. Tolba, Executive Director of the
UN Environment Program for their moral and financial support for the
participation of African scientists.

Michael H. Glantz   Boulder, Colorado
                                                     15 January 1986

FOREWORD*

When I was a junior member of the United States Congress in the early 1960s, there were probably not more than five or six members of the US House and Senate who were well informed about Africa. Now there is a cadre of more than 100 members of Congress who have an informed interest in African affairs. Regrettably, however, awareness of this vast, troubled, yet vital continent remains sparse among the American people at large.

Let me say at the outset that I quite agree with those who assert that national decision-makers and development practitioners have paid too little attention to drought in the past. If one of the results of this volume is to help ensure that climatological factors are regarded as significantly more than transient constraints on development, your efforts will be worthwile.

In saying this, I acknowledge that the United Nations Development Programme (UNDP), which I am privileged to head, has been guilty of this omission, even though from time to time we have recognized the importance of agro- and hydro-meteorology in development. It was UNDP that took the initiative with the Interstate Committee to Combat Drought in the Sahel (CILSS) in establishing the AGRHYMET Programme in the Sahel a decade ago. The aim of that initiative was to strengthen the national meteorological and hydrological services in each of the CILSS countries. AGRHYMET is the largest project ever executed by the World Meteorological Organization; 50% of the funding has been provided by UNDP and the balance by a consortium of bilateral donors. The program, with offices in the other CILSS countries, has its headquarters in Niamey (Niger) and is now managed almost entirely by nationals of those countries.

* Presentation by Bradford Morse (Administrator, United Nations Development Programme and Director, United Nations Office for Emergency Operations in Africa) at the Colloquium on Drought and Hunger in Africa: Denying Famine a Future, held at the National Center for Atmospheric Research, Boulder, Colorado, 14–16 August 1985.

AGRHYMET was a response to the African drought emergency of the early 1970s. Looking back on our collective response to that emergency, we can say that we demobilized too soon – that we did not give sufficient weight to drought-preparedness in African development programs. This is a lesson which must be learned by African governments, bilateral donors and international agencies. Indeed, the nature of the present emergency does not permit us to ignore it, since the emergency has been triggered by drought.

Of course, drought itself is not the fundamental problem in sub-Saharan Africa. After all, drought prevails in many parts of the world and, in affluent societies, need be no more than a nuisance. The real problem in Africa is poverty – the lack of development – the seeds of which lie in Africa's colonial past and in unwise policy choices made in the early days of independence by national governments and external aid donors. The present drought has, however, intensified the interaction of the factors impeding development in Africa; it has laid bare the African development crisis. This crisis has now reached such proportions that the overall Gross Domestic Product (GDP) of sub-Saharan Africa has been declining in the 1980s. Per capita income at this time is less than it was in the early 1970s.

In reviewing the obstacles to development in Africa, one certainly cannot overlook the very considerable colonial baggage which burdened many African countries as they came onto the world stage. This included unsettled frontiers, highly dependent market-economies keyed to the export of commodities of interest to the metropolitan powers, neglected food-production systems and imitative life-styles and institutions. African societies and economies were split between traditional/rural and modern/urban sectors, the latter dominating political choices before as well as after independence.

At the time they achieved independence, most African countries were short on physical infrastructure and institutional, administrative and managerial capacity. The mass of people were generally less exposed to modern education than those in other developing regions; many basic services – in health, for example – were scarce in relation to needs. Endemic human and animal diseases, including onchocerciasis and trypanosomiasis, made it virtually impossible to exploit large areas of potentially arable land in sub-Saharan Africa – about 40% even today. Yet these emerging countries were largely left to themselves to build the social and economic foundations on which self-generating and self-reliant development could be achieved. It was predictable that, in view of the lack of integration of African societies and African economies, development would be a very difficult and fragile process.

Yet many of the difficulties that African countries have encountered in this process have been within the sphere of influence of their national decision-making processes. Population growth, environmental degradation, and lack of incentives for agriculture are obstacles to which national policies hold the key.

With regard to population, the growth rate in Africa is higher than in any other region of the world, but even more disturbing is the fact that the growth rate itself is increasing. Demographic policies have been inadequate. Twenty years ago, the growth rate was roughly 2.5% per year; it was up to 2.7% in the 1970s; it is now slightly over 3% and still rising. One result of this growth is that many African countries have very young and vulnerable populations. In Kenya (with a population growth rate over 4% per year), for example, 52% of the population is under the age of 14 years.

Environmental degradation in Africa can usually be traced to human activity and is therefore related to the demographic phenomenon. In the 1960s, when the West African Sahel was enjoying a moist climate, a number of communities were established in areas previously regarded as inadequate to sustain agriculture. Ever-increasing human and animal populations have threatened traditional agricultural and livestock-raising practices. Over-cultivation and over-grazing have reduced the productivity of land. The widespread destruction of tree cover for fuelwood (encouraged by high prices of petroleum-based fuel) and construction has accelerated the degradation. Poor water management and the salinization of irrigation systems have also left their mark. The net result is that each year 1.5 million hectares are overwhelmed by a glacier of sand.

Most disturbingly, Africa's capacity to feed its growing population has been decreasing steadily. Even in 1975, 14 nations which made up over 30% of the area of sub-Saharan Africa, with 50% of the region's population, were not self-sufficient in home-grown food. At present, food imports are required to cover about one-fifth of domestic consumption. Part of this dependence is due to growing urban preferences for exotic wheat products.

Paradoxically, this growing dependence on food imports has often been accompanied by increases in agricultural production of non-food crops for export. In Burkina Faso, for example, during the 1960s, cotton production was roughly 2000 tons a year; in 1984 it was 75 000 tons. Even Chad, with all of its difficulties, had a record cotton harvest last year. The bias in favor of cash crops to generate foreign exchange prevails in many parts of Africa.

A legacy of the colonial economy, this bias is now explained by reference to comparative advantage and to the harsh realities of international markets, which do little to encourage the diversification of traditional trade patterns by weak participants.

The strategy of export-led growth was seen by most African leaders and their foreign advisers and financiers as the shortest route to the industrialized models beckoning from the North. Part of this strategy is the promotion of a modern, industrial urban economy, requiring cheap food supplies, requiring, in turn, the subsidization of food prices. This strategy is anti-rural in bias, depressing incentives for domestic food production and encouraging food imports. Although its shortcomings have become apparent, it remains at the root of Africa's food deficits.

Yet, while these policy deficiencies are increasingly recognized, it must also be acknowledged that African countries have been victimized by other forces over the last 10 or 15 years – forces external to them and over which they have little or no control.

The shock of the oil price increases in the early 1970s followed the steady increase in the prices of essential imports resulting from inflation in industrialized countries. Markets for African commodity exports had begun to weaken even before the worldwide recession set in. Available foreign exchange was reduced, resulting in the deterioration of equipment, infrastructure and institutions. At the same time, net capital inflows to sub-Saharan Africa, including Overseas Development Assistance (ODA), declined in real terms. Additional borrowing was inevitable to meet development needs and the debts of many African countries escalated, swelling a burden which has been aggravated by high interest rates. Although not as large in absolute terms as the Latin American debt, the African debt burden is heavier in relation to capacity to repay. The total external debt for the 39 sub-Saharan countries (excluding Nigeria) exceeded $56 billion in 1984, with debt service equivalent to 23.3% of the value of their exports. The annual interest on external debt of the African countries adversely affected by drought substantially exceeds the cost of the total unmet emergency needs for 1985 as determined by the United Nations Office for Emergency Operations in Africa. On top of these external economic factors, 17 years of the harshest drought in recent history have had a devastating effect upon the people and the economic and social systems of a score of African countries. It has wrenched the ecology and the environment of most of the continent. Successive crop failures have led to food scarcity, resulting in malnutrition, unchecked disease, the decimation of livestock herds and ultimately famine, with a staggering loss of human life. There has been enormous migration – within and between countries – and human suffering and hardship beyond the comprehension of those of us who have known affluence all of our lives. In addition, the drought has aggravated the economic crisis that almost all sub-Saharan countries have been experiencing. It has further depleted government revenues and foreign exchange, intensified unemployment and brought growth in major productive centers to a halt.

This tragic and pervasive situation triggered by drought is, as I have said, a crisis of development. One example will perhaps suffice to make the point: in one particular country, logistical problems have been critical for the past several weeks. Food aid has arrived in the port but it has not been possible to transport it in adequate quantities to the people in need in the distant hinterland. The railway system of that country, which should be the main carrier of relief food, used to have a capacity of some 3 million tons per annum, 20 years ago; it has declined to about 800 000 tons per annum today. The reversal of development (in this case transport development) combined with drought has produced famine.

A further tragic consequence of the drought-induced emergency is that it has disrupted development efforts already underway. The massive inflows of emergency food aid, necessary as they are, have disrupted transport systems and, in some cases, may disrupt the marketing of local produce. Some donors have diverted precious development funds to meet emergency needs. If this continues – if emergency funds are not truly additional funds – it will mean that many Africans who are fortunate enough to survive famine will be robbed of their future.

The tragedy of African famine, while calling forth an impressive humanitarian response, has also raised political and public awareness of the need for the massive mobilization of national and international resources and energies to revive the development process in Africa and make it sustainable. There is a growing convergence of views among African leaders and among donors as to the correct policy responses to the African development crisis. The food emergency has driven home the need to orient policies towards rural development and food production, although there are different ideas as to how to achieve the shift. All agree, however, that it is essential to offer greater incentives for food production by African farmers. There is a growing consensus on the need to devote more resources to the rehabilitation and maintenance of existing investments, instead of starting up new ones. Other points of consensus include the need for greater efficiency in the public sector, for the adaptation of education to African needs and for new, intensified efforts to reduce the rate of population growth.

The consensus is about objectives and, to a growing extent, policies. Given the difficulties in converting ideas into action, the process of achieving these objectives will be a very difficult one. It is clear that unless there is substantial support from the donor community, committed to supporting the policy prescriptions with necessary financial assistance, the prospect will be bleak indeed. Yet the response on this score to date has been inadequate. The World Bank's Special Facility to support policy reform in Africa has raised only $1.25 billion of the $2 billion target. The Bank itself has determined that Africa requires an additional $2 billion each year. Yet, as I noted, ODA has been declining in real terms.

There is no doubt that the scientific community has a great deal to offer to provide new hope. Certainly, the world has the know-how and the resources to eradicate famine from the face of the earth. This colloquium itself will be able to make a significant contribution in advising how to prepare for drought. I realize that there is a debate among climatologists on the nature of the drought phenomenon we are now witnessing in Africa. Is it a cyclical or an aperiodic phenomenon? Are we going through a climatic shift? Either way, African planners and external development practitioners must build drought into their thinking and their actions.

Much more research – valuable research – could be undertaken on Africa's traditional crops. The Consultative Group on International Agricultural Research (CGIAR), which is supported by the World Bank, UNDP and a number of bilateral donors, spends about 37% of its budget on Africa, but much of its activities relate to export commodities, rather than to staple foods. In 1983 one important bilateral donor spent twice as much in one African country on tobacco improvement than it did on food crop research.

The Green Revolution, which brought such great returns to Asia and Latin America, has not brought comparable benefits to sub-Saharan Africa. For a variety of reasons – not least, cost and cultural considerations – the techniques of the Green Revolution, which involved the modification of the environment to meet the needs of the plant, may be less relevant to Africa than a 'Gene Revolution' which would seek to modify the plant to suit the environment.

Perhaps of greatest importance, much more must be done to develop the capacity of Africa's human resources. Past emphasis on investment in physical infrastructure has resulted in the neglect of investment in human capital. It is increasingly clear that an expanded agricultural strategy to increase African food production must be centered on the small farmer at the village level where community organization exists and operates effectively.

As we look to the future, recognizing that effective development is necessary if we are to 'deny famine a future', we must acknowledge that much of the development assistance provided to sub-Saharan Africa in years past has generally not yielded the expected results. I think it is fair to say that one of the reasons is that development actions have not adequately taken into account the unique agro-climatic conditions of Africa nor the social and cultural frameworks of African societies.

The improvement of our development cooperation efforts will require hard decisions by peoples and nations who can provide the wherewithal to support the efforts of the African people to do the job, and those who have the skills to augment their efforts. In a word, we face a political problem to persuade those with influence in the industrialized countries to understand the vital importance of eradicating famine and of overcoming the obscene imbalances which characterize the world today. You and the institutions you represent can play an essential role in supporting the efforts of African countries to advance their economic and social development, but unless resources are made available by the industrialized countries, you will not have the opportunity to do so.

The basic questions remain. How do we make the affluent world understand? Do we cast the question in moral terms? Do we underline the dependence of northern industry on the commodities exported by developing countries? Do we emphasize the fact that two-thirds of the world's population represents a major market, with untold potential? Whatever the approach, whatever the arguments, there will be 'new' hope only when the affluent world understands its own stake in helping Africa to overcome the effects of this devastating drought, in helping Africa and other parts of the developing world in their quest for economic and social progress.

Development is about human beings like you and me. It is a process by which our less advantaged brothers and sisters can protect, enlarge and enrich their lives. We must study the lessons to be learned from the development catastrophe in Africa over the last several years. We must seek to understand where we have failed and how we can, on another day, succeed. We must understand that development is a multidisciplinary and multisectoral adventure in which all aspects must complement and support each other.

We have taken great heart from the massive generosity of people throughout the world in helping to meet the immediate needs of starving people. In one day, in the Federal Republic of Germany over DM110 million were raised. The *USA for Africa* recording, under the leadership of Harry

Belafonte and Ken Kragen, has mobilized some $50 million which has already been translated into effective relief for people in Ethiopia, Sudan, and other affected countries. The 'Live Aid' concert, stimulated and organized by Bob Geldof, Michael Mitchell and Kevin Jenden, has already raised some UK £50 million (about $70 million) which is also being transformed into meaningful support in many of the affected countries in Africa.

The challenge which we now face is to sustain that commitment and transform it into support for long-term development. The task will not be easy. When Maurice Strong thought of 'new hope', I am sure that he had in mind the need to inspire sustained support for the multidisciplinary development efforts which will be necessary if we are to 'deny famine a future'. If we fail to do so, if we fail to demonstrate the intimate linkages between the emergency and development, if we fail to persuade decision-makers that they must provide resources for development, we can be sure that famine will have a very healthy future.

We need not fail – we dare not fail.

PART I

Physical and social setting

1

Global climate change and variability: effects on drought and desertification in Africa

EUGENE M. RASMUSSON
Climate Analysis Center, National Weather Service, NOAA

Introduction

Annual rainfall over the African continent varies from near zero in the heart of the Sahara to over 1600 mm in the equatorial rainforests (Fig.1.1). Rainfall over the most arid regions is too scanty and unreliable to support a permanent population except around isolated water sources. In contrast, over the humid areas, the rainy seasons are relatively long and reliable. The most critical climate problems arise in the semiarid regions where rainfall amounts range between about 200 and 800 mm/y, and year-to-year variability is relatively large: typically 20–30% or more of the average annual value. The physical and socio-economic well-being of the sizable and rapidly growing population of these regions is extremely sensitive to the large year-to-year rainfall variations.

The semiarid areas of sub-Saharan Africa sweep out a huge crescent-shaped region, which extends eastward in a narrow band from Senegal and Mauritania to the Red Sea, then southward through parts of Ethiopia and Somalia and southwestward through marginally arid portions of East Africa before broadening in area to enclose the Kalahari and other arid regions of Southern Africa (Fig.1.1). Drought enveloped most of this area as well as parts of North Africa during 1982–84. This immense size of the region affected, together with the extreme severity of this drought, marks it as one of the most significant climatic events of modern times. The coincidence of drought over such a vast region was the result of complex climatic processes which are inadequately measured, difficult to describe and poorly understood. Some aspects of the problem are beginning to be untangled, but a comprehensive understanding of the basic causes of widespread African drought still eludes us. This chapter is devoted to a general description of the nature of African drought and to a review of the factors, both global and regional, which may play a significant role in this phenomenon.

African climatology

The mean circulation of the tropics is dominated by planetary-scale features, i.e. circulation systems which span distances comparable to the earth's radius. These circulations are a consequence of spatially varying surface boundary conditions, i.e. land, oceans and topographical features such as mountains, and the associated uneven distribution of the radiative heating of the earth (Shukla, 1984). Prominent among these features are the oceanic tradewind systems of both hemispheres, and the great cross-equatorial monsoon circulations, which evolve on a seasonal time scale. The low-level monsoon flow of the winter hemisphere acquires heat and moisture from the ocean surface as it sweeps across the equator. Reaching the warm continental and ocean areas of the summer hemisphere, the air rises to the upper troposphere in the towering rainclouds of the monsoon rainfall regions. It then flows outward from the convection region, is cooled by radiation and slowly sinks into the surface layer over the winter hemisphere continents or the cooler ocean areas of the tropics. These great atmospheric overturnings, which have both east–west and north–south components, are in effect direct thermal circulations (warm air rising, cold air sinking), which produce the energy for the great wind systems of the tropics. They evolve with the seasonal movement of the sun but, because of the huge

Fig. 1.1. Climatological mean annual rainfall (millimeters) over Africa (adapted from Nicholson, 1981).

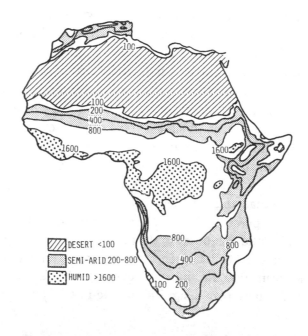

heat capacity and consequent thermal lag of the oceans, the peaks of the seasonal monsoons and the associated rainy seasons typically lag the solar season by a month or two. Since the northern hemisphere has by far the greater land mass, it exhibits the more intense monsoonal circulations.

The planetary-scale circulation features of the tropics play a fundamental role in determining the climatology of African rainfall. Large parts of the continent are influenced by the Atlantic or the more massive Indian Ocean monsoonal circulations. The zone of converging winds from the two hemispheres, loosely referred to as the Intertropical Convergence Zone (ITCZ), migrates northward and southward and changes its orientation with the waxing and waning of the seasonal monsoons (Fig.1.2). Over the Atlantic and West Africa, the ITCZ remains north of the equator throughout the year. Over the Indian Ocean and eastern Africa, it executes deep seasonal swings into each hemisphere.

The ITCZ normally marks a zone of rising, moist air and heavy rainfall. However, conditions over West Africa are profoundly modified during the northern summer because of the dry southwestward flow from the Sahara. Under these conditions, the zone of heavy rainfall is displaced well to the south of the position of the ITCZ, shown on Fig.1.2, into the region of the Atlantic Southwest Monsoon circulation (Lamb, 1980).

Although there is some regional variability, the annual cycle in African rainfall is largely a function of latitude (Griffiths, 1972; (Fig.1.2)). Within the tropics, the rainy season is tied to the high sun season. Two rainfall maxima are observed in the equatorial zone, associated with the twice-yearly solar passage. The subtropics experience a single high sun rainy season. Consequently, rainfall near the equator is distributed over many months while in the subtropics it is strongly concentrated in the summer season. For example, as one progresses northward through the Sahel, the rainy season shortens from 4 months to little more than a single month on the southern margins of the Sahara (Nicholson, 1980). With the shorter, more capricious rainy season comes less mean annual rainfall and greater year-to-year variability.

Africa spans a variety of temperate and tropical climatic zones, and the meteorological situation associated with precipitation events is quite different in different parts of the continent. In West Africa, monsoon winds bring a moist, low-level flow northeastward from the tropical Atlantic across the southwestern and southern coast of West Africa. Here, rainfall results primarily from atmospheric features known as West African Disturbance Lines or Ligne de Grains (LGDs) (Lamb, 1980). These are organized bands of thunderstorms which can extend up to 1000 km from north to south and

100 km from west to east. They derive much of their moisture from the low-level monsoon flow, but their movement is determined by the flow above the monsoon layer, which steers them westward at about 15 m/s (Burpee, 1972; Norquist, Recker & Reed, 1977). Some of these squall lines evolve into tropical cyclones, or even hurricanes, as they move westward across the tropical Atlantic.

The nature of rainfall events in the equatorial belt are not so well described or understood. They seem to be related to variations in the east–west component of the low-level wind. However, important questions remain to be resolved such as the role of organized planetary-scale fluctuations with time scales of 30–60 days (Weickmann, Lussky & Kutzbach, 1985), and

Fig. 1.2. Schematic diagram illustrating the mean annual cycle of rainfall and surface wind systems over Africa (adapted from Griffiths, 1972 and from Dhonneur, 1974).

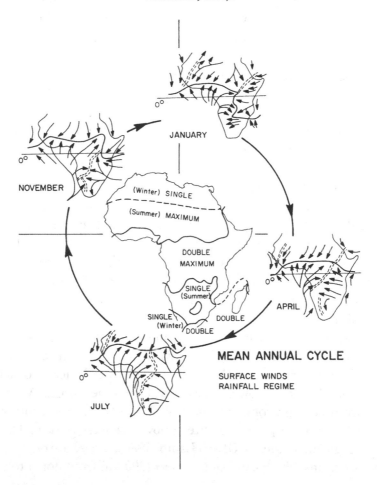

differences between the rainfall regimes of east Africa and the western equatorial belt.

There is a single wintertime rainfall maximum in the northern temperate latitudes, e.g. in Morocco and Tunisia, associated with middle-latitude synoptic weather systems. Although Southern Africa also extends into the extratropics, most of that region exhibits a summer rainfall maximum. Harrison (1983, 1984) and others have described South African summer rainfall episodes as being associated with diagonal cloud bands that extend southeastward from a tropical cyclonic vortex over the continent to a mid-latitude disturbance. Thus, they involve some degree of tropical–extratropical interaction. Dry periods over Southern Africa appear to be associated with a systematic eastward displacement of these features.

Interannual variability

The seasonally averaged circulation features of the tropics exhibit large year-to-year variability. Characteristic large-scale departure patterns include compensating east–west anomaly pairs in the plane of the equator (Walker type) and north–south-oriented changes in the vertical overturnings between the equatorial and subtropical belts (Hadley type) (Bjerknes, 1969). These patterns are associated with significant shifts in the position and intensity of the large-scale rainfall regimes of the tropics, and are often 'teleconnected' with circulation anomalies in the temperate latitudes. Such pronounced spatial coherence of anomalies over great distances is also a characteristic of African rainfall departure patterns.

The broad-scale characteristics of African rainfall variability have been extensively documented by Nicholson (1980, 1981, 1983, 1986) and Nicholson & Entekhabi (1986). Among the salient features is the very large spatial scale of many departure patterns (the strong east–west coherence of rainfall anomalies over the Sahel is a good example) and the tendency for departures of opposite sign over the equatorial zone and the higher tropical latitudes of both hemispheres. A significant fraction of the large-scale rainfall variability over the continent can be characterized in terms of a few basic anomaly patterns. Considering the entire continent, Nicholson (1986) identified six anomaly types which fall into two basic classes: (*a*) departures of similar sign over most of the continent, and (*b*) departures over the equatorial belt which are of opposite sign from those in the subtropics of both hemispheres. Of the 73 years examined, 29 were objectively classified into 1 of these 6 types. For the more restricted area north of 10° S, 43 years fell into 1 of 6 similar types.

The degree of persistence and the typical recurrence interval of African rainfall departures has been examined by a number of investigators. Quasi-periodicities have been documented in rainfall records from east Africa and many parts of southern Africa (see a review in Nicholson & Entekhabi, 1986). Extensive spectral analysis of rainfall data from equatorial and Southern Africa (Nicholson & Entekhabi, 1986) shows a predominance of short-term variability on time scales of 2–6 years.

The pattern of rainfall variability over the Sahel and Soudan is strikingly different from that observed to the south. Much longer, decadal-scale fluctuations associated with the long and persistent wet and dry spells of the region are the dominant mode of variability (Nicholson & Entekhabi, 1986). The difference between the Sahel and southern African rainfall regimes is strikingly reflected in the distribution of runs of consecutive dry years. In the Sahel, the frequency of long runs is significantly greater than would be expected by chance (several runs of 10–18 years are evident in the record), while for the semiarid regions analyzed in the southern hemisphere, the longest run was 5 years (Nicholson, 1983; Nicholson & Entekhabi, 1986).

African drought

Drought is a recurrent phenomenon in the semiarid regions of Africa. The first showers of the rainy season usually follow a long dry season, during which the soils become thoroughly dried out. The early rains signal the time for planting; consequently, an unexpected break in the rains may result in the withering of the emerging plants, even though subsequent rainfall is favorable for a good crop. If the total rainfall is inadequate, or if the rainy season terminates early, the crops may not mature and yields will be low. Thus, it is often the distribution, rather than the total amount of precipitation during the rainy season, that is the more relevant factor in food production (Wilhite & Glantz, 1985).

Many definitions have been offered for the term 'drought', but no single parameter (whether it be precipitation, runoff, evapotranspiration, temperature, soil moisture or crop yields) can serve as an adequate or comprehensive drought index. Drought implies an extended and significant negative departure in rainfall, relative to the regime around which society has stabilized. Thus, drought conditions in one region may be considered normal conditions in a more arid region, or during a more arid epoch. Ambiguities can arise in the definition of drought, because of the arbitrary definition of normal rainfall, particularly in an unstable climate regime such as the Sahel, when averages over periods of a few decades may vary substantially (Todorov, 1985).

Conceptually, the choice of the averaging period implies a time scale separation between the shorter time scales associated with 'climate variability' and the longer time scales associated with 'climate change'. It has become customary to use 30 year averages, updated every decade, to define the current climate (World Meteorological Organization, 1971). However, without an understanding of the climatic processes at work, this seems to be a rather arbitrary choice.

For the most part, we will assess rainfall deficiency by simply using normalized departures from average rainfall over the various lengths of the instrumental record. This is usually 80 years or less. The data are normalized either by dividing by the long-term standard deviation or by expressing the departures in terms of percentile rank, i.e. the driest year of record being 1 and the wettest year 100.

Recent rainfall fluctuations

The more notable large-scale African rainfall fluctuations during recent decades are: (*a*) wet conditions throughout much of Africa during the decade of the 1950s; (*b*) a reverse pattern during 1968–73 with abnormally dry conditions prevailing throughout Africa except for a narrow equatorial strip (Nicholson, 1985); and (*c*) easing of dry conditions during the remainder of 1970s, followed by a return to extreme drought conditions over much of the continent during 1982–84. Regionally, severe droughts occurred in West Africa during this century in the 1910s, 1940s, and again during the past two decades. In southern Africa, major widespread drought occurred during the 1910s, late 1920s, the late 1940s and the 1960s (Nicholson & Entekhabi, 1986), and again in the 1980s. In East Africa, since 1922, severe drought occurred over at least a portion of the region (Kenya, Tanzania, Uganda) during 1933–34, 1938–39, 1949–50, 1952, 1965, 1969, 1973–76, 1980 and during the first rainy season of 1984 (Ogallo & Nassib, 1984). Because of the inhomogeneous terrain, drought rarely encompasses the entire East African region. For example, the worst drought of the period occurred in 1949, but only 80% of East Africa was affected. East African droughts also exhibit low persistence. Locally, most last only 1 year, with the longest drought at an individual station, during the period analyzed by Ogallo and Nassib, being 6 years. Conversely, some part of the region experienced drought almost every year, with the whole of east Africa being drought-free only 21% of the time.

Short-term and long-term drought

A broad overview of these results suggests a two time scale conceptual framework for viewing African drought. Simply stated, the major African drought events tend to occur on two time scales; (*a*) relatively short, intense drought 'episodes' usually lasting no longer than 1–3 years, and (*b*) long dry 'regimes' of predominantly subnormal rainfall, spanning about a decade or more, which may include several intense drought 'episodes'. Both time scales appear in most rainfall records. The short-term, episodic pattern is the dominant regime in the semiarid regions of equatorial and Southern Africa. The Sahel–Soudan also experiences short-term drought episodes but the rainfall regime of this region is dominated by decadal-scale dry periods.

Relationship to sea surface temperature

The basic cause of short drought episodes may be quite different over different parts of the continent, but the short-term mode of variability can often be related to global-scale fluctuations in the atmosphere and ocean. Southeast Africa serves as a good example. Figure 1.3 depicts departures from the rainy season normal during the period 1875–1978, averaged for 16 stations scattered over the region from Zimbabwe–Mozambique southward through Botswana and northeast South Africa. The rainfall time series is clearly dominated by year-to-year variability. Although there are a few large back-to-back departures of the same sign, and a few generally wet

Fig. 1.3. Rainfall index derived from 16 stations over Southeastern Africa. The index is expressed as a departure from the long-term median (1875–1978) on a scale of +50 to −50. Shaded bars indicate years in which El Niño-Southern Oscillation (ENSO) episodes occurred.

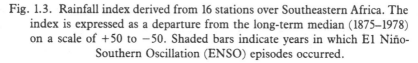

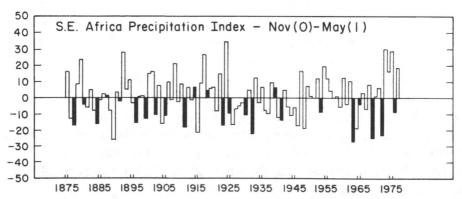

(e.g., 1949–62) or dry (e.g., 1925–46) periods, the long-term swings are not a prominent feature of the record.

This pattern of alternating wet and dry years has been clearly linked to the global El Niño/Southern Oscillation (ENSO) climate phenomenon, which is related to the warming of ocean surface temperatures half a world away in the eastern equatorial Pacific (Rasmusson & Wallace, 1983; Cane, 1983; Fig. 1.4). ENSO is associated with massive geographical shifts in the normal rainfall regime of the tropics, and clearly affects much of southern Africa (Nicholson & Entekhabi, 1986). The relationship with southeast African rainfall is particularly strong (Rasmusson, 1985). This is quite apparent from Fig. 1.3, which shows that the probability of subnormal rainfall is significantly enhanced during the warm ENSO episodes. Of the 27 moderate/strong warm episodes during the past 110 years, 22 were accompanied by below-normal rainfall, as measured by the index. Of the 20 driest years during the period shown on the diagram, 12 were warm-episode years, where only 5 would be expected by chance. In this and many other regions of the tropics, there is also a tendency for rainfall anomalies of reverse sign (above normal rainfall) during the opposite phase of the Southern Oscillation, when sea surface temperatures (SST) in the equatorial Pacific are below normal.

Over West Africa, year-to-year rainfall variability appears to be more closely related to atmospheric and SST fluctuations over the tropical Atlantic (Lamb, 1978*a*, *b*; Lough, 1986; Fig. 1.5). Above/below-normal SST in the southeast tropical Atlantic together with below/above-normal SST north of

Fig. 1.4. Sea surface temperature (SST) anomalies (degrees Celsius) during the El Niño-Southern Oscillation (ENSO) episode of 1982–83.

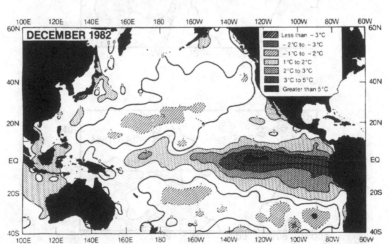

10° N, particularly over the Caribbean, is associated with below/above-normal rainfall over West Africa. This pattern of SST anomalies is also related, in the opposite sense, to rainfall fluctuations over northeast Brazil during their rainy season early in the year (Hastenrath & Heller, 1977; Moura & Shukla, 1981).

The observed pattern of Atlantic SST anomalies during 1984 was remarkably similar to that associated with West African drought (Fig. 1.5). Positive SST anomalies in the South Atlantic, among the largest of the century, first developed along the Angolan coast in early 1984. During the next several months, they spread gradually across the entire tropical Atlantic. The typical pattern of rainfall anomalies developed; intense summer drought over the western Sahel, and heavy rainfall over northeast Brazil during the rainy season of early 1985. The appearance of above-normal SST in the south Atlantic many months *before* the Sahel and northeast Brazil rainfall anomalies suggests that the seeds of these anomalies were present in the ocean/atmosphere climate system many months before they actually

Fig. 1.5. Top: July–September (JAS) Pattern of Atlantic SST anomaly (+ indicates above-normal, − below-normal temperatures) associated with drought in West Africa and above normal rainfall in northeast Brazil (from Lough, 1986). Bottom: Actual sea surface temperature anomalies (degrees Celsius) during July–September, 1984.

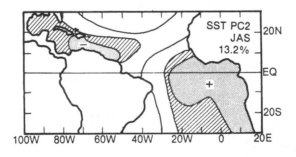

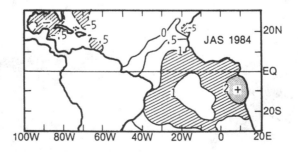

occurred. However, it is well to emphasize that these and many other correlations between climate variables do not necessarily indicate cause and effect relationships, even if they involve time-lag relationships. For example, the Atlantic anomalies do not appear to be entirely independent of El-Niño-related fluctuations in the equatorial Pacific. Thus, it is not clear to what degree Atlantic SST is a causal factor or to what extent both it and West African rainfall are responding to other forcing factors. What is clear is that on the year-to-year time scale, the pattern of large-scale African rainfall anomalies can often be related to global rather than simply to local patterns of climate variability.

Long-term Sahelian drought

The Sahel extends across the African continent in an east–west band a few hundred kilometers wide. It is defined in essentially phytogeographical terms, a zone of grassland, scrub and thornbush blending into the Sahara to the north and the savanna to the south (Grove, 1978). Rainfall amounts associated with the different sub-Saharan climate zones vary with author. Nicholson (1981) referred to the 100–400 mm/y rainfall zone as the Sahel and the 400–1200 mm/y zone as the Soudan. Grove (1978), whom we shall follow, used 200–600 mm/y for the Sahel. A difference of 100 mm in mean annual rainfall is of little importance in humid regions, but in the Sahel it is of great significance for the plant cover and usability of the land. Since 80% of the annual Sahelian rainfall occurs during the period July–September (Lamb, 1980), a deficient rainy season will handicap the region's agriculture and its socio-economic life for at least an entire year.

The Sahelian zone has been described by Grove (1978) and others. Where annual rainfall is less than 150 mm, the vegetation does not afford sufficient protection to prevent wind transport. North of the 250 mm isohyet, crops can scarcely be grown in most years without irrigation, and camels are better suited to the area than are cattle. To the south, regions with over 400 mm of rainfall are often well-wooded. South of the 600 mm isohyet is the more heavily settled sorghum and cattle country of Senegal, Burkina Faso, and northern Nigeria.

Twentieth century record

A rainfall index (Lamb, 1985) based on data from 20 stations west of 20° E between 11° N and 19° N is shown in Fig. 1.6. The time series is punctuated at irregular intervals by dry episodes, usually of longer extent, but similar in character to those observed over southeast Africa (Fig. 1.3). However, each succeeding dry episode since the early 1960s has been more extreme than the previous one, giving the series the appearance of downward trend in rainfall during the past two decades. This most recent long-term downturn in Sahelian rainfall is clearly the most severe of the twentieth century (Nicholson, 1986). The processes associated with the downturn are unclear, but they may be distinct from those associated with the shorter, more intense, dry episodes. Long-term trends do exist in other climate parameters, e.g., globally averaged surface temperature, but it has been difficult to forge convincing links with Sahelian rainfall. Recently, however, Folland, Parker & Parker (1985) have called attention to an apparent relationship with a global-scale change in the low-latitude north–south gradient of SST. The globally averaged pattern is similar to the change in the north–south gradient over the Atlantic Sector shown in Fig. 1.5 (warmer to the south/colder to the north) that has been related to shorter period drought episodes over West Africa.

The trend in rainfall over the Sudan and northern Ethiopia appears to have followed a pattern similar to that shown on Fig. 1.6, although data are inadequate to clearly define the year-to-year fluctuations. The decrease in Ethiopian rainfall has been reflected in the flow of the Atbara and Blue

Fig. 1.6. Rainfall index for 20 sub-Saharan stations in West Africa west of 10° E between 11° N and 19° N (developed by Lamb, 1985).

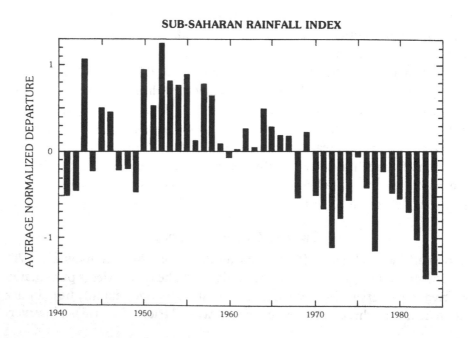

SUB-SAHARAN RAINFALL INDEX

Nile Rivers which are the major sources of the annual Nile flood. The flow at Aswan decreased sharply in the mid-1960s, and the flow into Lake Nasser, during 1984, was reported to be the lowest in the past 15 years (Starr, 1985). In February 1985, Starr (1985) reported the worst water crisis in Egypt since 1913, with the water stored behind the Aswan Dam barely sufficient to meet normal requirements for one more year.

Long-term reconstructions

Climate variations before the era of instrumental records must be deduced from qualitative as well as quantitative historical, archeological, and geological proxy sources. Although the reconstruction of these records is fraught with difficulty and uncertainty, it seems clear that at various times during the period extending from about 10 000 to 5000 BP (before the present time), the water levels were high in lakes across much of North Africa (Street & Grove, 1976). All closed basins of Africa for which there is information were occupied by lakes and it seems likely that rainfall in the Sahel was 50–100% greater than during the period preceding the most recent drought (Grove, 1978).

Lake Chad is a sensitive index of sub-Saharan rainfall fluctuations. Its drainage basin is centered near 15° N, 15° E and covers approximately 2.5 million km^2. The Chari and Logone Rivers, which drain the area to the south, where annual rainfall exceeds 1000 mm, are the principal sources of water for the lake. Around 9000 BP, the lake surface is estimated to have been 40 m higher than its present level, and its area around 350 000 km^2, about the size of the present Caspian Sea (Kutzbach, 1980). Climate model results (Kutzbach, 1981) indicate that this wet era resulted from variations in the earth's orbital parameters, i.e., Milankovitch forcing, which increased incoming solar radiation during the northern hemisphere summer, leading to a more vigorous summer monsoon circulation and heavier rainfall over the African–Eurasian land masses.

Attempts to reconstruct a record of African rainfall fluctuations during recent centuries have also been made. Conditions considerably wetter than those of the twentieth century seem to have lasted from the sixteenth through the eighteenth centuries (S. Nicholson, 1985, personal communication). Maley (1981) envisages Lake Chad as rising in the sixteenth century and remaining at a high level for much of the seventeenth century in response to plentiful rains in the headwaters of the Chari and Logone Rivers. However, several severe droughts must have occurred during the late seventeenth and during the eighteenth centuries (S. Nicholson, 1985, personal communication). A major decline in rainfall appears to have occurred near the beginning of the nineteenth century, leading to extreme droughts throughout much of the continent in the 1820s and 1830s (Nicholson, 1980). Lake levels, river flow and historical and geographical records suggest that considerably wetter conditions again prevailed throughout most of the period 1850–1900.

The most recent dry period has been dramatically reflected in the diminishing size of Lake Chad. In 1963, the lake covered 23 500 km². By 1973 the surface area had decreased to one-third this size and the lake had separated into two parts, with the northern section drying up seasonally. By early 1985, the lake had shrunk to about 2000 km², apparently its lowest level of the century (A.T. Grove, 1985, personal communication).

Theoretical considerations

African drought reflects the integrated effect of climate processes operating on a variety of temporal and spatial scales. The relative importance of global and regional influences, both natural and man-made, is yet to be untangled. In attacking such a complex problem, it seems essential first to obtain answers to a number of separate, more tractable questions. Each of these may involve quite different climatic processes.

Consider, first, the processes associated with monthly and seasonally averaged fluctuations. Conceptually, these fluctuations can be viewed as arising from two mechanisms (*a*) internal atmospheric dynamics and (*b*) boundary forcing (Shukla, 1984). The first category includes the day-to-day weather fluctuations, with which we are all familiar. These are manifestations of the growth, decay and propagation of meteorological disturbances which derive their energy from the internal structure of the atmosphere, i.e horizontal or vertical gradients of wind, temperature and moisture. Because of their unstable nature and the complexity of atmospheric dynamics, these fluctuations become unpredictable beyond a theoretical limit of a few weeks.

Consider now the second mechanism, boundary forcing. Complex interactions take place between the atmosphere and the lower boundary on all time and space scales. The changes at the lower boundary due to changes in SST, snow cover, sea ice, soil moisture, albedo (the percentage of incoming solar radiation reflected from the earth's surface) and vegetation cover affect the exchange of heat and moisture between the earth's surface and the atmosphere. This in turn, may result in significant changes in the time-averaged atmospheric circulation and associated climatic parameters, for example, rainfall and surface temperature.

Changes in surface boundary conditions normally occur at a much slower rate than atmospheric changes associated solely with internal dynamics. Thus, they represent an ocean or land surface layer memory mechanism for the atmosphere which is potentially of great importance for forecasting on seasonal or longer time scales. It may well be that the keys to the problem of African drought lie in the variation of regional and global boundary conditions, particularly SST, and African land surface conditions. Longer time scale changes in atmospheric aerosol concentrations, arising from volcanic eruptions or the introduction of combustion products (fossil fuels) and other man-made contaminants, must also be given consideration when examining decadal-scale trends.

It is usually difficult to distinguish between the effects of internal dynamics and surface boundary forcing, or to isolate the effect of various boundary forcings from data alone. An invaluable theoretical aid in the study of these processes is an atmospheric General Circulation Model (GCM). These numerical models are based on the mathematical representation of the physical laws governing large-scale atmospheric motions. They are similar to those used in day-to-day weather forecasting but are integrated, using large computers, for periods of 1 month or more of simulated time. The present state-of-the-art GCMs are capable of realistic simulation of the global climate. Such models have been coupled in an interactive manner with the land surface processes, including hydrology, and are now beginning to be coupled with companion oceanic GCMs.

Ocean-atmosphere interactions

Both observations and models clearly indicate that variations in tropical SST are strongly related to interannual variability of the global atmosphere, particularly in the tropics. The most notable example is the ENSO phenomenon, which owes its existence to coupled ocean-atmosphere interactions in the equatorial Pacific. ENSO has a pronounced effect on year-to-year variations over southeast Africa, as previously documented. Such interactions between the tropical oceans and global atmosphere are now recognized as a major element in year-to-year global climate variability, and are being addressed by a major research element of the World Climate Research Program (WCRP) called TOGA (Tropical Ocean Global Atmosphere; Rasmusson, 1985).

Because of the relatively slow response of the extratropical oceans, effective dynamic coupling of the ocean and atmosphere on relatively short interannual time scales can take place most effectively in the deep tropics.

Elsewhere, the predominant large-scale dynamic response of the ocean appears to occur on decadal rather than interannual time scales (Philander, 1979). This suggests the possibility of global ocean involvement in the longer time scale fluctuations of climate in the Sahel.

Land-surface–atmosphere interactions

The unusually strong persistence of drought conditions in the Sahel-Soudan belt invites speculation as to the possibility of a natural feedback between surface conditions and the drought-sustaining atmospheric circulation. There are, in addition, widely expressed views, but as yet no real proof, that man-induced environmental degradation (deforestation and desertification) is leading to climatic deterioration in this region. In this regard, it should be noted that the reconstruction of eighteenth–twentieth century rainfall variations suggests that conditions of the past two decades are indeed rare; however, they may not yet be outside the limits of the natural variability experienced during the past few centuries.

Charney (1975) first discussed the possibility of feedback between the West African land surface and the overlying atmospheric conditions. He hypothesized that changes in surface albedo induced by overgrazing may have been a factor in the drought in the early 1970s. It is now recognized that the problem of land-surface–atmospheric interaction is quite complex, possibly involving surface hydrology and roughness changes as well. For a realistic model simulation, evapotranspiration must also be allowed to vary with ground temperature and soil-moisture conditions, and the resulting change in cloudiness must be taken into consideration. Furthermore, the problem is not simply one of local heat/moisture balance considerations. The integrated effect of surface anomalies over a large region may produce changes in the large-scale atmospheric circulation and associated transport of moisture across the boundaries of the region. It is not obvious whether or when such changes reinforce, reduce, or totally override the direct effects of the local surface anomalies (Sud & Fennessey, 1984).

The GCM experiments of the past decade have yet to produce definitive answers to these questions for two reasons. First, there is a problem of observations. To date, direct observations of surface conditions are so inadequate that it is not possible to ascertain how realistic the climatological values and variability of surface parameters used in the GCM experiments really are (Rasool, 1984). Efforts are now being made to develop a new generation of satellite data products which may improve this situation (see, for example, Tucker, Townshend & Goff, 1985). The second problem is associated with the models themselves which, to date, have lacked the sensitivity needed to evaluate the effect of realistic anomalies. Thus far, most of the experiments have been in the nature of sensitivity studies, in which the prescribed anomalies are unrealistically large in magnitude and spatial extent (e.g., Shukla & Mintz, 1982). Although a consensus on the importance of land-surface feedback is yet to emerge, the numerical experiments to date clearly suggest the possibility of a significant influence on climatic variability from land surface processes.

Climate prediction

While it will likely be many years before GCMs evolve from purely a research tool to routine use in seasonal climate prediction, there is some optimism that in the interim modest predictive skill can be attained under some conditions, using empirical/statistical techniques. The most promising applications at present are for seasonal predictions a few months in advance in regions strongly influenced by the ENSO phenomenon or, more generally, for regions where relationships can be established with slowly varying parameters such as Atlantic SST. Even then, the best to be expected is a prediction of seasonally averaged conditions over a relatively large region, such as Southeast Africa. While a skillful forecast of this type is of marginal use to the individual farmer, it would be of significant value to national planners and decisionmakers.

The variance maxima in the spectral analysis of African rainfall which appear at periods between 2 and 6 years can be related to the Southern Oscillation but, in themselves, they are mainly of diagnostic and theoretical interest and have little direct value for prediction. Use of longer period trends and 'cycles' is indeed hazardous, unless there is some understanding of the underlying processes of work, whether natural or man-made. The fact that a fluctuation with a particular period has repeated two or three times during the length of record may provide an estimate of natural variability, but in and of itself it implies little else. The meteorological literature is replete with descriptions of cycles found *a posteriori*, but which failed miserably when used predictively.

Concluding remarks

An understanding of the climate processes responsible for African drought will be difficult to attain without a significant improvement in the network of routine surface and upper-air meteorological observations as well as a better description of surface conditions over the African continent. These data are necessary ingredients for the extensive program of empirical and theoretical studies needed to resolve this problem. Unfortunately, this data collection network suffers during budget stringencies in various countries in sub-Saharan Africa. Yet, without the data provided through such networks, it will be increasingly difficult for scientists to understand better the origins of African droughts.

Clearly, amelioration of the impact of African drought is a problem which requires the cooperation of scientists in a variety of disciplines. While drought will always be a recurrent phenomenon in Africa and the degree to which it is predictable has yet to be established, we must make better use of our present and future knowledge of the behavior of the climate system in a multidisciplinary effort to deny famine a future on this continent.

Acknowledgements

I am deeply indebted to A. T. Grove (African Studies Center, Cambridge), for providing information on the Sahel and Lake Chad; to P. J. Lamb, Illinois State Water Survey; and to S. E. Nicholson, Florida State University for generously providing information on African rainfall variability.

References

Bjerknes, J. (1969). Atmospheric teleconvection from the equatorial Pacific. *Monthly Weather Review,* 97, 164–72.

Burpee, R. W. (1972). The origin and structure of easterly waves in the lower troposphere of North Africa. *Journal of Atmospheric Sciences,* 29, 77–90.

Cane, M. A. (1983). Oceanographic events during El Niño. *Science,* 222, 1189–95.

Charney, J. G. (1975). Dynamics of deserts and drought in the Sahel. *Quarterly Journal of the Royal Meteorological Society,* 101, 193–202.

Dhonneur, G. (1974) *Nouvelle Approche Des Réalités Météorologiques de l'Afrique Occidentale et Centrale.* ASECNA-Tome 1 (358p.), Tome 2 (470p.). Dakar, Senegal: Université de Dakar.

Folland, C. K., Parker, D.E. & Palmer, T. N. (1985). Sahel drought and worldwide sea surface temperatures. *Proceedings of the First WMO Workshop on the Diagnosis and Prediction of Monthly and Seasonal Atmospheric Variations over the Globe.* Geneva, Switzerland: World Meteorological Organization.

Griffiths, J. F. (1972). *Climates of Africa: World Survey of Climatology,* Vol. 10. Amsterdam: Elsevier.

Grove, A. T. (1978). Geographical Introduction to the Sahel. *Geographical Journal,* 144, 407–15.

Grove, A. T. (1985). The environmental setting. *The Niger and its Neighbors.* Rotterdam: Balkema.

Harrison, M. S. J. (1983). The Southern Oscillation, zonal equatorial circulation cells and South African rainfall. *Preprints of the 1st International Conference on Southern Hemisphere Meteorology* ed. American Meteorological Society, pp.302–5. Boston: American Meteorological Society.

Harrison, M. S. J. (1984). The annual rainfall cycle over the central interior of Africa. *South African Geographical Journal*, 66, 47–64.

Hastenrath, S. & Heller, L. (1977). Dynamics of climatic hazards in northeast Brazil. *Quarterly Journal of the Royal Meteorological Society*, 106, 447–62.

Kutzbach, J. E. (1980). Estimates of past climate at Paleolake Chad, North Africa, based on a hydrological and energy-balance model. *Quarternary Research*, 14, 210–23.

Kutzbach, J. E. (1981). Monsoon climate of the early Holocene: climate experiment with the earth's orbital parameters 9000 years ago. *Science*, 214, 59–61.

Lamb, P. J. (1978a). Large-scale tropical Atlantic surface circulation patterns associated with Subsaharan weather anomalies. *Tellus*, 30, 240–51.

Lamb, P. J. (1978b). Case studies of tropical Atlantic surface circulation patterns during recent sub-Saharan weather anomalies: 1967 and 1978. *Monthly Weather Review*, 106, 482–92.

Lamb, P. J. (1980). Sahelian drought. *New Zealand Journal of Geography*, 68, 12–16.

Lamb, P. J. (1985). Rainfall in Subsaharan West Africa during 1941–83. *Zeitschrift für Gletscher kunde und Glazialgeologie*, 21, 131–9.

Lough, J. M. (1986). Tropical Atlantic sea surface temperatures and rainfall variations in Subsaharan Africa and northeast Brazil. *Monthly Weather Review*, 114, 561–70.

Maley, J. (1981). *Etudes palynologiques dans le bassin du Tehad et paleoclimatologie de l'Afrique nord-tropicale de 30 000 ans à l'époque actuelle*. Paris: ORSTROM.

Moura, A. D. & Shukla, J. (1981). On the dynamics of droughts in northeast Brazil: observations, theory and numerical experiments with a general circulation model. *Journal of Atmospheric Sciences*, 38, 2653–75.

Nicholson, S. E. (1980). The nature of rainfall fluctuations in subtropical West Africa. *Monthly Weather Review*, 108, 473–87.

Nicholson, S. E. (1981). Rainfall and atmospheric circulation during drought and wetter periods in West Africa. *Monthly Weather Review*, 109, 2191–208.

Nicholson, S. E. (1983). Sub-Saharan rainfall in the years 1976–1980: evidence of continued drought. *Monthly Weather Review*, 111, 1646–54.

Nicholson, S. E. (1986). The spatial coherence of African rainfall anomalies – interhemispheric teleconnections. Submitted to the *Journal of Climate and Applied Meteorology*.

Nicholson, S. E. & Entekhabi, D. (1986), The quasi-periodic behavior of rainfall variability in Africa and its relationship to the Southern Oscillation. *Archiv für Meteor. Geophys. Bioclimo. Scr. A. Meteorology and Atmospheric Physics*, 34, 311–48.

Norquist, D. C., Recker, E.C. & Reed, R. J. (1977). The energetics of African wave disturbances as observed during Phase III of GATE. *Monthly Weather Review*, 105, 334–42.

Ogallo, L. A. J. & Nassib, I. R. (1984). Drought patterns and famines in East Africa during 1922–83. *Extended Abstracts of Papers Presented at the Second WMO Symposium on Meteorological Aspects of Tropical Droughts*, Fortaleza, Brazil, 24–8 September, 1984, pp.41–4. Geneva: World Meteorological Organization.

Philander, S. G. F. (1979). Variability of the tropical oceans. *Dynamics of Atmospheres and Oceans*, 3, 191–208.

Rasmusson, E. M. (1985). El Niño and variations in climate. *American Scientist*, 73, 168–78.

Rasmusson, E. M. & Wallace, J. M. (1983). Meteorological Aspects of the El Niño/Southern Oscillation. *Science*, 222, 1195–202.

Rasool, S. I. (1984). On dynamics of deserts and climate. In *The Global Climate*, ed. J. T. Houghton, pp.107–20. Cambridge University Press.

Shukla, J. (1984). Predictability of time averages, Part II: the influence of the boundary forcing. In *Problems and Prospects in Long and Medium Range Weather Forecasting*, eds R. M. Burridge & E. Kallen, pp. 155–206. London: Springer-Verlag.

Shukla, J. & Mintz, Y. (1982). Influence of land-surface evapotranspiration on the earth's climate. *Science*, **215**, 1498–501.

Starr, J. (1985). Growing anxieties along the Nile. *Washington Times*, 22 Feb. 1985.

Street, F. A. & Grove, A. T. (1976). Environmental and climatic implications of late Quaternary lake-level fluctuations in Africa. *Nature*, **261**, 385–90.

Sud, Y. C. & Fennessy, M. J. (1984). Influence of evaporation in semi-arid regions on the July circulation: A numerical study. *Journal of Climatology*, **4**, 383–98.

Todorov, A. V. (1985). Sahel: the changing rainfall regime and the 'normals' used in its assessment. *Journal of Climate and Applied Meteorology*, **24**, 97–107.

Tucker, C. J., Townshend, J. R. G. & Goff, T. E. (1985). African land-cover classification using satellite data. *Science*, **227**, 369–75.

Weickmann, K. M., Lussky, G. R. & Kutzbach, J. E. (1985). Intraseasonal (30–60 day) fluctuations of outgoing longwave radiation and 250 mb streamfunction during northern winter. *Monthly Weather Review*, **113**, 941–61.

Wilhite, D. A. & Glantz, M. H. (1985). Understanding the drought phenomenon: the role of definitions. *Water International*, **10**, 111–20.

World Meteorological Organization (1971). *Climatological Normals (CLINO) for Climate and Climate Stations for the period 1931–1960*. WMO 117, TP52. Geneva: World Meteorological Organization.

2

Some aspects of meteorological drought in Ethiopia

WORKINEH DEGEFU

Ethiopian National Meteorological Services Agency

Some meteorological facts about Ethiopia

Rainfall amount and distribution

Ethiopia, situated in the Horn of Africa, lies between 3° 25′N and 18° N latitude, and 33° E and 48° E longitude. The sources of moisture that account for almost all rains in the country are the Indian and Atlantic Oceans. Southeasterly winds during the months of February to May carry moisture from the Indian Ocean into most of the country, while southwesterly as well as southeasterly winds bring moisture during June to September, the main rainy season. During the main rainy season, moisture gradually penetrates into the country from the southwest, as the Inter-Tropical Convergence Zone (ITCZ) progresses northward with the equatorial trough. As a general rule, rainfall should decrease as one moves from the south to the north of the country. However, this situation is somewhat modified by the topography of the country; rainfall maxima are found in the southwest and minima in the northeast and southeast of the country (Fig. 2.1).

Seasons in Ethiopia

As a result of its geographical location and topography, the spatial and temporal distribution of rainfall in the country has given rise to three main seasons, unlike the two seasons normally found in most tropical areas. These seasons are locally known as *Kiremt* (main rains), *Belg* (small rains), and *Bega* (dry season). This classification does not encompass the southern and southeastern lowlands of the country which have a bimodal rainfall distribution with rainfall periods from March to May and from September to October. The southwestern part of the country also does not follow the three-season pattern, because there it rains from February to November.

For other parts of the country, *Kiremt* is the main rainy season in which about 85–95% of the food crop of the country is produced. The *Kiremt* rain, which begins around the end of May, overtaking the *Belg* rain, in the southwestern part of the country, gradually moves in the north and northeasterly direction until by mid-July it approaches the northern tip of the country. The *Kiremt* rain ceases in the north around the end of August, with the retreat of the ITCZ, which gradually moves southward until by early October, most of the country comes under the influence of the north-easterly trade winds.

The *Kiremt* season is followed by the *Bega* season (October–January) which is mostly dry. Although October and November are classified as dry months, there is rainfall in the southern and southwestern part of the country during these 2 months. In the south, the retreat of the ITCZ produces rain up to early November while, in the southwest, the rainfall continues up to the end of November.

Agriculturally, the *Bega* season is a harvest season. However, occasional rainfall caused by the interaction of the northern hemisphere winter frontal systems with the tropical systems, causes widespread rain over Ethiopia. The other sources of rain during this time of the year are the tropical

Fig. 2.1. Mean annual rainfall (in mm).

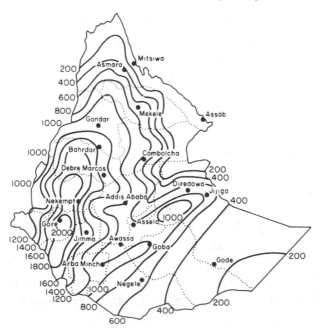

depressions of the Indian Ocean. The rainfall during the *Bega* season is
significant hydrologically, as well as for the availability of grasses for cattle.
The absence of this rainfall, during the *Bega* season of 1983, was the
beginning of the catastrophic Ethiopian drought of 1984.

The *Bega* season is then followed by the *Belg* season (February–May).
most parts of the country benefit from rainfall during this period. In the
case of southern Ethiopia, the rains continue up to the end of May. For
that part of the country, *Belg* rain is the main rainy season.

Belg rain is extremely important from an agricultural as well as from a
hydrologic point of view. *Belg* crops, which account for 5–15% of the national
food crop, are produced with this rain. Long-season crops, like maize and
sorghum, which constitute major food crops of the country, are planted
during this season. Small rivers, ponds and water reservoirs, which usually
dry up during the *Bega* season, will start to regenerate. A delay of the *Belg*
rain usually means the absence of water and pasture which, in turn, results
in the death of thousands of animals.

The three seasonal rainfall amounts expressed as a percentage of the
annual total are shown in Figs. 2.2–2.4. Rainfall regimes in Ethiopia,
showing periods and amounts of rainfall, are shown in Fig. 2.5.

Fig. 2.2. Percentage of *Kiremt* rain to that of mean annual total.

Fig. 2.3. Percentage of *Bega* rain to that of mean annual total.

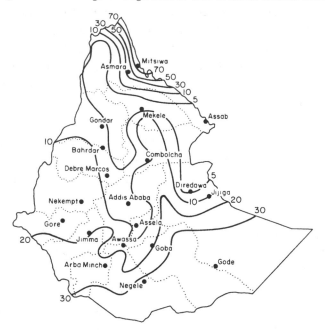

Fig. 2.4. Percentage of *Belg* rain to that of mean annual total.

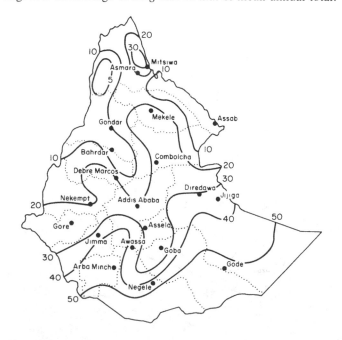

Rainfall variability in Ethiopia

Rainfall is the most important climatic element that influences Ethiopian agriculture. The entire agricultural activity of the country is associated with the behavior of the rainfall. For example, crops fail primarily because the rains are late, the rainfall season is too short, or the amount of rainfall received is insufficient for good crop growth. Another factor to be considered is referred to as effective precipitation, that is, the amount of water which actually becomes available for crop growth. This is basically the rainfall less the portion lost to evaporation and runoff. These losses are particularly large in regions where rainfall occurs in violent and infrequent episodes, and are much smaller where rains are moderate and steady.

An important characteristic of Ethiopian rainfall is that it exhibits high variability in time and space. This variation is largely due to orographic effects and to other weather extremes affecting the country. The coefficient of variability (CV) of rainfall, a measure of the variation from the mean of the observed rainfall, is shown in Fig. 2.6.

Meteorological drought in Ethiopia

In Ethiopia, climatic variability, including the occurrence of drought, is not unusual. During the last two decades, however, the frequency of

Fig. 2.5. Rainfall regimes.

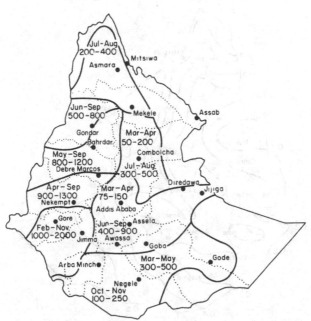

recurrence and intensity of drought has increased. Unfortunately, droughts in Ethiopia and their impacts have not as yet been systematically documented. According to Wood (1977), the history of famines, which in many cases were suspected to be caused by drought, goes back to the eleventh century. Oladipo takes these dates further back to 253 BC, as quoted from the works of Shove (1977; see also Nicholson, 1979). Table 2.1 lists drought and famine years in Ethiopia, as quoted in these sources and from information obtained from the Ethiopian Relief and Rehabilitation Commission.

As seen from the chronology of Table 2.1, it is evident that there is no pattern (trend, periodicity, or cycle) for Ethiopian drought episodes. This may be largely the result of the absence of proper documentation. Drought events, since meterological records began in Ethiopia in the early 1950s, exhibit some kind of pattern. Since then, droughts of different intensities have occurred in different parts of the country, incurring losses of life and property. There have been six notable drought years, of which the 1972 and the most recent one of 1983–84 have been the most catastrophic.

In 1957 the failure of the *Belg* rain, compounded by the outbreak of locust and epidemics, brought famine and suffering to the people, particularly in the provinces of Wello and Tigrai. From 1965 through 1966

Fig. 2.6. Coefficient of rainfall variability.

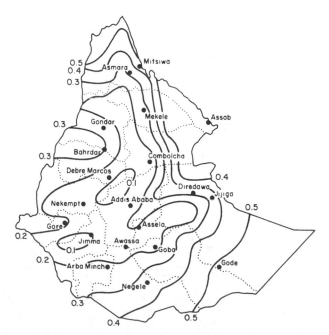

there were widespread rainfall deficiencies that brought the national food-crop production to a lower level. Famine was observed in many parts of the country in these years.

In 1972 the most devastating drought, since meteorological recording

Table 2.1 *Occurrences of drought and famine in Ethiopia*

Year	Affected area	Severity
253–242 BC	Ethiopia	Mainly deduced from the chronology of low Nile River levels.
1066–72	Ethiopia and Egypt	Low Nile River levels
1252	Refers to Ethiopia as a whole	Noted as famine years
1272–73	Refers to Ethiopia as a whole	Noted as famine years
1274–75	Refers to Ethiopia as a whole	Noted as famine years
1435–36	Refers to Ethiopia as a whole	Noted as famine years
1454–68	Refers to Ethiopia as a whole	Noted as famine years
1543–62	Especially in Harar region	For three years following the killing of Emperor Glaudios, there was no rainfall, especially in Harar (eastern Ethiopia)
1800	Refers to Ethiopia as a whole	Both people and horses died of famine
1826–27	Refers to Ethiopia as a whole	There was great failure of both cotton and grain crops and many cattle died
1829	Shoa region	Crop failure occurred and cattle disease epidemic followed
1835	Shoa (central Ethiopia) and western Eritrea regions	Many people of Shoa (central Ethiopia) died following failure of rain. If that was the same drought that the people of western Eritrea (northwestern Ethiopia) remembered as 'The Great Starvation' when 'rain disappeared from the earth and famine came over men and beasts', as Pankhurst(1968) suggests, it would have been a major widespread drought.
1836–37	Northern provinces	A holocaust of drought, famine, cattle disease, epidemic, and cholera. In southwestern Eritrea, it was a continuation of the 1835 famine that was referred to as 'the year of stagnations'.

Table 2.1. *Occurrences of drought and famine in Ethiopia (continued)*

Year	Affected area	Severity
1888–92	Whole of Ethiopia, highlands and lowlands included	It was one of the most serious droughts experienced in Ethiopia and was known in Ethiopian history as the 'Kifu Ken' (the harsh days). Both the big and small rains failed. The entire period was hot and dry and the effects of drought were magnified by the catastrophic rinderpest epidemic (which killed 90% of the cattle) and invasion of locusts, caterpillars and rats. Suicide and cannibalism occurred and wild animals attacked people. About 1/3 of the population perished.
1895–96	Refers to Ethiopia as a whole	A minor drought occurred that year, due to the failure of the winter and spring rains. Yet, many people and cattle died.
1899–1900	Refers to Ethiopia as a whole	An unrecorded drought was manifested through a fall in the level of Lake Rudolf. The Nile flood was also abnormally low.
1913–14	Ethiopia (northern part, especially Tigrai region)	Very low Nile flood (the lowest since 1695). In the northern part of Ethiopia, the price of grain increased 30-fold and there was great starvation in Tigrai.
1921–22	Ethiopia as a whole	Similar drought to that of 1895–96. According to the recollection of a long-time English resident of Ethiopia, there was a complete failure of the rains from Oct. 1920–May 1921.
1932–34	Ethiopia as a whole	The level of Lake Rudolf dropped, implying a serious decrease in rainfall in southern Ethiopia. A drought was recorded in northern Kenya and in 1934 a relief camp was set up in British Somaliland to aid drought victims.
1953	Wello and Tigrai region	Another undocumented drought in Wello and Tigrai.
1957–58	Wello and Tigrai region	The complete failure of rain in the 1957–58 period brought, together with the outbreak of locusts and epidemics, famine, of which the worst year was said to be 1957, which did not have more than 10 rainy days. More than 100 000 people died. Locust plagues in September and October 1958 had a devastating effect in all the *awrajas* of Tigrai and probably in the neighboring regions as well.

began, started. That year, drought affected many parts of Africa, but the impact was more pronounced in Ethiopia.

After the 1972–73 drought, climatic conditions began to normalize. However, droughts of minor intensities continued to appear in some parts of the country. In 1980 there was a failure of the *Belg* rain in southern Ethiopia, and as a result thousands of cattle died. In 1982 the onset of the

Table 2.1. *Occurrences of drought and famine in Ethiopia (continued)*

Year	Affected area	Severity
1964–65	Ethiopia as a whole	A virtually undocumented drought said to be more widespread in Ethiopia than that of 1973–75.
1965–66	Wello and Tigrai regions	The failure of the spring and summer rains and the high temperatures that accompanied the drought affected five of the eight *awrajas* of Tigrai and eight of the twelve *awrajas* of Wello. The impact on human lives had not been reported, but the number of cattle, pack animals, sheep and goats lost is estimated to be above 297 350 for Tigrai alone.
1969	Eritrea region	Severe drought affected about 1.7 million people
1971–78	Northern, southeastern and eastern part of Ethiopia, but particularly Tigrai and Wello region	Complete failure of the spring (*Belg*) rain. RRC in its publication of December 1982 gives the number of dead to be about 200 000 for Tigrai, Wello, and northern Shoa. Other estimates give 400 000 to 1 million for Tigrai and more than 100 000 for Wello. 80% of cattle, 50% of sheep, and 30% of goats perished.
1975–76	Wello and Tigrai regions	The four *awrajas* of western Wello and parts of Tigrai have collected poor harvests in 1975–76 and 1976–77, while at least 75% of the main harvest of 1977–78 was destroyed due to 'unfavorable climatic conditions'. In Wello alone, about 1.2 million people were affected. For both Wello and Tigrai estimates of affected people go up to 2–3 million people.
1978–79	Southern Ethiopia	Failure of spring rain resulted in drought
1982	Northern Ethiopia	Delay of monsoon rain by two months

ECA, 1984.

Kiremt rains was delayed by 2 months and, as a result, there was crop failure in northern Ethiopia, particularly in parts of the Gondar and Wello regions.

Rainfall activity, during the 1983 growing season, has been one of the most favorable ones since the beginning of the decade. The onset and withdrawal, as well as the amount and distribution, of precipitation, during both the *Belg* and *Kiremt* seasons, were generally good. However, the cumulative rainfall values in northern Ethiopia, particularly in parts of the Wello and northern Shoa regions were just below normal.

The *Bega* season of 1983 started with the unusually wet month of October. The rainfall ceased as of the second 10-day period in October, and the dry spell prevailed throughout the country. The intensive heating, as a result of a cloudless sky, gradually diminished the availability of surface and ground water. By the end of January 1984, the *Bega* season's aridity had already caused widespread movement and dislocation of rural populations, particularly the nomadic community, in search of grass and water. By the end of February 1984, synoptic conditions for the onset of the *Belg* rain did not look favorable. The persistence of unfavorable meteorological conditions continued into the months of March and April and the *Belg* rain of 1984 also failed.

Thousands of domestic, as well as wild, animals perished due to lack of water. Forest fires broke out at various regions of the country, causing incalculable damage to the forest-based economy of the country. Large-scale population migration took place from the arid north to the west and southwest direction.

Rainfall was widespread in May 1984. However, it was too late not only to grow *Belg* crops, but also to plant long-season crops such as maize and sorghum, the country's major food crops. Another effect of the drought on the ensuing *Kiremt* season's farming activities was that farm animals had died or were in weakened conditions in most parts of the country. As a result, farmers' activities were greatly handicapped.

By June 1984 prolonged aridity during the *Bega* season of 1983 and the subsequent failure of the *Belg* rains of 1984 had already caused widespread human dislocation and suffering. Rainfall, which came to many parts of the country in May, established itself as *Kiremt* rain. The short dry spell that normally exists from mid-May to mid-June was now replaced by a wet spell. This unusual climatic pattern confused the farmers, and, as a result, they were unable to perform their usual agricultural activities.

The *Kiremt* rain, which started ahead of the normal onset period, continued throughout June and July, with decreased amounts and with poor spatial and temporal distribution. The second half of July is normally the time when the *Kiremt* rain approaches its peak. However, in 1984 the conditions were reversed: the dry spell continued until the end of the month. It was the time when most crops were at the flowering stage. That was yet another adverse effect on the year's crop production.

The rainfall time-series for the period 1951–1984, in drought-prone areas of northern Ethiopia, is shown in Fig. 2.7. Figures 2.8 and 2.9 depict rainfall departure from normal for the month of April 1984 (i.e. rainfall for the 1984 *Belg* and the *Kiremt* seasons, respectively).

Anomalous meteorological features of 1984

The Belg *season*

The surface meteorological conditions were characterized by the following features: (*a*) the ridge of the Arabian anticyclone extended mainly along the Horn of Africa; normally it lies over the Arabian Sea; (*b*) the confluence zone of the winds from the south Atlantic Ocean and the South Indian Ocean anticyclones which coincide with the trough of the low over central Africa, and which normally lies over eastern Africa (including Ethiopia),

Fig. 2.7 Northern Ethiopia annual rainfall departures.

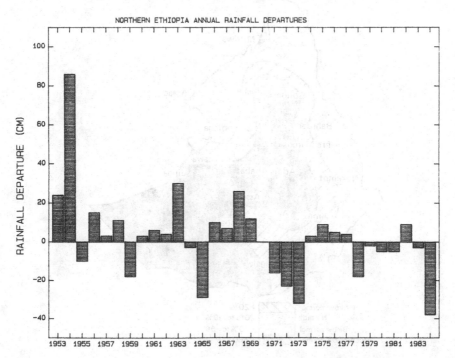

shifted far to the west of this position; (c) the Saharan anticyclone had an east–west extension and usually mixed with the anticyclone over Arabia, as a result of which the interaction between the tropical and mid-latitude lows was hindered.

The middle-level atmospheric conditions had the following features: (a) there was a frequent blocking high along 0°–10° E over the 700 and 500 mb levels, this hindered the west–east movement of the middle latitude waves; (b) the flow pattern was mainly zonal with less intensity in wind speed; (c) late in the *Belg* period (May), the effect of the waves over the easterlies was one of the rain-producing components in the southern parts of Ethiopia. During this period these waves were observed many times and the rainfall during the period was better than it had been during the earlier part of the *Belg* season (February, March and April).

Conditions at the upper tropospheric levels were as follows: (a) the mid-latitude waves were not deep, the flow pattern was zonal and the penetration of the subtropical jet stream to the tropics was less frequent, wind speed was also weak; (b) the formation of cutoff low and the penetration of mid-latitude troughs to the tropics were also minimum.

Fig. 2.8. Rainfall deviation from the mean in percentage of the month for April 1984.

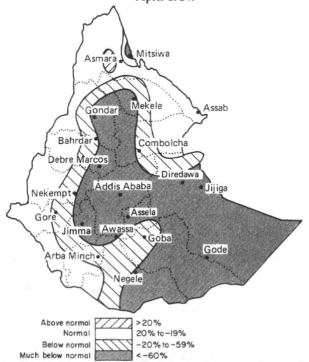

Above normal	///	>20%
Normal		20% to −19%
Below normal	\\\	−20% to −59%
Much below normal	▓	< −60%

The Kiremt *season*

This season was characterized by the following features. (*a*) The tropical easterly jet stream was weak and less frequent over the Horn of Africa. Usually, it appeared north of its normal position. (*b*) The monsoon depressions of the Indian subcontinent were very weak. (*c*) Drift flow from the southern hemisphere was not strong. (*d*) Fewer than normal easterly waves were observed.

Editor's Note

This chapter reinforces the need to supplement the relatively widespread coverage in the news media and in the research literature of drought, food-production problems and famines in Ethiopia with meteorological information. Prolonged droughts throughout various parts of Ethiopia have clearly contributed to its food crises. Recognizing this fact, the Ethiopian Relief and Rehabilitation Commission (RRC) incorporates such information into its early-warning system for food shortages. This information, along with other relevant data such as changes in grain and livestock prices in

Fig. 2.9. Rainfall deviation from the mean in percentage for the *kiremt* season of 1984.

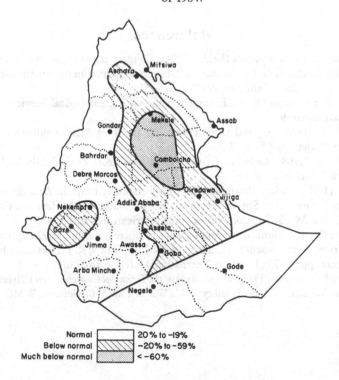

Normal		20% to −19%
Below normal		−20% to −59%
Much below normal		< −60%

the marketplace, nutritional status of various segments of the population, livestock conditions, rangeland conditions and the like, are used to monitor food production and food availability for the purpose of early response to identified local or regional shortfalls in agricultural and livestock production. The early-warning system also serves to alert governmental and non-governmental donors to impending food shortages.

As with other African countries, there is a need to maintain, if not strengthen, national meteorological services so that more accurate information about the probability of drought recurrences can be collected and preparation for drought mitigation can take place. Strengthening meteorological services may be important not only for the purpose of early famine detection but for development purposes as well. Climate should not only be taken as a hazard to society but should also be viewed as a resource to be more effectively exploited.

Acknowledgment

I wish to express my gratitude to Mr Taffesse Olkeba and Mr Bekuretsion Kasahun of the National Meteorological Services Agency, Ethiopia, for providing me with relevant materials for the preparation of this chapter.

References

Economic Commission for Africa (ECA) (1984). Country paper presented to the ECA's Scientific Round Table Conference on 'Drought and the climatic situation in Africa', Addis Ababa, 20–3 February 1984.

Farming Weather Bulletin (1984). Ethiopian National Meteorological Services Agency, Publication No. 9.

Nicholson, S. E. (1979). Revised rainfall series for the West African sub-tropics. *Monthly Weather Review*, **107**(5), 620–3.

Pankhurst, R. K. (1968). *Economic History of Ethiopia, 1800–1935*. Addis Ababa: Haile Sellassie I University Press.

Shove, D. J. (1977). African droughts and the spectrum of time. In *Drought in Africa – 2*, African Environment Special Report 6, ed. D. Dalby, R. J. Harrison Church & F. Bezzaz, pp. 38–53. London: International African Institute.

Wood, C. A. (1977). A preliminary chronology of Ethiopian droughts. In *Drought in Africa – 2*, African Environment Special Report 6, ed. D. Dalby, R. J. Harrison Church & F. Bezzaz, pp.68–73. London: International African Institute.

World Climate Programme (1982). *Proceedings of the Technical Conference on Climate for Africa*, Arusha, Tanzania, 25–30 January 1982, WMO No. 596. Geneva: WMO.

3

Drought and economic development in sub-Saharan Africa

MICHAEL H. GLANTZ

Environmental and Societal Impacts Group, National Center for Atmospheric Research*

Introduction

The purpose of this chapter is to remind those concerned with economic development issues how drought can, and often does, affect the process of development in sub-Saharan Africa. This reminder might seem unnecessary because of what now seems to be a widespread awareness of the devastating impact of drought in developing countries. Yet, there are many troublesome signs that constantly reappear suggesting that such a reminder is clearly warranted.

A review of how the West African Sahel, plagued by drought between 1968 and 1973, and Ethiopia, affected by drought in 1972–74, were treated by scholars, policymakers, governmental and non-governmental development agencies, and of course the media, once those droughts and their impacts had seemingly ended, raises concern about how drought is viewed over the long term. While a drought is in progress, it is on everyone's mind. Once it ends, however, the interest in, as well as the perceived importance of, drought rapidly disappears.

The latest surge in awareness of African droughts and their societal impacts seems to be a direct result of recent news accounts of deaths and suffering of humans and livestock and of the stark photographs of a degraded African environment. This was clearly the case in the early 1980s for Ethiopia (e.g. BBC film aired in October 1984), as it was in the early 1970s both for the West African Sahel (e.g. Morentz, 1980) and for Ethiopia (e.g. Shepherd, 1975). These accounts dramatize some of the more visible impacts of climate variability (e.g. drought) on society and on the environment in sub-Saharan Africa's arid and semiarid areas.

The drought-ridden 17-year period in the West African Sahel that began

* The National Center for Atmospheric Research is sponsored by the National Science Foundation.

in 1968 (actually, after the 1967 harvest) prompted the publication of scores of books and hundreds, if not thousands, of articles on various physical aspects of drought and on their impacts on various economic and social activities. The focus of these publications encompasses various levels of social organization from herder and farmer to households and tribal groups, to the state, to the international community. A similar situation arose 40 years ago as a response to drought in the US Great Plains; Tannehill (1947, 18) noted that 'each time there is a serious drought millions of words are written on crop failures, misuse of the land, overpopulation, rainfall records . . . '.[1]

Yet, even with all these words during the past few years, many concerned with the study of drought have been left with an uneasy feeling that drought is still generally viewed as either an 'idiosyncratic' occurrence, a transient event, or a 'temporary climatic aberration'. Evidence shows that such views are misleading and that in some areas, meteorological drought is a recurring but aperiodic phenomenon; it is a part of climate and not apart from it. As such, drought is closely related to the problem of achieving sustained agricultural production in sub-Saharan Africa. It should no longer be ignored in development planning.

African climate since independence

Post-independence African leaders have had to cope with innumerable problems of nation-building in a world rent by political, economic, and ideological cleavages. At the same time several of them have been forced to cope with the environmental and societal impacts of one of their worst and most prolonged drought episodes in recent times.

Peter Lamb has developed a regional rainfall index for an area somewhat broader than the West African Sahel (Lamb, 1982, 1983; Kerr, 1985). The rainfall record used by Lamb encompasses the years 1941–84. As Fig. 3.1 shows, rainfall in this area declined drastically within a relatively short time. By coincidence, this decline followed political independence of the countries whose rainfall stations are included in the index.

Changes in rainfall amounts at the local level are also quite graphic, as shown in the following figure for Gao, Mali. Figure 3.2 shows rainfall box plots in terms of the annual rainfall quartile ranges, medians, and highs and lows. It also compares the rainfall record before independence, after independence, and for the total time series. When one combines the post-independence years with the rest of the time series, the recent decline in rainfall becomes overshadowed by the longer (of the two) time series segments. Clearly, rainfall at this particular location has been considerably lower since independence than in the pre-independence period. However, this does not mean that the current downward trend in annual rainfall will continue, as the above-average rainfall years of the 1950s and early 1960s (see Fig. 3.1) did not mean that the then-prevailing wet conditions would continue.

Although many sub-Saharan African countries, like Mali, have been plagued by drought-related constraints on domestic food production during a large part of their post-independence period, there is no simple correlation between meteorological drought and declines in agricultural production. As the box plots for Gao and Lamb's regional rainfall index, however, do suggest, drought conditions in the post-independence period should receive more attention than they have in the past; they should be included as one of the several main factors contributing to declining per capita agricultural production in many African countries. Drought should also be seen not only for its direct societal and environmental impacts but for the varying degrees to which it exacerbates other related problems in Africa such as balance of payments, debt repayment, food imports and urbanization.

To better understand how climate, drought and development issues should be considered in the future, it is useful to look at how they have been considered in the past.

Fig. 3.1. Rainfall index for 20 sub-Saharan stations in West Africa west of the 10° E between 11° N–19° N developed by Lamb (1985).

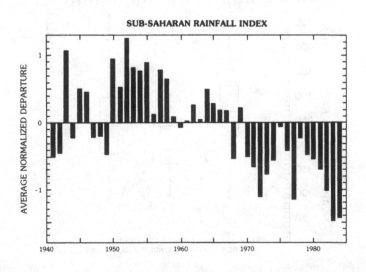

SUB-SAHARAN RAINFALL INDEX

Climate and development: a review of the literature

Until the most recent Sahelian and Ethiopian droughts, climate has been presented in the scientific and popular literature primarily as a boundary condition, that is, as a fixed and relatively unchanging precondition affecting society's development. Scientists and policymakers who perceive climate as a set of boundary conditions suggest that there is little, if anything, one might do to alter the climate or its impacts in a large-scale way, except for technological fixes such as those suggested during the past century for various parts of Africa (Glantz, 1977).

Many discussions that assess the effects of climate on human activities begin with comments on a book by Ellsworth Huntington, *Climate and Civilization* (1915), in which he hypothesized about the effects of climate on levels of development of different cultures. He concluded that there was an ideal temperature and degree of 'storminess' that made possible the development of industrialized societies. Brooks noted that, according to Huntington, a

> certain type of climate, now found mainly in Britain, France and neighbouring parts of Europe, and in the eastern United States, is favourable to a high level of civilization. This climate is characterized by a moderate temperature, and by the passage of

Fig. 3.2. Box plots of pre-independence (1920–1960), post-independence (1961–1984), and combined (1920–1984) wet season rainfall at Gao, Mali. These plots give the minimum (indicated by an 'x'), lower quartile (the bottom of the box), median (the horizontal line within the box), upper quartile (the top of the box), and maximum (indicated by an 'x') of the probability distribution.

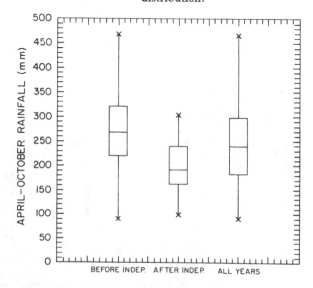

frequent barometric depressions, which give a sufficient rainfall
and changeable stimulating weather. (*Brooks, 1926, 292*)
Huntington viewed inhabitants in the Northern temperate regions as
industrious and energetic, as a result of climatic factors (especially
temperature), while those in the tropical areas were subjected to enervating
climatic conditions that sapped them of the desire and energy to undertake
productive work.

Huntington's views represent the thinking of a group of geographic
determinists who tend to reduce the ills of the developing world (as well
as the successes of the industrialized world) to climatic differences between
these two regions.[2] Several books followed Huntington's *Climate and
Civilization* including such titles as *Climate through the Ages* (Brooks, 1926),
Climate and the Energy of Nations (Markham, 1944), and *World Power and
Shifting Climates* (Mills, 1963). While some authors later adopted and
modified some of the stances taken by Huntington, many challenged them
as being at best ethnocentric and at worst racist. Still others felt that he
had overplayed the importance to society and culture of natural climatic
factors.

After World War II, attention on the relationship between developing
countries and climate shifted from an explanation of why the levels of
development of the tropical countries were so different from those in the
temperate regions to discussions of how to develop tropical countries, given
their climatic boundary conditions.[3] Paul Streeten observed that after World
War II, a new optimism had emerged, suggesting that development could
be brought about by simply supplying some missing economic elements:

> It is part of the stages-of-growth mythology that all countries
> tread inexorably the same path to eventual 'take-off' and self-
> sustained economic growth; that the speed of this march is
> determined by savings ratios, investment ratios, and capital-
> output ratios; and the role of rich countries is to supply missing
> components, like foreign exchange or skills. (*Streeten, 1976, xi*)

According to Streeten (1976, xi) 'the neglect of the role of climate fitted
well into the new optimism'.

In the early 1950s, the US Council on Foreign Relations established a
study group on climate and economic development in the tropics. Their
concern for convening such a study group was raised by the following
observation: 'By any rational definition of 'under-developed country' most
of them lie entirely or partially in the tropics. Is climate the common factor
that keeps them under-developed?' (Lee, 1957, vii). The study group's
report discussed climate in terms of its effects on soil fertility, on human

and livestock health, on plants, on storage and handling of crops, and so forth. Nowhere in the report, however, was either climate variability or drought mentioned.

In 1969 Bernard Oury, an economist with the World Bank, lamented the lack of interest that economists, in their consideration of the development process, show in the effects on societies of weather and climate. He noted that 'From the earliest times, the weather and climate have influenced, if not controlled, the progress of civilization. Yet, in spite of its obvious importance and often dramatic impact, professional economists generally take little interest in the weather as a prime factor in economic development' (Oury, 1969, 24). This article appears to be one of the first from the World Bank to consider weather and development.

In the 1970s another World Bank economist, Kamarck, discussed climate and economic development in the tropics (Kamarck, 1976). Kamarck, unlike the geographic determinists writing earlier this century, did not reduce all ills of countries in the midst of economic development to climatic factors. He noted that

> None of this is to claim that climate has a mechanical one-to-one relation to economic development, nor that climate with its effects is the only ruling constraint on economic development, nor that if the effects of climate were removed as a constraint in today's poor countries development would be unbounded. Rather, in today's poor countries climatic factors have hampered economic development through their impact on agriculture. . . . These effects need to be better understood. (*Kamarck, 1976, 11*)[4]

Although Kamarck commented on the neglect of climate factors by earlier students of economic development, he provided little additional information about how drought or, more generally, climate variability might affect development. Yet, the importance of considering the impacts of drought, as an element of climate, on the development process should not be underestimated.

Biswas (1984), like Kamarck, presented a cursory review of the works of economists who made references to climate and development, [such as Galbraith (1951), Lewis (1955), Myrdal (1968) and Tosi (1975)]. Biswas (1984, 7) raised an interesting and important issue: 'One can, however, ask if development economists have failed miserably to consider climate as an important factor for development planning, why have not the climatologists ensured that such a neglect is not allowed to continue'. He then suggested that 'Much though the climatologists know about climate, they have not ventured out of their own discipline: they have tended to remain isolated within their own field. . . .Accordingly one is indeed hard pressed to name more than a handful of climatologists who are even active in the fringe areas of development'. Further, he suggested that it was no surprise that at the World Climate Conference (WMO, 1979, 7) convened by the World Meteorological Organization, 'not a single paper analyzing the relations between climate and overall development' was presented.

One such recent attempt at a 'new' approach of understanding the relationship of climatic characteristics and development was the seminar convened in 1978 by the Institute of Development Studies at the University of Sussex, in which seasonality and its relationship to various aspects of rural poverty were addressed (Chambers, Longhurst & Pacey, 1981). Underscoring the relevance to development of seasonality (as a characteristic of climate), Chambers *et al.* (1981, xv) succinctly stated the case, as follows:

> most of the very poor people in the world live in tropical areas with marked wet and dry seasons. Especially for the poorer people, women and children, the wet season before the harvest is usually the most critical time of year. At that time adverse factors often overlap and interact: food is short and food prices high; physical energy is needed for agricultural work; sickness is prevalent, especially malaria, diarrhoea and skin infections; child care, family hygiene, and cooking are neglected by women overburdened with work; and late pregnancy is common, with births peaking near harvest. This is a time of year marked by loss of body weight, low birth weights, high neonatal mortality, malnutrition, and indebtedness. It is the hungry season and the sick season. It is the time of year when poor people are at their poorest and most vulnerable to becoming poorer.

Thus, a more effective way to assess the impacts of climate on the development process is to move away from considerations of climate as a boundary condition and to focus on seasonality and other characteristics of climate, such as variability, changes in climate regimes, and droughts and their impacts on the environment and on human activities. To do so means that we can move on to a new phase in the consideration of climatic factors by integrating them into the long list of factors that constrain the development process.

Climate variability and development

Understanding drought

There are several major difficulties in dealing with drought. One is that it is a creeping phenomenon. Its onset as well as its end are often difficult to identify, because they lack a sharp distinction from non-drought dry spells.

Tannehill (1947, 2), for example, suggested that 'The first rainless day in a spell of fine weather contributes as much to the drought as the last day, but no one knows precisely how serious it will be until the last dry day has gone and the rains have come again'.

Another difficulty is that drought is generally viewed as a transient phenomenon. As a result, it is usually not taken seriously, once the rains have returned. Yet, drought is a major disruptive force with which policymakers must reckon. The return time for droughts in various parts of sub-Saharan Africa is on the order of decades, while the length of tenure in office for policymakers is most often on the order of years. How does one get these decisionmakers to keep in mind events that recur on a decadal scale when structuring development policies for their countries?

Yet another problem relates to identifying the impacts of drought on human activities. Impacts of drought are pervasive; while there are some obvious effects (e.g. withering crops, dry watering points, reduced forage for livestock), we often are less aware of second- and third-order effects (e.g. price increases, increased food imports, surges in rural-to-urban migration rates). Therefore, many of the impacts that might be attributable to drought are difficult to identify.

The view of drought as an idiosyncratic event that does not need to be taken seriously once it has passed is subtly reinforced by the fact that in many instances drought is not the sole factor responsible for a variety of social dislocations.[5] Yet, drought can, and often does, exacerbate existing sociopolitical, economic, or cultural factors that vary from one country to another and from one point in time to the next.

As another example of how drought exacerbates existing societal conditions, a US Department of Agriculture (USDA) situation report for sub-Saharan Africa noted that 'the continuing economic crisis in the region was worsened by the drought'; that 'drought in addition to aged plantation trees and disease reduced the output of cocoa in Cameroon'; that 'drought and continued weak finances among farmers adversely affected crop yields by reducing the need for, and use of, fertilizers'; that 'the combination of guerrilla disruption and drought has had a disastrous impact on food production [in Mozambique]' (USDA, 1984, 17); and that 'during the last year the drought and import restrictions caused severe distortions in the corn price' (USDA, 1984, 9). Along similar lines, Lester Brown (1985, 71) argued that 'Three forces are acting in concert to put Africa on the skids in terms of food supplies (population growth, widespread soil erosion and government neglect of agriculture) . . . Only now has this situation been brought into sharp focus by severe drought'. And finally, with respect to famines, many authors have explained why droughts need not result in famine and famines do not necessarily have their origins in drought (Sen, 1981; Watts, 1983; Torry, 1984; Bush, 1985).

Thus, while drought may not be solely or even directly responsible for many of the societal disruptions that occur during drought episodes, its combination with other factors specific to a country at a given time can make a bad situation worse and can be devastating to the development process.

Another difficulty in dealing with drought is that drought means different things to different people, depending on their specific interest in, or need for, rainfall. Here, only meteorological, agricultural, and hydrological droughts are discussed (for a more detailed discussion on drought see, for example, Tannehill, 1947; Palmer, 1965; Wilhite & Glantz, 1985).

Meteorological drought can be defined, for example, as a 25% reduction of the long-term average rainfall in a given region. There are scores of variations of this definition. A meteorological drought is sometimes difficult to identify with any degree of reliability, in part because of the nature of the phenomenon and in part because meteorological and climatological information in many African countries has only been available for relatively short time periods or is of relatively poor quality.[6]

It is not always the case, however, that rainfall information by itself is of immediate, direct, or prime use to policymakers and agricultural planners. Agricultural drought occurs when there is not enough moisture available at the right time for the growth and development of crops. As a result, yields and/or absolute production declines. Many people now realize that the timing of precipitation throughout the growing season is as important as the absolute amount of seasonal or monthly precipitation. What appears to have been adequate seasonal rainfall (in terms of amounts) may have been poorly distributed throughout the season (see, for example, Palutikof, Farmer & Wigley, 1982). Crops have varying moisture needs throughout their growth and development cycles, and thus the timing of rainfall is crucial in rainfed agricultural regions in determining whether there will be a good harvest or a poor one. Dennett, Elston & Rodgers (1985) have recently shown that there has been a change in the seasonal distribution of Sahelian rains, primarily a reduction in August rainfall. Such a change is detrimental to agricultural development plans but can only be detected (as a trend) in retrospect. Also, different crops require different amounts of moisture. Thus, many argue, drought must be defined in terms of the water requirements of specific crops.

Hydrologic drought has been defined as one in which streamflow falls below some predetermined level (Dracup, Lee & Paulson, 1980). Most often, that level is defined in terms of a reduction in streamflow that interrupts the successful undertaking of human activities, such as irrigated farming. In the West African context, an added dimension to the impact of such a drought would be the inability to cultivate land along the edge of a river or a body of water, i.e. flood recession farming, during periods of extremely low flow.

When discussions of drought take place, there is often confusion between meteorological and agricultural droughts. There has been a growing number of references to the '17-year drought situation' in sub-Saharan Africa (Nicholson, 1983; Kerr, 1985; Winstanley, 1985). Lamb's index, shown earlier, suggests that meteorological drought has occurred in the West African Sahel for this length of time. Winstanley (1985) suggests that, based on *his* interpretation of yet unpublished data, 'the current 17-year drought in sub-Saharan Africa has a 1 in 125 000 probability of occurrence'. Aside from the issue of whether the actual meteorological situation can accurately be called a 17-year drought or whether such a probability statement can be justified, the agricultural production situation does not necessarily reflect the same adverse conditions for these 17 years. In 1974, for example, observers were convinced that the Sahelian drought had 'broken'. Meteorological drought (or at least its devastating impacts) was absent in 1975 and, to some observers, in 1976 as well. Nicholson (1983, 1946) noted that

> Rainfall in 1974 and 1975 was still 15–20% below normal . . . but compared to the previous years [1968–73], conditions had dramatically improved and drought was generally presumed to have ended. The apparent return of 'normal' economic and human conditions supported this assumption.

On a local scale, the spatial variability of rainfall was (as usual) quite large. Campbell (1977, 178) noted that, for example, 'Tahoua (Niger) had below average precipitation in 1969, 1971, and 1972 and above average precipitation in 1968 and 1970. A few hundred kilometers away, however, at Ingall (Niger), a run of dry years began in 1968'.

More recently, the 1984 USDA situation report for sub-Saharan Africa referred to increases in agricultural production in the early 1980s, despite meteorological drought. It mentioned, for example, that in 1983 'only marginal areas of Mali were affected by drought, so national cereal production increased slightly . . . ' (USDA, 1984, 2). More generally, the 1984 situation report noted that 'The drought of 1983 followed two years of average to above-average harvests, which contribute to the economic

gains made by most of these [Sahelian] countries during 1982 and 1983'
(USDA, 1984, 2).

A similar situation occurred in southern Africa, as noted by Bratton (this
volume) who, commenting on the resourcefulness of peasant farmers in
Zimbabwe during their worst drought this century, wrote the following:

> How, in the face of drought, can we account for the fact that
> peasant farmers grew and sold more maize in the first five years
> of independence? What factors were at work to counteract the
> failure of the rains? We recognize that surplus production
> emanated principally from the areas of highest potential where
> drought was least severe. Nonetheless, expansion of food crop
> output is most unusual under drought conditions.

Thus, although there have been interannual and intraseasonal reductions
in regional rainfall, such reductions did not necessarily translate into reduced
agricultural production for each drought year in various locations.

Perceptions about drought and development

It is not necessary to search long and hard for reasons why one might feel
uneasy about how drought in sub-Saharan Africa is perceived, as it relates
to development prospects. All too often the pervasive impacts of drought
as a recurrent phenomenon receive too little sustained attention. For
example, a recent issue of *Time Magazine* (*Time*, 1984), devoted to Africa,
identified on its cover Africa's woes as coups, corruption, and conflict.
Only a few sentences in that issue mentioned drought. Yet, that issue
appeared in mid-January 1984, a time when tens of thousands of Ethiopians,
among other Africans, were dying from drought-related food shortages.

Recurrent drought has also been ignored by development planners. A
vice-president of the World Bank commented that

> Despite all our achievements, I think it is fair to say we have
> failed in Africa along with everybody else. . . . We have not fully
> understood the problems. We have not always designed our
> projects to fit the agroclimatic conditions of Africa, and the social,
> cultural, and political framework of the African countries. (*Walsh,
> 1984, 22*)

Bradford Morse (this volume) also called attention to this problem, when
he commented: 'I quite agree with those who assert that national decision
makers and development practitioners have paid too little attention to
drought in the past'.

Perhaps one of the most alarming examples (to this author) of how drought is perceived by those who have an influence on development planning in the Third World is a 1981 report prepared by the US Department of Agriculture (USDA, 1981) for the US Agency for International Development discussing factors that affect the food balance situation in sub-Saharan Africa. Figure 3.3 depicts the way in which those who prepared this particular report viewed the role of weather (and, therefore, drought) in food production, an activity in which a large majority of the population of sub-Saharan Africa engages.

This figure suggests that the most important and direct impact of weather is on crop yields. Yet, weather (especially drought) also affects to varying degrees several other factors shown on the schematic, such as migration, labor supply (urban and rural), acreage planted, on-farm grain storage, home consumption, food imports, export crop acreage (in some instances), and land quality. While the importance of the direct and indirect impacts of drought on the food-balance situation will vary greatly from one location

Fig. 3.3 Interaction among food balance factors, sub-Saharan Africa (USDA).

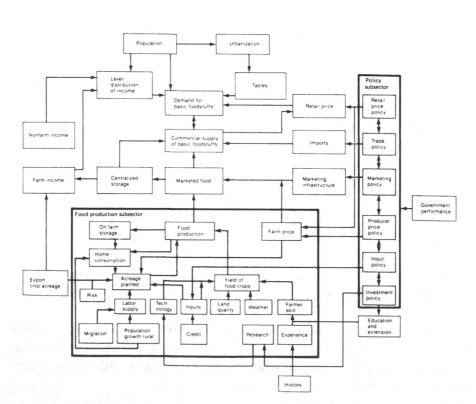

to another and from one time to another, and while drought is but one factor among many others that require consideration, its importance must not be underestimated. Some of the factors that have not been well represented in the chart or in the supporting text of the USDA report are briefly discussed in the following paragraphs. As these are not mutually exclusive factors, the discussions within the following sections overlap.

Land quality

In the years before the outbreak of the Sahelian drought in the late 1960s, when rainfall in the region was more abundant, wetter prevailing conditions masked the adverse effects on land quality of inappropriate land use practices. Wetter conditions also prompted governments and farmers to encroach on and to cultivate parts of the seemingly seldom-used rangelands. In addition to depriving the pastoralists of their sorely needed dry- as well as wet-season pastures, this action increased the vulnerability of farmers, since these former rangelands were in the long run unable to sustain agriculture, as a direct result of an inevitable return to drier conditions in that semiarid or arid region.

With the onset of the multiyear drought in the late 1960s, many of the impacts of inappropriate land use practices became visible. The slow process of environmental degradation became accelerated by prevailing drought conditions. Drought not only exposes and accelerates existing land quality problems, it can also initiate new ones. The cultivation of lands subject to a high degree of rainfall variability makes the land extremely susceptible to wind erosion (and desertification) during prolonged drought episodes, as the bare soils lack the density of the vegetative cover necessary to minimize the effects of aeolian processes. Desertification has become a catch-all term for environmental degradation in arid and semiarid lands. It can result from natural or human causes, or a combination of the two. The term now encompasses soil erosion due to wind and water, soil compaction due to trampling, firewood gathering, reduced fallow time, salinization and waterlogging, and so forth.

As the fertility of the land and crop yields decline, farmers (and their governments) search for new land to cultivate. Assuming that the best rainfed agricultural land is already in production, farmers are forced to cultivate lands considered increasingly marginal from the standpoint of soil quality, terrain slope, and rainfall (see Glantz *et al.*, 1986). Thus, newly cultivated lands are high risk areas in the long run for rainfed agricultural activities: high risk from the standpoint of soil fertility and reduced fallow time and from the standpoint of their susceptibility to the adverse effects of prolonged droughts. The relationship among drought, environmental deterioration, and agricultural stagnation is too often ignored in the formulation of agricultural development strategies.

Acreage planted

In most years, farmers are prepared to make several attempts at planting. If rain does not fall within a certain time following the sowing of seed, the seeds die or germinate and then die. Each time the farmers reseed, they draw down their grain reserves. Successive planting failures reduce the reserves and seed becomes scarce. The farmer must then decide whether to plant the remaining seed or to use it for consumption. Eventually, there is nothing left for planting and little for consumption.

The farmer is then obliged to borrow, offering his labor or perhaps a portion of a future harvest as payment for the loan. In the event of a prolonged drought, the farmer may drift to the urban centers in search of work or the entire family may abandon their land in search of emergency food supplies at famine relief centers. Those who remain to farm the land in case the rains return are operating under conditions where there will be fewer and weaker family members to till the land and where most, if not all, of the necessary draft animals might have perished (McCann, this volume). Clearly, drought conditions affect the amount of planted acreage.

The amount of acreage planted to food crops is also affected by land quality. The poorer the land, the lower the yields and, therefore, the more land that must be brought into production to maintain a certain required absolute level of production. To maintain the same level of production, there is considerable pressure to bring even more land into production, forcing farmers to cultivate climatically marginal lands. As a result of the movement into these areas, human activities become vulnerable to an increase in the probability that a drought will affect their activities, not because the climate of the newly cultivated area will have changed but because people will have undertaken activities more appropriate to higher rainfall areas.

Export crop acreage

One activity generally considered to be insulated from the impacts of drought is the production of agricultural commodities for export. Governments usually give these commodities favored treatment, allowing them to be grown in the relatively better-watered and most-favored agricultural areas. They receive costly but necessary inputs such as better seed varieties, technology, water resources (including irrigation and well-drilling technology), fertilizer, and so forth. They earn foreign exchange and, as a result, are perceived to be essential for long-term development. They are considered so important that the export of cash crops continued or increased during the droughts and famines in the Sahel and Ethiopia in the early 1970s and 1980s (see Lofchie, 1975; Shepherd, 1975; Hancock, 1985), while the production of food crops declined sharply. Much criticism has been leveled against this practice, especially in the midst of famine.

In many instances the record shows that foreign exchange earned from agricultural exports does not necessarily feed back into the development process and, more particularly, into agricultural development. It is often diverted to support other non-development-oriented government programs, such as the military, or it is used to subsidize the food demands of urban populations by importing non-traditional food supplies for their consumption.

Some studies do suggest, however, that prolonged drought can affect cash crop production. de Wilde (1984, 8), for example, notes that 'Widespread, repeated droughts may reverse, for some time, a trend toward the growing of cash crops, particularly if no effective steps have been taken to provide food relief, even if the prices of cash crops appear attractive'. Campbell (1977, 130) too, has noted that

> . . . as food production decreases so the amount available at the market declines more rapidly due to the increasing proportion of the harvest used for consumption at the farm. In the event of a prolonged drought the amount of food available at the market would decline and its price rise and farmers might attempt to sow food crops . . . in place of cash crops, or increase their consumption of edible cash crops.

Thus, even cash crops grown in relatively fertile and better-watered areas might not be immune from the indirect effects of prolonged droughts.

Migration, labor supply and urbanization

As climate variability (or, more specifically, drought) is a common feature of arid and semiarid parts of sub-Saharan Africa, so too is migration. Migration has been resorted to as a mechanism for survival for centuries (Baier, 1980) by regional inhabitants, herders and farmers alike, on a seasonal basis (Rempel, 1981, 210–14) as well as during prolonged drought episodes (Caldwell, 1975).

Baier (1980, 139) noted, for example, that during the 1910–14 drought in this region, 'migration to the south was an age-old reaction to climatic downturn'. Migration tactics and strategies depend to a great extent on the duration and intensity of meteorological drought and on the degree of its impacts on food production processes.

With the advent of a major drought, men often leave their villages in search of income-producing labor. As Campbell (1977, 177) noted, about the 1968–73 Sahelian drought,

> A large proportion of the strategies for dealing with drought . . . involve the movement of people away from their homes in search of means for minimizing the possibility of famine arising as a consequence of drought. For example, once the Hausa realized that a food shortage was imminent, greater numbers of men began to move in search of employment, and later, as the problems became more severe, whole families abandoned their homes and moved to other areas in search of sustenance.

This robs the rural areas of some of its labor supply, placing the total burden of agricultural activities on the shoulders of the women, children, and the elderly. Drought, therefore, accelerates a process already underway, even in normal rainfall years, when some segments of the population migrate from rural to urban areas seeking wage labor to repay money borrowed for seed and, perhaps, food supplies required in order to get through the 'seasonal hunger' that precedes annual harvests (Watts, 1983).

In addition, as individuals and households migrate from their villages in search of food, an increasing burden falls on those agricultural areas less directly (or less harshly) affected by drought conditions. With respect to the recent situation in Ethiopia, its Relief and Rehabilitation Commission (RRC) noted that:

> In recent years, particularly since 1980, prolonged rain failures (mostly drought but also inappropriate and untimely precipitation) have caused widespread crop and pasture failures. Over the same period, cultivators and pastoralists. . .have been forced to migrate in masses in search of relief and resettlement. (RRC, 1984, 7)

Urban centers in sub-Saharan Africa are relatively small, compared to cities in Asia and Latin America. Yet, their growth rates have been quite high, even in the absence of droughts. As noted in a recent report on Africa: 'The rate of urban growth, at 5 to 7 percent per year (2.7% growth rate results in a doubling of population in about 10 years), is the highest in the world. . . . Immigrants to African cities often come from impoverished rural areas and their movements add to the destabilizing effects of rapid urbanization' (USOTA, 1984, 19). During extended drought, existing migration from rural to urban areas becomes accelerated (Campbell, 1977,

180; Caldwell, 1975, 23–31). Many who go to the cities in either their own or neighboring countries do not return to the rural areas. As urban populations expand in pulses during drought episodes, the rural areas become increasingly stressed.

> There is at present in Ethiopia a catastrophic displacement of population which is putting enormous pressure on agriculturally favoured areas of the country . . . the burden of feeding the affected population is being undertaken by local people (some of whom may themselves be drought affected) and by the RRC.
> (*RRC, 1984, p. 7*)

In sum, many have written about the rural-urban conflict over scarce resources within developing countries and have criticized the bias of national governments toward the urban populations at the expense of those in the rural areas (Lipton, 1977). Drought episodes serve not only to highlight such disparities but also exacerbate them.

Food imports

During extended droughts, food imports, either as aid or as trade, are required to sustain the nutritional needs of urban and rural populations. Even when food imports are primarily geared to the needs of the urban populations, drought also comes into play because, during such periods, there are influxes of migrants into urban areas in search of food or employment. Increased rates or urbanization not only deprive the rural areas of agricultural producers, as noted earlier, but increase the need for imported food for the expanding urban populations.

Because urban areas are the centers of power for the ruling elites, they are supplied with inexpensive, often subsidized, food either drawn from the rural areas or imported from abroad. The artificially low prices for agricultural products act as a disincentive to produce greater quantities of food and discriminate against the rural population.

Food imports for urban populations are usually easily prepared convenience foods such as wheat and rice, purchased with scarce foreign exchange. These inexpensive, subsidized food imports also alter the tastes and, subsequently, the food preferences of the urban population away from the traditional crops grown in the countryside, such as millet, sorghum and cassava. As noted earlier, the urban population is constantly expanding as a result of the influx of migrants from rural areas. As the urban population swells and individuals develop a preference for, and a dependence on, wheat and rice, the situation takes on the appearance of a vicious circle. In response to this particular situation, some governments have resorted to import substitution by growing wheat and rice in their own country. To do so is not without cost and sacrifice in many cases, however, as it requires the displacement of traditional crops from certain cultivated areas, as well as costly inputs of fertilizers and irrigation equipment.

Rural poverty

Rural poverty exists in sub-Saharan Africa, even in years of rainfall favorable for agriculture and livestock raising. In the villages there are both rich and poor farmers. Frequently, the poorer farmers borrow from those more well-to-do, repaying them with labor or with a share of the crop they produced. During times of drought stress, the disparities between the rich and the poor increase, as the former are in a position to pay low prices for livestock as well as exact high interest for grain they lend. More often than not the relatively richer peasants are in a position of having grain in reserve that they can sell in the marketplace when grain prices rise sharply, as they do during drought periods. Some of the poorer peasant farmers might eventually sell their land for food and become part of a migrant labor force of landless peasants (see, for example, Colvin *et al.*, 1981; Chambers *et al.*, 1981).

Conclusions

These examples identify some ways that drought affects components of the food balance system and, hence, the development process. They are by no means exhaustive. Clearly, many publications now exist that discuss how drought has affected a local community, the environment, water resources, livestock, a sector of an economy, international trade, foreign assistance, and so forth. These publications constitute a large body of case-specific information, not only for sub-Saharan Africa but for other developing and developed countries as well. Yet, the literature related to economic development has focused little attention on drought (a recurring, aperiodic phenomenon) as a constraint to economic development.

To gain a proper understanding of why societies that are partially or wholly in the arid and semiarid tropics and sub-tropics have had great difficulty in developing their economies, one must consider, along with other relevant factors, the implications of recurrent prolonged meteorological, agricultural, and hydrologic drought. Rephrasing the adage that war is too important to be left to the generals, I would assert that drought considerations are too important to be left only to the meteorologists.

This chapter does not claim that all constraints on development in African societies are the result of droughts (or, more broadly, climate variability). It does claim, however, that drought must be integrated into the already lengthy list of natural and societal influences that can adversely affect the economic development in sub-Saharan Africa.

Notes

1. Drought is not a problem for developing countries alone. Coping with recurrent drought episodes has also been a formidable task for leaders of industrialized countries, as was the case, for example, in the 1930s and 1950s in the US Great Plains and in the mid-1950s in Soviet Central Asia (Brezhnev, 1978). There is considerable speculation among scientists today about whether and how society in the United States could cope with a return of a 1930s-like US drought in, say, the 1990s (e.g. Bernard, 1980; Bowden *et al.*, 1981; Warrick, 1984).

2. In fact, the belief that the climate of the tropics is a primary factor in underdevelopment and the climate of the temperate regions is a factor in industrialization persists today, not only among 'northerners' but among some 'southerners' as well. For example, Bandyopadhyaya (1983, vi) wrote that

   In India and other tropical countries I have noted farmers, industrial labourers, and in fact all kinds of manual and office workers working in slow rhythm with long and frequent rest pauses. But in the temperate zone I have noticed the same classes of people working in quick rhythm with great vigour and energy, and with very few rest pauses. I have known from personal experience and the experience of other tropical peoples in the temperate zone that this spectacular difference in working energy and efficiency could not be due entirely or even mainly to different levels of nutrition. I had no doubt at all in my mind that the principal interpretation lay in the differences in temperature and humidity between the two climatic zones.

3. Streeten (1976) referred to these boundary conditions as initial conditions. By initial conditions he meant those conditions that the migrants (in the North American case, the immigrants) discovered upon arrival; for the developing countries he apparently referred to the existing conditions with which they had to cope at the time of independence. He noted the reluctance of writers 'to admit the vast differences in initial conditions with which today's poor countries are faced compared with the pre-industrial phase of the more advanced countries' (1976, p. 9).

4. Although Kamarck referred only in passing to the adverse impacts of the Sahelian drought of the early 1970s, he was one of the first to draw attention to drought as having an adverse effect on the development process. In reference to drought in northeast Brazil he noted that 'Although these droughts [in Brazil] are a major obstacle – perhaps the biggest one to economic development – development approaches heretofore have tended to invest mainly in fixed capital of various kinds rather than in necessary research to find out how best to handle the droughts' (1976, 20).

5. Torry's (1986) reference to a distinction between underlying (ultimate) and catalytic (proximate) causes of famine are of direct relevance to discussions of drought and its societal impacts. As Torry (1986, 8) noted,

   Proximate causes are situational and originate shortly prior to or during an emergency. Ultimate causes can be construed as predisposing conditions transforming proximate causes into famine distresses. . . . In fact

proximate causes (e.g. drought) can land a household in the clutches of
famine with or without the involvement of ultimate causes.
However, while a specific drought may be considered a proximate cause of famine,
droughts as recurrent phenomena can be considered an underlying cause.

6. In the past most African countries have put a relatively low value on their meteorological
services. Such services were considered of value mostly for aviation, not only during
colonial times but during post-colonial periods as well. Recently, there has been a
growing interest in, and support for, many of the national and regional (e.g.,
AGRHYMET in Niamey, Niger) meteorological services in sub-Saharan Africa. Some
of these services have become more active in assisting decisionmakers by identifying
how meteorological information might improve the value of their decisions.
Meteorological services are increasingly being called upon to provide input into
decisionmaking processes regarding agricultural development and into early famine-
warning systems.

References

Baier, S. (1980). *An Economic History of Central Niger*. Oxford: Oxford University Press.

Bandyopadhyaya, J. (1983). *Climate and World Order: An Inquiry into the Natural Cause of
Underdevelopment*. New Delhi: South Asian Publishers.

Bernard, H. W., Jr (1980). *The Greenhouse Effect*. Cambridge: Ballinger Publishing Co.

Biswas, A. K. (1984). Climate and development. In *Climate and Development*, ed. A. K.
Biswas, Natural Resources and the Environment Series, Vol. 13. Dublin: Tycooly
International Publishing.

Bowden, M. J., Kates, R. W., Kay, P. A., Riebsame, W. E., Warrick, R. A., Johnson, D.
L., Gould, H. A. & Weiner, D. (1981). The effect of climate fluctuations on human
populations: two hypotheses. In *Climate and History: Studies in Past Climates and their
Impact on Man*, ed. T. M. L. Wigley, M. J. Ingram & G. Farmer, pp. 479–513.
Cambridge University Press.

Brezhnev, L. (1978). *The Virgin Lands*. Moscow: Progress Publishers.

Brooks, C. E. P. (1926). *Climate through the Ages*. Second revised edition, published in 1970.
New York: Dover Publications, Inc.

Brown, L. (1985). 'Human element' not drought causes famine. *U.S. News and World Report*
(February 25), 71–2.

Bush, R. (1985). Drought and famines. *Review of African Political Economy*, Vol. 33, pp.
59–64.

Caldwell, J. (1975). *The Sahelian Drought and its Demographic Implications*. Overseas Liaison
Committee Paper 8. Washington: American Council on Education.

Campbell, D. J. (1977). Strategies for coping with drought in the Sahel: A study of recent
population movements in the department of Maradi, Niger. Ph.D. dissertation, Clark
University, 270 p.

Chambers, R., Longhurst, R. & Pacey, A. eds (1981). *Seasonal Dimensions to Rural Poverty*.
Totowa, New Jersey: Allenheld, Osmun & Co.

Colvin, L. G., Ba, C., Barry, B., Faye, J., Hamer, A., Soumah, M. & Sow, F., eds (1981).
The Uprooted of the Western Sahel: Migrants' Quest for Cash in the Senegambia. New
York: Praeger.

Dennett, M. D., Elston, J. & Rodgers, J. A. (1985). A reappraisal of rainfall trends in the
Sahel. *Journal of Climatology*, 5, 353–61.

de Wilde, J. C. (1984). *Agriculture, Marketing, and Pricing in Sub-Saharan Africa.* Los Angeles: University of California African Studies Center and African Studies Association.

Dracup, J. A., Lee, K. S. & Paulson, E. G., Jr. (1980). On the definition of droughts. *Water Resources Research*, 16(2), 297–302.

Galbraith, J. K. (1951). Conditions for economic change in underdeveloped countries. *American Journal of Farm Economics*, 33, 689–96.

Glantz, M. H. (1977). Climate and weather modification in and around arid lands in Africa. In *Desertification: Environmental Degradation in and around Arid Lands*, ed. M. H. Glantz, pp. 307–31. Boulder, Co: Westview Press.

Glantz, M. H., Katz, R. W., Magalhaes, A. R. & Ogallo, L. (1986). Drought follows the plow. Boulder, Co: National Center for Atmospheric Research, draft mimeo.

Hancock, G. (1985). *Ethiopia: The Challenge of Hunger.* London: Victor Gollancz Ltd.

Huntington, E. (1915). *Civilization and Climate*, reprinted 1971. Hamden, Connecticut: The Shoe String Press.

Kamarck, A. M. (1976). *The Tropics and Economic Development: A Provocative Inquiry into the Poverty of Nations.* Baltimore: The Johns Hopkins University Press, for The World Bank.

Kerr, R. A. (1985). Fifteen years of African drought. *Science*, 227, 1453–4.

Lamb, P. (1982). Persistence of Subsaharan drought. *Nature*, 299, 46–8.

Lamb, P. (1983). Subsaharan Rainfall update for 1982: continued drought. *Journal of Climatology*, 3, 419–22.

Lamb, P. J. (1985). Rainfall in Subsaharan West Africa during 1941–83. *Zeitschrift für Gletscherkunde und Glazialgeologie*, 21, 131–9.

Lewis, W. A. (1955). *Theory of Economic Growth.* London: Allen & Unwin.

Lee, D. H. K. (1957). *Climate and Economic Development in the Tropics.* New York: Harper & Brothers.

Lipton, M. (1977). *Why Poor People Stay Poor: Urban Bias in World Development.* London: Temple Smith.

Lofchie, M. F. (1975). Political and economic origins of African hunger. *Journal of Modern African Studies*, XIII(4), 551–67.

Markham, S. F. (1944). *Climate and Energy of Nations.* London: Oxford University Press.

Mills, C. A. (1963). *World Power Amid Shifting Climate.* Boston: The Christopher Publishing House.

Morentz, J. W. (1980). Communications in the Sahel drought: comparing the mass media with other channels of international communication. In *Disasters and the Mass Media*, Proceedings of the Committee on Disasters and the Mass Media Workshop, February 1979. Washington: National Academy of Sciences.

Myrdal, G. (1968). *Asian Drama: An Inquiry into the Poverty of Nations*, Vol. III, pp. 2121–38. New York: Pantheon.

Nicholson, S. E. (1983). Sub-Saharan rainfall in the years 1976–80: evidence of continued drought. *Monthly Weather Review*, 111(Aug), 1646–54.

Oury, B. (1969). Weather and economic development. *Finance and Development*, 6, 24–9.

Palmer, W. C. (1965). Meteorological drought. *Research Paper No. 45.* Wahington: US Weather Bureau.

Palutikof, J. P., Farmer, G. & Wigley, T. M. L. (1982). Strategies for the amelioration of agricultural drought in Africa. In *Proceedings of the Technical Conference on Climate: Africa*, 25–30 January 1982. Geneva: WMO Secretariat.

Rempel, H. (1981). Seasonal outmigration and rural poverty. In *Seasonal Dimensions to Rural Poverty*, ed. R. Chambers, R. Longhurst & A. Pacey, pp. 210–14. Totawa, NJ : Allenheld, Osmun Publishers.

RRC (Relief and Rehabilitation Commission, Ethiopia) (1984). *Project Proposal for the Strengthening of the Food and Nutrition Surveillance Programme.* Addis Ababa: RRC.

Sen, A. (1981). *Poverty and Famines: An Essay on Entitlement and Deprivation*. Oxford: Oxford University Press.

Shepherd, J. (1975). *The Politics of Starvation*. Washington: Carnegie Endowment for International Peace.

Streeten, P. (1976). Foreword. In *The Tropics and Economic Development: A Provocative Inquiry into the Poverty of Nations*, A. M. Kamarck, pp. ix–xii. Baltimore: The Johns Hopkins University Press, for The World Bank.

Tannehill, I. R. (1947). *Drought: Its Causes and Effects*. Princeton: Princeton University Press.

Time (1984). Africa's woes: coups, conflict, corruption. Vol. 123, No. 3, January 16.

Torry, W. I. (1984). Social science research on famine: A critical evaluation. *Human Ecology*, **12**(3), 227–52.

Torry, W. I. (1986). Economic development, drought and famines: some limitations of dependency explanations. *GeoJournal*, **12**, 5–8.

Tosi, J. (1975). Some relationships of climate to economic development in the tropics. In *The Use of Ecological Guidelines for Development in the American Tropics*, New Series No. 31. Morges, Switzerland: International Union for Conservation of Nature.

US Department of Agriculture (1981). *Food Problems and Prospects in Sub-Saharan Africa*. Washington: Government Printing Office.

US Department of Agriculture (1984). *Sub-Saharan Africa: Outlook and Situation Report*. Washington: Economic Research Service, USDA.

US Office of Technology Assessment (1984). *Africa Tomorrow: Issues in Technology, Agriculture, and U.S. Foreign Aid*. Washington: Office of Technology Assessment.

Walsh, J. (1984). Hunger in West Africa: A crisis in development. *Technology Review*, **87**(6), 22–3.

Warrick, R. A. (1984). The possible impacts on wheat production of a recurrence of the 1930s drought in the US Great Plains. *Climatic Change*, **6**(1), 5–26.

Watts, M. (1983). *Silent Violence: Food, Famine and Peasantry in Northern Nigeria*. Berkeley: University of California Press.

Wilhite, D. A. & Glantz, M. H. (1985). Understanding the drought phenomenon: the role of definitions. *Water International*, **10**, 111–20.

Winstanley, D. (1985). Africa in drought: a change of climate? *Weatherwise*, **38**, 75–81.

World Meteorological Organization (1979). *Proceedings of the World Climate Conference*, 12–23 February. Geneva: WMO.

4

African pastoralism and poverty: some implications for drought and famine

MICHAEL M. HOROWITZ and PETER D. LITTLE
State University of New York at Binghamton and
Institute for Development Anthropology

In this chapter, we examine the complex relationship between pastoral poverty and ecology, with emphasis on its implications for drought and famine. We neither challenge the claim that African rangelands are being degraded nor examine the evidence in any detail. We do note a rising sense – if not consensus – among environmentalists and ecologists that the identification of long-term secular degradation on rangelands is not easy, and there are very few areas where longitudinal data sets of sufficient reliability exist to support either the claim of desertification or, what is more important for policy, the contribution of pastoral production systems to its creation. Blaikie, in his recent examination of the political economy of environmental degradation, notes considerable uncertainty around the concept of erosion itself:

> [Uncertainty] arises from the difficulty of obtaining accurate and
> widespread measurement of environmental deterioration through
> a long enough time period to indicate trends. Although there
> may have been many recent improvements in measurement and
> monitoring, long established and reliable data sets are few and
> far between. Secondly, it is often difficult to single out the effect
> of humans on soil erosion and sedimentation rates, from other
> effects such as climatic change, and ongoing 'natural' erosion
> processes. (*Blaikie, 1984, 1*)

Blaikie is echoing the conclusions reached by Warren & Maizels (1977, 1) at the 1977 UN Conference on Desertification who noted that 'only over periods greater than a decade can desertification be clearly distinguished from the less lasting effects of drought'. On what appears to be the other hand, a recent World Bank document insists that low rainfall, even over an extended period, does not alone affect the long-term biological viability of a habitat:

> . . . droughts, no matter how severe, are ephemeral occurrences; when the rains return, the land's inherent productivity is fully restored. If desertification occurs, however, a return to normal rainfall can never fully restore the land's productivity. If the desertification is severe, the land may remain unproductive for many human generations, unless costly remedial measures are taken. While drought can trigger rapid desertification and can make its effects more keenly felt by those living in the affected area, most scientists agree that changes in climate are not responsible for the vast areas of semi-arid land going out of production each year. (*Kirchner* et al., *1985, 73*)

During the past decade – the years since the 1968–74 Sahelian drought largely set the agenda for donor activities in the African livestock 'sector' – the simple assertion of herder-induced degradation, often bracketed to the assurance that grazing replaces 'palatable perennials with unpalatable annuals', has been sufficient to elaborate projects calling for fundamental changes in pastoral practice. These projects are remarkable in their almost universal lack of success. Despite some half-billion plus dollars devoted to the sector (Eicher, unpublished paper presented to World Bank conference 25 February – 1 March 1985, Bellagio, Italy), there is a monotony of evaluations attesting to their failure (AID, 1985a, b; IBRD, 1985). Productivity is not increased; producer income and 'quality of life' are not improved; anticipated financial rates of return are not achieved; and the retardation or reversal of environmental decline is not demonstrated. It is clear that more rigorous standards of proof need be applied to the claims made in these projects, especially where the planned interventions involve sedentarization, stock reduction, privatization of pasture, and other actions that are interpreted by herders as threatening their very ability to reproduce their households, herds and ways of life.

In this chapter we examine some of the conditions under which African pastoralism operates today, and confront thereby a number of the recurrent claims and assumptions about pastoralism that inform development thinking and interventions. Rather than concentrating exclusively on the microlevel presentation of the pastoral group and its production system, as is customary anthropologically, we shall explore the political-economic context within which pastoral production systems operate, and indicate the relevance of political-economic analysis for understanding the process of desertification, its impacts on local communities and regional systems, and some of the more fruitful directions to be considered in its mitigation.

Marginalization and social differentiation

A political-economic analysis illuminates the twin processes of increasing **marginalization** and **social differentiation** that characterize societies dependent on pastoral production systems in Africa's arid and semiarid rangelands. These processes affect other African peasants, of course, but are exceptionally intense among pastoralists. Marginalization here refers to the compaction of ruminant herding in areas of low biological productivity, usually areas not yet experiencing agriculture. In other words, pastoralism retreats as agriculture and commercial ranching expand. Social differentiation refers to growing inequality between pastoralists and other segments of regional and national economies, and among pastoralists themselves. In simple terms, most herders are becoming poorer, despite the relatively high value placed upon meat, while a small number of livestock owners from pastoral communities are becoming rich and powerful and a perhaps larger number of rich and powerful men who are not from pastoral communities are becoming livestock owners. An appreciation of the processes of marginalization and social differentiation is essential to a full understanding of environmental degradation and famine.

Marginalization is fairly straightforward. Pastoralism retreats to areas of low biological productivity in the face of the appropriation of rangelands by other users. Huge areas of East Africa – both rangelands and farmlands – were appropriated by conservationists for game/tourist parks and by European farmers during the colonial period. In both East and West Africa, the removal of land from food-crop production, under the notion of a 'comparative economic advantage' in the production of export crops, forced an expansion of rainfed food cultivation onto former rangelands (Little & Horowitz, 1986). Large-scale irrigated schemes, again mainly in export-crop production, claim important tracts of riverine land previously used as dry-season grazing reserves and for recessional cultivation of cereals (Horowitz & Badi, 1981; Salem-Murdock, 1984). Urban settlements place great demands on the arboreal pasture for fuelwood, essentially depriving herders of a critical component of their dry-season browse. Finally, rangeland areas are diverted from communal to private or highly restricted use, increasing the charge on the pasture that remains available.

The migration of rainfed cultivation onto traditional rangelands – a rapidly expanding cause of pastoral marginalization – is itself a consequence of the impoverishment of agrarian peasantries. Ibrahim describes the process for the western Sudan:

. . . the sound, traditional system of shifting cultivation turned into land misuse, and a chain of processes of deterioration of land productivity was begun: population increase led to excessive cultivation, which, in turn, led to enhanced soil erosion and soil impoverishment. This resulted in the decrease of millet yields per ha in the Sudan by half in the last 15 years. To make amends for this, the population, which is constantly increasing at an annual rate of 2.5%, had to increase the area cultivated with millet, from 392 000 ha in 1960 to 1 055 000 ha in 1975. This expansion of cultivation meant a fresh wave of desertification. . . . The increasing persistence of the inhabitants in tilling the land despite lack of sufficient rainfall proves that they are not able to keep pace with the natural fluctuations any more. Instead of shifting southwards [i.e. away from the pastoral zone] the peasants try to enlarge the area cultivated to be able to exist. This expansion of cultivation to counteract the decrease of rainfall works as a catalyst for the processes of desertification. (*Ibrahim, 1984, 110–18*)

The case of Niger is also instructive. During the colonial regime and continuing through the early independence period, the administration officially imposed a division of the country into a *zone nomade* and a *zone sédentaire*, for herding and for rainfed farming, respectively. Agriculture was discouraged north of the delimitation, and farmers had no legal recourse for crop losses due to animal incursions. In return, herds were not permitted to graze south of the line until the grain harvest was complete. The post-harvest period saw farmers enticing animals onto the cropped fields with gifts to the herders of money, sugar and tea. Farmers were known to dig wells on their fields to attract post-harvest grazing, and to consign a few animals to the care of a particular herder in the hopes that he would lead the herd onto the owner's field. Farmers profited from high-quality manuring, especially in the early weeks following the harvest when the stubble was relatively rich in protein. Such manuring, along with rotational farming and some intercropping (especially leguminous cowpeas with millet), constituted the sole available techniques for restoring soil fertility.

Relationships between farmers and herders were occasionally strained (Horowitz, 1972, 1973, 1975; Horowitz *et al.*, 1983), as when a desiccation of northern pastures forced a southward retreat of herds before the harvest was complete, with resultant conflict between the two groups. In general, however, land use was complementary rather than competitive. The marked increase in export-crop production in the agricultural zone – in Niger this refers mainly to groundnuts – led to a decrease in the surface of the *zone sédentaire* committed to food. The cereals shortfall was supposed to be recouped by earnings from groundnut sales. The marketing of groundnuts, however, was monopolized by a government corporation, SONARA, which, along with a class of large traders, appropriated the bulk of the earnings, with little benefit for small-scale growers.

There were two main consequences of this impoverishment of the peasantry through export-crop production. First, households were unable to reproduce themselves solely in agriculture and large numbers of persons, especially young men whose labor was sorely missed in production, were forced to enter the wage labor economies of towns and of coastal West African countries. While much has been made of remittances earned in wage labor as being, on balance, beneficial to labor-exporting households, there is considerable recent evidence that the migration served mainly to shift survival-maintenance requirements from the household to the individual, without any significant surplus generated to be reinvested at home (Painter, 1985). The second effect was an extension of cereals cultivation northward into the *zone nomade*, traversing what Ibrahim (1978) in the Sudan labels the 'agronomic dry boundary'. Unable to earn enough on their farms to purchase grain, and lacking both the capital and the labor to intensify production, farmers attempted to meet their consumptive requirements by bringing more land under the hoe. These new fields had been rangelands. The extension of farming into the pastoral area seems to cause a progressive deterioration of soil fertility on the newly cultivated lands, with their exceptionally thin top soils subject to rapid aeolian and pluvial erosion; and a further compaction of herding with a concomitant increased charge on pasture resources. As farmers appropriate rangelands, relationships with herders – competing with farmers for the same space – become increasingly antagonistic.

Preference for agriculture over pastoralism

For reasons that are hardly difficult to discern, both states and financing organizations favor agriculture over herding; little, if anything, is done to retard the expansion of cultivation, despite its untoward ecological consequences. First, donors have tended recently to be less willing to invest in the livestock sector, having more (though perhaps not a great deal more) confidence in their technical packages for farmers. That is, apart from veterinary interventions, there is not much persuasive cost-effective technology at hand that will improve productivity among ruminant livestock on arid and semiarid ranges without further assaulting the fertility and stability of the land.

Secondly, with the exceptions of Somalia and Mauritania, African administrators tend to be drawn from ethnic groups whose roots are in farming rather than herding, and whose understandings and sympathies are biased toward sedentary life. Since pastoral groups have few representatives in the halls of power, it is quite a bit easier for governments to turn deaf ears to their needs than to the demands of farmers for more land. Rather than shift from export to food-crop production on existing rainfed and irrigated lands, with a consequent loss of foreign exchange, governments allow the movement of grain cultivation onto pasturelands. A recent examination of desertification in Niger notes:

> The 1977 report noted that the northward movement of cultivated fields, briefly interrupted during the drought, had started up again two years ago. This migration continues today, and we saw an almost unbroken chain of fields along the road from Tahoua to Agadez as far as Abalak, that is some 80 km north of the officially defined limit for rainfed agriculture. Along the Tassara road we saw fields with their earthen granaries 25 km north of Tchin Salatin, which is 100 km above the line. These millet fields in constant expansion seem hardly to interest the Administration. (*Fauck, Bernus, & Peyre de Fabrègues 1983, 51*)

Finally, donors as well as governments share in an 'anti-nomad morality', which provides ready, if false, rationalizations to restrict herding. This morality predates modern development writings by at least 600 years. The fourteenth century historian, Ibn Khaldun, elaborated the basic attack on pastoralism which has remained essentially unchanged to the present. Ibn Khaldun, a great advocate of sedentary life, wrote:

> . . . civilization always collapsed in places where the Arabs took over and conquered, and . . . such settlements were depopulated and the very earth there turned into something that was no longer earth. . . . When the Banu Hilal and the Banu Salaym pushed through from their homeland to Ifriqiyah [Tunisia] and the Maghrib in the beginning of the Fifth century [eleventh CE] and struggled there for 350 years, they attached themselves to the country, and the flat territory in the Maghrib was completely ruined. Formerly, the whole region between the Sudan and the Mediterranean had been settled. This fact is attested by the relics of civilization there. (*Ibn Khaldun, 1967, 304–5*)

These accusations were echoed by E. H. Palmer, writing about the Sinai Peninsula and the Negev in the late nineteenth century:

> wherever [the Bedouin] goes, he brings with him ruin, violence, and neglect. To call him a 'son of the desert' is a misnomer; half the desert owes its existence to him. . . . The soil he owns deteriorates, and his neighbors are either driven away or reduced to beggary by his raids and depredations. If the military authorities were to make more systematic expeditions against these tribes, and take from them every camel and sheep which they possess, they would no longer be able to roam over the deserts, but would be compelled to settle down to agricultural pursuits or starve. . . . They might thus be tamed and turned into useful members of the community. (*Palmer, 1977, 297, 299–300*)

The intellectual tradition of anti-nomadism has found support even among twentieth century anthropologists (Herskovits, 1926; Murdock, 1959; Lomax & Arensberg, 1977), although rarely among those with first-hand field work in a pastoral community. If so many distinguished scholars claim that herders are hell-bent on destroying the environment, it is almost *infra dig* to ask for evidence. Yet it seems clear that in place of data and analysis, one author just quotes another, or simply invokes 'common knowledge'. For example,

> It is *generally agreed* that overstocking and the lack of managed grazing patterns in the Sahel are the most important causes of desertification in the region and that desertification is a symptom of more fundamental problems of rapid population growth and the inability of individuals and communities to adopt *known* land management and conservation technologies. (*Ferguson 1977, 7, unpublished emphasis added*)

Lamprey (1983, 656) states categorically that the evidence pointing to overgrazing as the cause of 'widespread damage to semiarid and arid zone grasslands . . . [is] overwhelming', but neglects to share that overwhelming evidence with the reader. The FAO escalates the argument to an assault on the very character of pastoralists, '. . . caring for nothing, disdaining manual labor, balking at paying taxes, and being unwilling to sell their animals. . . ; they do not make the economic contribution to their countries that is *rightfully* expected of them' (FAO, 1973, 14, emphasis added). More recently FAO (1980, 56) continued to attribute degradation on rangelands to mismanagement by pastoralists. 'It is basically a problem of the misuse of land . . . particularly in pastoral areas, much of the problem results from the customs, value systems and attitudes of the people concerning grazing lands and livestock, together with the lack of government mechanisms for effective control'. Finding the 'cause' of environmental degradation in the attitudes of herders rather than in the conditions under which pastoralism operates may be a temporary comfort, but it is a false comfort since it cannot but lead to inappropriate action.

With such rhetoric defining the discourse on pastoralism, it is little wonder that international donor organizations and African governments can easily dismiss those with more positive views.

Carrying capacity

The mainstream development paradigm today (often, though not invariably) recasts the accusatory language of anti-pastoralism in a more temperate diction, but its import is little different from that of earlier critics: herders destroy the range, and they have got to be stopped for the general as well as for their own good. The modern argument was given form by Garrett Hardin (1968) in his famous paper in *Science* on the 'tragedy of the commons'. It is elegant in its simplicity, resting on two assumptions. The first is that rangelands are 'communally' owned, public property; the second is that livestock are privately owned. This disjunction of ownership generates a situation in which rational action to increase individual benefit conflicts with behaviors beneficial to the larger group. The costs of each additional animal placed on the common range are shared by all users, while the benefits of herd increase pertain exclusively to the individual owner. A follower of Hardin rephrases the position as follows:

A conservation ethic oriented toward the long-term preservation of the ecological resource does not exist among the pastoral cultures of the Sahel . . . [because] the benefits of conservation are delayed and tend to be shared. Thus, an individual herdsman realizes only a fraction of the rewards of his efforts, which he then discounts heavily. His benefits are shared because, since property is held in common, he cannot prevent others from grazing on pastures for which he has limited his own herds. The limitation of his own herds is an immediate cost which he alone bears. Thus, conservation has always been negatively reinforced by the same . . . system that positively reinforces herd maximization behavior. (*Picardi, 1975, 164–5*)

Since each herd owner seeks to maximize his number of animals, charge rapidly exceeds the carrying capacity of the range. At first the effect may be contained within the herd itself, as a given liveweight is distributed among a larger number of animals. But soon the costs are transferred to the range in the form of overgrazing, erosion, declining biological productivity, and desertification. It is a persuasive scenario. Yet it is vulnerable at a number of points, some of which have already been mentioned.

In the first place, there is little evidence that, controlling for environment, private ranges are better managed than public ones. The Borana material (Cossins, 1985), which is presented in some detail below, argues to the contrary. Sandford (1983, 119–20) notes that 'overstocking and environmental deterioration appear to be just as common and serious in areas of rangeland where, as in parts of the USA and Australia, both land and livestock are individually owned'.

Secondly, the very notion of 'ownership' is left unclarified. Almost all of the anthropological literature on pastoralism points to restrictions on both access and use of 'communal' ranges, and widespread claims on the use and disposition of 'private' livestock. In other words, pastoralists, no less than anyone else, live in communities, and these communities have moral bases which do not allow for unchecked personal aggrandizement at the expense of one's fellows. To be sure, the transition from more egalitarian to more differentiated communities is clearly associated with a decline in the commitment to general morality (cf. Taussig, 1980); but this transition is normally accompanied by the very privatization of public resources which Hardin and others, who accept the logic of the tragedy-of-the-commons argument, favor.

An instructive example can be seen in the AID-financed Lesotho Grazing Lands Management Project. An unpublished evaluation carried out in 1985 notes that the project area was removed from common use, and given to a newly formed grazing association, the members of which (affluent herd owners) have exclusive rights to run animals on the range. State power is invoked to keep others, including former users, off the range by impounding trespassing animals. The irony of the situation is that environmental improvement is a project objective. Yet members of the association, affluent and powerful, ignore grazing regulations with impunity: 'by virtue of their stature, [they] may demand exemption . . .'. These upper class herd owners, in other words, enjoying reduced competition on the range from a large number of small herders, are unconstrained by rules designed to maintain productivity and soil fertility.

The notion of 'overgrazing', so central to discussions of desertification of rangelands, requires close consideration, for it assumes that there is some maximum weight of animals that can be sustained on a given range more or less indefinitely. When that gross weight is exceeded, one of several alternative scenarios is supposed to happen. Either a number of animals is removed from the range – through starvation, forced migration, or sale – and the 'proper ratio' between charge and carrying capacity restored; or the capacity of the range to sustain that weight is permanently impaired. The first notion, that assumes a fluctuating number of animals about a fixed carrying capacity, is illustrated by an **equilibrium model**. The second notion, in which the carrying capacity of the range deteriorates, is illustrated by a **degradation model**. A third alternative, illustrated by a **resiliency model** (Hjort, 1981, 173–4), sees both stocking rate and carrying capacity in some kind of association, such that temporary increases in load cause short-term declines in the carrying capacity, but that capacity is restored over time as the charge pressure is reduced.

A great deal of discussion in the Sahel and in other arid and semiarid pastoral regions turns on the appropriateness of the particular model held by the various discussants, without their being made explicit. Anthropologists of the structural school, that was most prominent in the study of pastoralism until a few years ago, tended to impose an equilibrium model on their analyses. American and British range scientists, for the most part, historically identified with the degradation model, and that is the one most commonly invoked in documents regarding livestock sector development. Increasingly, persons with extensive field research in pastoral areas show considerable dissatisfaction with both equilibrium and degradation models, and find the resiliency model to accord better with the empirical situation. Resiliency is implicit in Sandford's exploration of range exploitation strategies, which he terms 'opportunistic' and 'conservative'.

> An *opportunistic* pastoral strategy is defined as one which varies the number of livestock in accordance with the current availability of forage. Such a strategy enables the extra forage available in good years to be converted directly into economic output (milk, meat) or into productive capital in the form of a bigger breeding herd. The economic output may be immediately consumed or it may be exchanged for easily storable wealth, such as money or jewelry, that can be re-exchanged for consumables when needed. In bad years livestock numbers are reduced. In most cases where an opportunistic strategy is actually attempted livestock numbers

in bad times are reduced too little and too late and as a consequence ecological degradation may occur. One can, therefore, distinguish an *efficient* opportunistic strategy as one where livestock numbers are varied at the *appropriate* time. If one thinks in terms of a 'proper use factor' (a range management term for the proportion of forage produced which can safely be consumed by livestock. . .), then an efficient opportunistic strategy is one which ensures that the proper use is adhered to.

A *conservative* strategy is defined as one which maintains a population of grazing animals at a relatively constant level, without overgrazing, through good and bad years alike. A conservative strategy implies that during good years livestock numbers are not allowed to increase to utilize all the additional forage available. (*Sandford, 1982, 62*)

The notion of 'carrying capacity' is appealing, because it provides a quantifiable index, and development planners and range scientists tend understandably to favor conservative pastoral strategies. But how is the 'carrying capacity' to be calculated? A salient feature of arid and semiarid rangelands is the enormous and unpredictable variation in the quantity, frequency, and distribution of rainfall and therefore of terrestrial graze. Holding constant the nutrient profile of the soil, forage production varies almost linearly with rainfall (Diarra & Breman, 1975, cited in Sandford, 1982, 65). Since the amount of forage is as unpredictable as the amount of rainfall, and in some years will be reduced to near zero, the selection of a conservative figure for carrying capacity will inevitably mean that in most years the range could sustain far more animals than actually utilize it. Sandford (1982) refers to this as **understocking**, and asks, over the long run, which strategy is economically sounder: one that habitually overstocks or one that habitually understocks? Livestock development rhetoric tends to favor the latter strategy, rendering it on a collision course with the more opportunistic use of rangelands practiced by African pastoralists. (It is important to note that on ranges like the Sahel, dominated by annual species, unconsumed forage cannot be saved from one year to the next, except, perhaps, by haying.)

Some recent findings by range biologists also question the appropriateness of the degradation model of carrying capacity for designing ecologically sound livestock sector interventions. The ecology team on AID's Niger Range and Livestock Project discovered that, contrary to conventional wisdom, moderate grazing in the early rainy season (which, in fact, is the practice of Nigerien pastoralists) provides for greater residual moisture, a soil moisture reserve, than does the lighter grazing pressure more often recommended by pasture management specialists. 'Increased soil moisture at the end of the rainy season affects the quality of the vegetation by decreasing percent dry matter, extending the 'green feed' period, and by increasing perennial regrowth, thus possibly allowing additional animal gains' (Swift, 1984, 758). The ecologists, led by Robert Bement, himself a rancher, discovered that many pastures are actually *under*utilized, and recommended harvesting those grasses for hay and silage to provide high-protein feed supplements in the dry season.

Can 'overgrazing' improve pasture? There is clear evidence that undergrazing may be detrimental. Conant (1982) documents the invasion and colonization by non-palatable thorny acacias of a rangeland in northern Kenya removed from pastoralism for several years. Others have cited the importance of grazing in reducing the canopy habitat required by *Glossinae* ('tsetse' fly) for their reproduction. A recent paper on gregarious grazing, while specific to wild ruminants, is suggestive for livestock:

> Continual grazing clearly presents a very strong selection pressure, and so this direct, short-term reaction is eventually translated to an evolutionary response, with a shift in the genetic and phenotypic characteristics of the surviving plants. There is, therefore, a coevolution between grazers and their 'prey'. . . : for the plants there is an effect on their morphology, and for the animals, an effect on behaviour. (*Lewin, 1985, 567*)

Lewin does not speculate on the implications for domesticated stock, but they are obvious. Grazing animals, rather than necessarily degrading their environment may, over time, transform it by favoring the reproduction of those species which better respond to their consumptive requirements. By increasing the biomass per unit volume – instead of the usual measure per unit area – gregariousness increases the efficiency of grazing.

> The high biomass concentration in grazed, short grasses is the result of a more densely packed foliage within the canopy volume, which becomes ecologically significant for grazers. The most important property in their food source is energy intake per bite, not the amount of standing biomass. It is possible for a herbivore to starve in the midst of apparent plenty, if the quantity of food culled in each tongue-swing is of a low concentration.
>
> Stobbs calculated that for a cow-size grazer, a bite size of about 0.3 gram of usable nutrients was necessary for survival, a figure

that translates to 0.8 milligrams per milliliter biomass concentration. McNaughton's data from the Serengeti show that vegetation taller than 40 centimeters would be deficient in support of such an animal. An animal grazing on a 10-centimeter greensward would be reaping rich rewards in terms of available energy per bite. Moreover, plants cropped at this level are in a more juvenile state, and therefore offer higher protein content and greater digestibility. (*Lewin, 1985, 568*)

While findings such as these do not negate the theoretical utility of the 'carrying capacity' concept, they should at the least have us regard any specific projections (cf., Lamprey, 1983) of the number/weight of animals that can be sustained on the range with a good deal of skepticism.

The productivity of pastoral production

The evidence to support the notion of low pastoral herd productivity, another assumption that has informed development programs, has been equally flimsy. In fact, recent studies demonstrate that livestock productivity in pastoral areas has been underestimated and that, on many criteria, pastoral systems are actually more productive than commercial ranching enterprises. In southern Ethiopia, for example, Cossins (1985, 10) found that the pastoral 'Borana system is very productive; compared with Australian commercial ranches in a very similar climatic environment, the Borana produce nearly four times as much protein and six times as much food energy from each hectare'. Rather than looking at productivity per animal, which is a measure used in commercial ranching areas, he uses a per land unit measure, which is of greater significance to pastoral producers. In terms of productivity per unit of cost (labor and other), the Borana system also has an advantage over commercial ranches. The amount of US dollar investment required to produce one kilogram of animal protein is 0.14–0.28 for the Borana system, 2.01 for commercial ranches in Laikipia, Kenya, and 1.93–3.89 for Australian ranches. If production of food energy (milk and meat) is used as an indicator, the difference in benefits from pastoral production is even more outstanding.

In a similar analysis, but based on data from Botswana, de Ridder & Wagenaar (1984, 6–7) reveal that the country's pastoral production systems are '95% more productive in terms of liveweight production equivalents than ranching systems on a per hectare basis'. While they do not use as many production indicators as does Cossins, they do show that with regard to one important criterion – kilograms of liveweight per hectare per year –

notions of pastoral inefficiency are unwarranted. What is even more significant about the research is that it raises important questions about optimal levels of stocking:

> Because of uncertainty over stocking rates in traditional systems, a sensitivity analysis was carried out to estimate the stocking rate at which productivity in traditional systems would fall to that of ranching, assuming that other production parameters were constant. This estimated stocking rate was 9.5 ha LSU-1 (one livestock unit per 9.5 hectares), compared with the observed stocking rate in traditional systems of around 6.0 ha LSU-1. (*de Ridder & Wagenaar, 1984, 7*)

The findings have profound implications for programs of controlled stocking which, as indicated earlier in the chapter, tend to favor 'understocking' the range. In the Botswana case, lowering the stocking rate ('increasing the land available per livestock unit') diminishes overall productivity per land unit, as well as reduces the number of pastoralists who can be effectively supported in the area.

The materials about Ethiopia and Botswana are discussed here to illustrate the difficulties of defining an appropriate measure of productivity in pastoral regions. In terms of productivity per land, investment, energy, and animal protein units, the evidence favors pastoralism over commercial ranching; in calculations of productivity per animal, commercial ranching is superior. It is unfortunate that the type of analyses carried out by Cossins and de Ridder & Wagenaar rarely inform the design of pastoral development programs.

The changing context of pastoralism

Several processes that are related to the larger political economy have affected contemporary pastoralism in Africa and complicate discussions of ecology and productivity. They include the centralization of resource-related decision-making by the state; the diversification of pastoral economies, including an expanded involvement of herders in wage employment and in farming; and the privatization of productive resources, such as land and water. Each of these phenomena has contributed to increased pastoral marginalization and greater differentiation at community, regional and state levels. In addition, they have, to some extent, led to the general impoverishment of pastoralists, to the inability of herders to respond to climatic fluctuations without devastating human and livestock losses, and to certain ecological problems. Donor and state efforts to improve the pastoralists' condition have often worsened the situation, and thus, directly contributed to the impoverishment and alienation of herders.

Centralization of decision-making

The state may attempt to centralize resource management decision-making directly, by imposing public land boards or laws – as in the case of Botswana (Devitt, 1982) or in parts of the Sudan (Teitelbaum, 1984) – removing decision-making from the community to the state, or indirectly, by weakening the power base of local institutions and leaders. In the latter case, the presence of competing sources of power at the local level, such as government-appointed leaders who may hold ultimate authority over land-use decisions, diminishes the effectiveness of local institutions to regulate water and range use. In such cases, no attempt is made directly to intervene in pastoral tenure and resource management systems, but the effect on local decision-making can be equally discouraging.

The usurpation of power by governments has been incomplete in most African states, creating for herders what Runge (1981) calls a problem of 'assurance'. In this situation, producers lack confidence in the capacity of either state or local institutions to regulate resource use. In the eastern African context, it is characterized by considerable ambiguity over who has legal access to range and water, and who has the authority to regulate the use of these resources (Little & Brokensha, 1985). Under these conditions, such outsiders as agriculturalists, civil servants, and urban-based businessmen are able to encroach onto pastoral lands, in some cases fencing off areas for private farming or livestock raising (Little, 1985). In other cases, they develop private water sources, which can give them *de facto* control over surrounding range (Peters, 1984) and further constrict the area available to local herders. The ecological implications of these infringements include localized overgrazing around water sources, and soil and wind erosion caused by the cultivation of rangelands (as discussed earlier). In the most extreme cases, as apparently occur in parts of Botswana, uncertainties over rights to resources can lead to a virtual 'free-for-all' assault on the range.

Diversification of pastoral economies

The causes of economic diversification in pastoral areas are complex. As was noted earlier in the chapter, involvement in non-pastoral activities may be a response to poverty, with the lack of livestock assets, as well as the capital to re-build herds after losses, compelling herders to seek wage employment. The extent of herder involvement in wage employment has only recently been recorded, but indications are that it is considerable, with up to 30% of adolescent and adult males of Kenyan pastoral areas engaged in it (Institute of African Studies, 1982). Such employment tends to be unskilled and not financially lucrative, providing only bare subsistence reproduction costs with little surplus for re-investing in livestock. While urban-based employment is most prevalent, impoverished pastoralists have also been observed remaining in rural areas as hired herders for wealthy businessmen and civil servants (Swift, 1984; Little, 1985), or as casual laborers on agricultural development schemes. The concentration of impoverished herders around irrigation schemes in eastern Africa requires substantial imports of famine relief, even for those who have acquired employment or irrigation plots.

Diversification on the part of wealthy pastoralists also has taken place in the last 10–15 years, but it is strikingly different from that described above. The large stockowners, in some cases, have educated their children for higher-paying employment, invested in irrigated agriculture, and/or purchased trading businesses. While for the poorer herders diversification is a question of survival, for the richer stock owners it is an investment strategy, allowing them to spread financial risks and rebuild their herds rapidly after droughts (Little, 1983). Social and economic differentiation accelerates, since wealthy herders and non-pastoralists can purchase devalued stock from poor herders during times of drought, and can use profits from other activities to buy stock after droughts. The irony of the diversification process is that while certain range areas are subject to overexploitation (e.g. in proximity to settlements and irrigation schemes), other areas are actually underutilized due to the transfer of herding labor into other sectors (Hogg, 1984; O'Leary, 1984).

Pastoral investment in agriculture has accelerated in many areas, and again its prevalence defies a simplistic explanation. In some cases, cultivation by herders is related to the inadequacies of grain markets; that is, pastoralists turn to farming because the market no longer is a reliable source of grain (Little, 1983). In other areas, the loss of livestock has forced herders to cultivate, either on a supplemental or full-time basis. (This is, of course, an option only in areas where agriculture is technically feasible.) In still other regions, farming may be a response to insecure tenure conditions where herders cultivate to acquire recognition of land rights (Sandford, 1983). Herders may farm to gain this official acknowledgment (which may be denied to pastoral users), as well as to keep 'migrant' farmers from cultivating and making their own claims.

The diversification process has strong implications for drought, famine and environmental decline. We noted above the effects that loss of herding labor can have on local management practices and on ecology, since the uneven utilization of range may facilitate the encroachment of less palatable plant species (cf., Conant, 1982). Poorer herders are especially affected because often, if labor is scarce, they must herd animals near settlements where vegetation is depleted. Wealthy herders, in contrast, have the domestic labor and/or resources to hire others to take advantage of distant grazing. In times of drought the concentration of stock around settlements is even greater, as impoverished pastoralists send family members out to seek employment and then settle near centers to acquire grain (famine relief). Management and maintenance of water points, activities requiring considerable work and skill, also suffer from the exodus of labor.

Privatization

The loss of grazing areas to state development schemes (including tourist parks), private farmers, and ranchers, has restricted herder options during drought. In the case of Kenya, encroachment on pastoral lands dates to the early colonial period when large tracts of grazing land were handed over to European settlers (Little, 1984). With independence and the ensuing political dominance of agriculturalists, such as the Kikuyu, cultivators were permitted to settle in the higher rainfall, dry season pastoral areas. Further loss of pastoral lands in Kenya is attributed to the expansion of the tourist industry and national parks. Most of Kenya's tourist sector, its second largest earner of foreign exchange, focuses on wildlife resources, which are concentrated in the country's range areas. Indeed, two of the most important wild game areas in Kenya (Amboseli and Maasai Mara) are in Maasai regions. Both of these, particularly the swamp-grazing area of Amboseli, were significant grazing areas for the Maasai. Recent legislation in Kenya, however, transformed Amboseli from a game reserve (which allows some pastoral use of the area) into a national park that excludes pastoralists altogether (Galaty, 1980, 170).

It is during drought years that herders most severely feel the loss of strategic water points and reserve/dry-season grazing areas. Without access to traditional highland and swamp-grazing areas, pastoral groups like the Maasai of Kenya and Tanzania must concentrate their animals on already-depleted lowland-range areas. Under these conditions livestock losses are numerous, with up to 50% of cattle dying during recent droughts in certain districts in Kenya (Little, 1984, 47). Carr vividly describes what the loss of grazing lands has meant to the Dasanetch, whose homeland straddles the Kenya/Ethiopian border:

> In sum, the input of territorial restriction to the system has resulted in a self-perpetuating (or runaway) deterioration within the system, especially along the lines of: (1) environmental breakdown in the plains in the form of reduction of total plant cover, invasion of bush and unpalatable plants, and disruption of natural faunal assemblages, soil erosion, and (2) economic breakdown within the major production activity, stock raising, through reduction of stock vigor, increase in disease and death, and reduced milk yields. These have generated strong pressure on production units to diversify their production activities to horticulture, fishing, gathering and hunting. (*Carr, 1977, 226*)

The environmental problems that are apparent in the Dasanetch region are not caused by pastoral practices, but rather are the result of a political decision that removed much of their land from pastoral use.

The role of the state and of donors

The migration of agriculture, the privatization of range resources, and the impoverishment of pastoralists, processes that have accelerated in recent years, receive both direct and indirect support from the state. These processes are also favored by donor organizations who interpret sedentarization and privatization as 'progressive'. Yet not only have these processes led to greater impoverishment of the majority of herders, they also have had negative environmental and productive impacts.

What types of policy interventions by states and donors would be beneficial to pastoralists and appropriate to ecological realities? What can be done to ameliorate the hardships of drought and famine that increasingly affect herder populations? Policy changes in at least three areas could improve the conditions under which pastoral production takes place.

Land tenure and decentralized management

The first area of improvement has to come from better informed policies regarding land tenure and use. While it is impractical to conceive of herders recovering previous lands, current policy should recognize pastoralist rights to land and promote appropriate institutions to oversee land use and management. In some cases, indigenous organizations, which already exist for these purposes, could be granted greater decision-making powers and legal recognition by the state. In other areas, new institutions, such as herder associations, are required. The work of Swift on the Niger Range and Livestock Project is instructive on this point:

> The policy should seek to allocate real control over rangelands and pastoral water sources to the traditional pastoral users and to create institutions by which discrete areas of range can be administered and managed communally by those users. The steps involved are: (a) the mapping of socio-ecological units covering viable pasture areas used principally by identifiable pastoral groups; (b) the definition, on the basis of kinship and historical presence, of principal rights' holders in these areas, as well as a clear definition of all other pastoral groups which might legitimately claim historical pasture rights in the same areas, and a list of other areas sometimes used by the principal rights' holders; (c) the enactment of a legal text specifying the rights and obligations of these herders' associations. There is no reason why the associations should have exclusive pasture rights in the first place, and the rights of other herders who have traditionally used that area must be safeguarded, perhaps through agreements on reciprocal access. (*Swift, 1982, 174*)

We would be naive to think that the type of power devolution recommended here would not be stringently opposed by most states. The 'localization', to use Thomson's (1985) term, of tenure and management decisions will be seen as a threat to many states. We would argue, however, that until the present ambiguities over land rights and the role of state and local institutions are clarified, most development interventions will be peripheral to the plight of herders. In most cases, the remedy calls for real decentralization on the part of the state.

Marketing policy

Marketing is another area where previous policies have been detrimental to herders, and where donors and states can play a constructive role in the future. Most herder groups are in an especially vulnerable position in the marketplace. Their isolation from urban centers and market infrastructure often means that they pay premium prices for imports (e.g. grain) and receive low prices for cattle and livestock product exports. They often reside in zones that are severely grain deficient, and must compete for scarce grain supplies with other national grain-deficient areas (namely, the urban centers). State policy is usually directed toward ensuring that cereal supplies move from the country's surplus areas to cities or to export markets (as in the case of the Sudan), with little concern for rural deficit areas. During times of acute food shortages, herders rely on famine relief or move to the larger urban centers where supplies of grain are more secure.

In many countries, regional trade between agriculture and livestock sectors has been discouraged by those policies that advocate strong, vertical market linkages to the large urban centers. This spatial orientation restricts trade between rural surplus and rural deficit zones. While retail price controls on staple food commodities may be strictly enforced in the urban sector, controls are often ineffective in remote areas, where herders can pay as much as 200% over official prices during drought years (Sen, 1981).

It is not feasible to expect governments to channel foods away from politically sensitive urban areas to remote regions of the country. Yet, there are practical measures that could ameliorate some of the marketing constraints in pastoral areas. First, a program to establish regional and district grain stores would facilitate herder access to food supplies, and would reduce inefficiencies and costs associated with the transport of grain from distant sources. Under AID's Niger Range and Livestock Project, the decentralization of grain distribution was accompanied by a credit program, which allowed herders to purchase grain at times of the year when prices were low rather than high (Swift, 1984).

On the production side, the pursuit of an 'opportunistic' stocking strategy would benefit from more flexible livestock marketing policies, especially those that would facilitate the sale of excess stock during drought years. Reform is needed at several levels in the marketing chain. For example, where there are urban price controls, they should be relaxed during drought years. In these years, the number of animals that come on the market is very high, causing producer prices to fall. While it becomes a buyer's market, state-imposed consumer prices are kept up in the cities, rather than allowed to fluctuate according to supply and demand conditions. This inflexibility greatly limits the absorptive capacity of the domestic meat market (which tends to be concentrated in urban areas), and allows private traders to reap considerable profits by buying cheaply and selling dearly in the urban marketplace. None of this value added is received by the producer. In Kenya, urban prices were not changed during the 1984 drought, although producer prices in the range areas dropped by as much as 65%. Even then herders had difficulty finding outlets for their surplus stock. The result was that up to 50% of cattle died in certain regions, where markets could not be found. The figure would have been higher had not the government increased its purchases of meat for the country's canning facilities.

Finally, market policies that strengthen regional trade, particularly between pastoral and agricultural areas, would benefit herders in certain cases (cf. Ibrahim, 1984; Swift, 1982). The vertical, urban orientation of grain marketing has at times redirected food supplies from those indigenous regional markets that in the past served pastoral areas. Herders, who formerly could obtain grain from these regional markets, are forced to depend on a national grain-distribution system that may be unreliable, or function only at the discretion of Western donors who funnel massive amounts of grain through it during drought years. In some countries, the state could facilitate regional trade by being less restrictive on trader licensing, by lifting bans on the movement of grains between regions, and by permitting peasants to sell more of their grain through private channels.

Programs for returning herders to livestock production

It is questionable whether irrigation or other non-pastoral programs are the appropriate vehicles for herders, who wish eventually to return to pastoralism. While forced sedentarization is rare in Africa, governments have encouraged pastoralists to adopt non-pastoral livelihoods and to settle, in the hopes of avoiding the worst effects of drought. Governments, assuming that the risks of famine are reduced through settlement, have tried to facilitate sedentarization by establishing irrigation, fishing and other schemes in pastoral areas. Impoverished herders move to these schemes, often with hopes of returning to pastoralism, but low returns from these schemes result in a more permanent form of pauperization. They find themselves in a 'Catch-22' situation where, in order to meet reproduction costs, they are required to further deplete their assets (livestock) to purchase needed food (Little, 1983). In many cases, the likelihood of famine is not further reduced, since production on these schemes can involve even greater risks: irrigation pumps can fail, needed inputs can be delayed and product markets can disappear. The herders who are engaged in low-return agricultural and wage activities have little chance to rebuild their herds, and are 'increasingly vulnerable to drought' (Hogg, 1984, 4).

The preceding discussion supports the premise that Africa's arid and semiarid zones are mainly suited to extensive livestock production. Programs of sedentarization and agricultural development, while appropriate in some cases, are not generally beneficial to either the herder population or the environment. In post-drought periods, such as the one we presently find ourselves in, more attention should be directed to restocking pastoral herds and to improving pastoral production systems, as opposed to replacing them. In the short and medium term, alternatives to pastoral production in Africa's range areas are not likely to be economically, socially, or ecologically cost effective.

References

AID (1985a). *Need to Redesign the Niger Integrated Livestock Production Project*. Audit Report No. 7-683-85-4, 28 February, 1985. Washington, DC: Agency for International Development.

AID (1985b). *Progress and Problems in Managing the Mali Livestock Sector II Project*. Audit Report No. 7-688-85-5, 27 March, 1985. Washington, DC: Agency for International Development.

Blaikie, P. (1984). *The Political Economy of Soil Erosion in Developing Countries*. London & New York: Longmans.

Carr, C. J. (1977). *Pastoralism in Crisis: The Dasanetch and their Ethiopian Lands*. Research Paper No. 180, Department of Geography. Chicago: University of Chicago.

Conant, F. P. (1982). Thorns paired, sharply recurved: cultural controls and rangeland quality in East Africa. In *Desertification and Development: Dryland Ecology in Social Perspective*, eds B. Spooner & H. S. Mann, pp. 111–22. New York: Academic Press.

Cossins, N. J. (1985). The productivity and potential of pastoral systems. *International Livestock Centre for Africa (ILCA) Bulletin*, **21**, 10–15.

de Ridder, N. & Wagenaar, K. T. (1984). A comparison between the productivity of traditional livestock systems and ranching in eastern Botswana. *ILCA Newsletter*, 3(3), 5–7.

Devitt, P. (1982). *The Management of Communal Grazing in Botswana*. Pastoral Network Paper 14d. London: Overseas Development Institute (ODI).

Diarra, L. & Breman, H. (1975). Influence of rainfall on the productivity of grasslands. In *Evaluation and Mapping of Tropical Rangelands*, Addis Ababa: International Livestock Centre for Africa.

FAO (1973). *Propositions Préliminaires pour une Approche Integrée du Développement à Long Terme de la Zone Sahélienne de l'Afrique de l'Ouest*. Doc. de Travail WS/D7404. Rome: Food and Agriculture Organization of the United Nations.

FAO (1980). *Natural Resources and the Human Environment for Food and Agriculture*. Environment Paper No. 1. Rome: Food and Agricultural Organization of the United Nations.

Fauck, R., Bernus, E. & Peyre de Fabrègues, B. (1983). *Mise à jour de l'étude de cas sur la désertification et renforcement de la stratégie nationale en matière de lutte contre la désertification*. Paris: UNESCO/UNSO.

Galaty, J. G. (1980). The Maasai group-ranch: politics and development in an African pastoral society. In *When Nomads Settle*, ed. P. Salzman, pp. 157–72. New York: J. F. Bergen.

Hardin, G. J. (1968). The tragedy of the commons. *Science* **162**, 1243–8.

Herskovits, M. J. (1926). The cattle complex in East Africa. *American Anthropologist*, **28**, 230–72, 361–80, 494–528, 633–64.

Hjort, A. (1981). A critique of 'ecological' models of pastoral land use. *Ethnos*, **46**(3–4), 171–89.

Hogg, R. (1984). *Re-Stocking Pastoralists in Kenya: A Strategy for Relief and Rehabilitation*. Pastoral Network Paper 19c. London: Overseas Development Institute.

Horowitz, M. M. (1972). Ethnic boundary maintenance among pastoralists and farmers in the western Sudan (Niger). *Journal of Asian and African Studies*, 7(1,2), 105–14.

Horowitz, M. M. (1973). Relations entre pasteurs et fermiers: compétition et complémentarité. *Notes et Documents Voltaiques*, **6**(3), 42–5.

Horowitz, M. M. (1975). Herdsman and husbandman in Niger: values and strategies. *Pastoralism in Tropical Africa*, ed. T. Monod, pp. 387–405. Oxford: Oxford University Press for International African Institute.

Horowitz, M. M., Arnould, E., Charlick, R., Eriksen, J. Faulkingham, R., Grimm, C., Little, P., Painter, M., Painter, T., Saenz, C., Salem-Murdock, M. & Saunders, M. (1983). *Niger: a Social and Institutional Profile*. Binghamton, NY: Institute for Development Anthropology.

Horowitz, M. M. & Badi, K. H. (1981). *Sudan: Introduction of Forestry in Grazing Systems*. Forestry for Local Community Development Programme. GCP/INT/347/SWE. Rome: Food and Agriculture Organization of the United Nations.

Ibn Khaldun (1967). *The Magaddimah*, vol.1, trans. F. Rosenthal. Princeton: Princeton University Press.

Ibrahim, F. N. (1978). *The Problem of Desertification in the Republic of the Sudan with Special Reference to Northern Darfur Province*. Khartoum: Khartoum University Press.

Ibrahim, F. N. (1984). *Ecological Imbalance in the Republic of the Sudan – with Reference to Desertification in Darfur*. Bayreuth: Druckhaus Bayreuth Verlagsgesellschaft mbH.

IBRD (International Bank for Reconstruction and Development) (1985). *Expérience Acquise par le Projet Elevage Centre Est et Reflexions sur l'Avenir du Développement de l'Elevage au Niger*. Paper prepared for the Niger National Debate on Livestock Herding, Tahoua, April 2–10, 1985. Washington, DC: The World Bank.

Institute of African Studies (1982). *Social/Cultural Profile of Samburu District*. Nairobi: Ministry of Finance and Planning.

Kirchner, J. W., Ledec, G., Goodland, R. J. A. & Drake, J. M. (1985). Carrying capacity, population growth, and sustainable development. In *Rapid Population Growth and Human Carrying Capacity: Two Perspectives*, ed. D. J. Mahar, pp. 45–89. World Bank Staff Working Paper No. 690. Washington, DC: The World Bank.

Lamprey, H. F. (1983). Pastoralism yesterday and today: the over-grazing problem. In *Ecosystems of the World 13: Tropical Savannas*, ed. F. Bourlière, pp. 643–66. Amsterdam: Elsevier Scientific Publishing Co.

Lewin, R. (1985). Gregarious grazers eat better. *Science*, **228**(4699), 567–8.

Little, P. D. (1983). The livestock-grain connection in Northern Kenya: an analysis of pastoral economics and semi-arid land development. *Rural Africana*, **15/16**, 91–108.

Little, P. D. (1984). Land and pastoralists. *Cultural Survival Quarterly*, **8**(1), 46–7.

Little, P. D. (1985). Absentee herdowners and part-time pastoralists: the political economy of resource use in Northern Kenya. *Human Ecology*, **13**(2), 131–51.

Little, P. D. & Brokensha, D. W. (1985). *Local Institutions, Common Property, and Resource Management in East Africa*. Paper presented at the Conference on 'The Scramble for Resources in Africa, 1884–1984', Cambridge University, England, 19–20 April 1985.

Little, P. D. & Horowitz, M. M. (1986). Subsistence crops are cash crops. *Human Organization*, forthcoming.

Lomax, A. & Arensberg, C. M. (1977). A worldwide evolutionary classification of culture by subsistence systems. *Current Anthropology*, **18**(4), 659–701.

Murdock, G. P. (1959). *Africa, its Peoples and their Culture History*. New York: McGraw-Hill Book Company, Inc.

O'Leary, M. (1984). Ecological villains or economic victims: the case of the Rendille of Northern Kenya. *UNEP Desertification Control Bulletin*, **11**, 17–21.

Painter, T. (1985). *Peasants, Migrants, Petty Traders, and Precarious Reproduction: The Squeeze on Simple Reproduction among Zarma Peasant Smallholders in Niger*. Paper prepared for the Annual Meeting of the African Studies Association, 23–6 November 1985, New Orleans, LA.

Palmer, E. H. (1977). *The Desert of the Exodus*, vols I. and II. New York: Arno Press.

Peters, P. E. (1984). Struggles over water, struggles over meaning: cattle, water and the state in Botswana. *Africa*, **54**, 29–49.

Picardi, A. C. (1975). *A Systems Analysis of Pastoralism in the West African Sahel. Framework for Evaluating Long-term Strategies for the Development of the Sahel–Sudan Region.* Annex 5. Center for Policy Alternatives. Cambridge, MA: Massachusetts Institute of Technology.

Runge, C. F. (1981). Common property externalities: isolation, assurance and resource depletion in a traditional grazing context. *American Journal of Agricultural Economics*, **63**(4), 595–606.

Salem-Murdock, M. (1984). *Nubian Farmers and Arab Herders in Irrigated Agriculture in the Sudan: From Domestic to Commodity Production.* Doctoral Dissertation in Anthropology, State University of New York at Binghamton.

Sandford, S. (1982). Pastoral strategies and desertification: opportunism and conservatism in dry lands. In *Desertification and Development: Dryland Ecology in Social Perspective.* eds B. Spooner & H. S. Mann, pp. 61–80. London & New York: Academic Press.

Sandford, S. (1983). *Management of Pastoral Development in the Third World.* Chichester, England: John Wiley & Sons.

Sen, A. (1981). *Poverty and Famines: An Essay on Entitlement and Deprivation.* Oxford: Oxford University Press.

Swift, J. (1982). The future of African hunter-gatherer and pastoral peoples. *Development and Change*, **13**, 159–81.

Swift, J., ed. (1984). *Pastoral Development in Central Niger: Report of the Niger Range and Livestock Project.* Niamey, Niger: United States Agency for International Development.

Taussig, M. (1980). *The Devil and Commodity Fetishism in South America.* Chapel Hill: University of North Carolina Press.

Teitelbaum, J. M. (1984). The transhuman production system and change among Hawazma nomads of the Kordofan Region, Western Sudan. *Nomadic Peoples*, **15**, 51–66.

Thomson, J. T. (1985). The politics of desertification in marginal environments: the Sahelian case. In *Divesting Nature's Capital*, ed. H. J. Leonard, pp. 227–62. New York: Holmes & Meier.

Warren, A. & Maizels, J. K. (1977). *Ecological Change and Desertification.* UN Conference on Desertification. Paper No. A/CONF. 74/7.

PART II

Internal–external perspectives

5

The decline of African agriculture: an internalist perspective

MICHAEL F. LOFCHIE

African Studies Center, University of California, Los Angeles

Introduction

Chronic food deficits have become the most visible symbol of inadequate agricultural performance in a large and growing number of African countries. Nearly half of sub-Saharan Africa's 45 independent countries face serious food emergencies. In at least six of these (Chad, Ethiopia, Mali, Mozambique, Niger and Sudan) food shortages are so severe that only concerted action by the international community can avert widespread starvation. The Food and Agriculture Organization of the United Nations estimated that countries affected by food shortages might need to import as much as 7.0 million tons of food during 1984–85, and this figure would have been substantially greater, had not an adequate pattern of rainfall returned to East Africa (Food and Agriculture Organization of the United Nations, 1985). The portentous feature of this situation is that it reflects a pattern of long-term agricultural decline: Africa's food shortages can no longer be understood as the outcome of short-term episodic events such as droughts or crop blights but, instead, must be analyzed as the consequence of fundamental structural properties of the international economic system and inappropriate agricultural policies on the part of African nation-states.

Current trends are such that Africa can be expected to remain a food-importing region for the foreseeable future. As early as 1981, a major study by the Africa/Middle East Branch of the US Department of Agriculture (USDA) documented a pattern of diminishing per capita food availability for a large and growing number of African countries. Of 25 countries surveyed, about two-thirds (17) had experienced a measurable decline in per capita food production during the decade of the 1970s, and in 7 of these (Mali, Senegal, Burkina Faso, Guinea, Angola, Ethiopia and Uganda), domestic per capita food production had fallen by as much as 30% or more during this period (US Department of Agriculture, 1981). Africa had emerged dramatically as the only major world region to have suffered a decline in per capita food production since the early 1960s (see Fig.5.1; chart provided by USDA). This pattern has ominous implications not only for human nutrition but, because of the high cost of food imports relative to national resources, for overall development as well. The need for food imports is not a matter of great concern in industrially developed countries such as Japan which can easily afford to enter the world market to purchase their requirements. It can be disastrous, however, for poor Third World countries whose economies are marginal and which lack the financial reserves to provide for their populations.

The beginnings of Africa's agrarian crisis were visible as early as the immediate post-independence period in the 1960s. The continent had already beome a net importer of grains to make up for shortfalls in domestic production. However, as the 1981 USDA Report notes, this was not a matter of urgent concern, because import volumes were relatively low; adequate, low-cost supplies were readily available from donor countries prepared to make grain shipments on concessional terms, and foreign exchange reserves did not pose a constraint on the import of other necessities (US Department of Agriculture, 1981).

During the decade of the 1960s, however, food imports grew to become a matter of utmost concern. Not only did the volume of food imports double during this period but, because of increasing prices, the cost of these imports

Fig. 5.1.

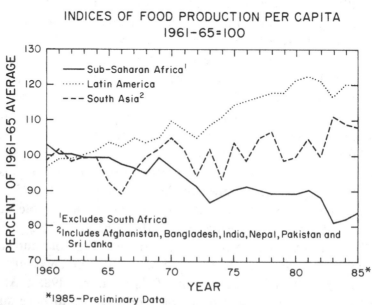

INDICES OF FOOD PRODUCTION PER CAPITA
1961–65=100

*1985–Preliminary Data

increased nearly six-fold. Even before the Sahelian drought of the early 1970s, it had become clear that the poor performance of the agricultural sectors of a large number of African countries had become the root cause of a broader and more diffuse economic crisis. A large number of countries had embarked upon costly programs of industrialization. The need to allocate larger and larger amounts of hard currency for food imports competed directly with the capital goods and raw materials required by the new industrial sector.

The Sahelian drought of the early 1970s focused attention dramatically on the underlying weaknesses in the agricultural systems of West African states. It demonstrated that there was a margin of reserve capacity with which to respond to a temporary deterioration of climatic conditions. It also demonstrated that these countries lacked the financial capacity, on their own, to import sufficient volumes of foodstuffs to avert starvation. Although extreme drought was confined principally to western Africa and Ethiopia, the agrarian weaknesses it exposed could also be discerned in eastern African countries such as Tanzania, which also experienced severe food shortfalls during this period. Although the West African drought exposed the weaknesses of the region's agrarian economies, it could by no means be held accountable for them, for the continent's need for food imports has continued to escalate in periods of both favorable and adverse weather conditions. According to one World Bank report, Africa's food imports had climbed to nearly 10 million tons per year by the early 1980s, an amount adequate for virtually the entire urban population (World Bank, 1984). If current production trends continue, the need for food imports could well double during the present decade, and even this increase might not be sufficient to prevent a further erosion of nutritional standards or occasional episodes of famine in countries most affected by any serious disruption, including prolonged drought, plant disease, insect infestation and internal war.

This scenario has the most ominous implications since the international donor community, whose patience has already become strained by poor agricultural policies, may become increasingly reluctant to provide concessional or humanitarian aid. Nor is it likely that African countries will be able to provide for this level of imports out of their own financial resources, since export-oriented agriculture, which would provide the principal basis of foreign exchange earnings necessary to finance food imports, has also been subject to stagnating production. Indeed, Africa's world market share of major agricultural exports such as coffee, tea, cotton, bananas and oil seeds has been falling steadily since the early 1970s. Since world demand for these commodities is projected to grow only slowly at best between now and the end of this century, Africa's prospects of developing an adequate foreign exchange earning capacity must necessarily lie in the introduction of agricultural policies that will enable the continent to recapture its former share of world trade in these commodities. This would require a major and politically difficult diversion of national resources away from urban industries and services to the countryside.

The faltering performance of export agriculture has already posed a serious constraint on the performance of Africa's urban industries. Since much of sub-Saharan Africa's industrial sector is based on import substitution, it depends heavily on the hard currency earnings of other economic sectors, principally agriculture, to finance the import of needed inputs. As the earnings from export agriculture have stagnated, and as diminishing reserves of foreign exchange have been required to finance the acquisition of other necessary imports such as food and energy, Africa's industries have begun to suffer from the lack of replacement capital goods, spare parts and raw materials. The most conspicuous symptom of these shortages has been a falling rate of capacity utilization, already as low as 25% to 30% in some countries. Other symptoms of the problem are to be discerned in such phenomena as high rates of inflation, caused by the scarcity of consumer goods; increasing rates of urban unemployment, caused by industrial closures; and a falling real wage rate, caused (in part) by the fall in labor productivity.

There is an urgent need for African countries to create the basis for an expansion of smallholder agriculture. The continent's cities already seem increasingly unable to provide economic opportunity for displaced members of peasant society who migrate to urban centers in a desperate quest for livelihood. With the absolute decline of industrial activity in numerous countries, rates of urban unemployment of 40% or more are by no means uncommon. This figure does not include those persons whom economists designate technically as 'underemployed'. Even Africa's better-off countries already confront a social crisis in which the number of those leaving school exceeds the number of new jobs being created and, in societies where stagnation of export agriculture has caused a particularly serious industrial decline, annual job creation in the modern sector is negative. Nor is it likely that Africa's informal economies, vast as they are, can provide employment opportunity on the necessary scale, for their buoyancy is heavily dependent upon the purchasing power of urban middle and working classes, and the capacity of these groups to act as consumers for goods produced in informal industries has been sharply eroded by inflationary trends. Despite the fact that agriculture holds the major prospect of employment generation, few African countries have made systematic efforts to develop smallholder-based systems of agricultural production.

Food deficits, industrial decline and urban unemployment are only the most conspicuous outcomes of Africa's agrarian malaise. Africa has also become a continent of generalized political and social instability – phenomena that are, at one and the same time, both cause and effect of agricultural deterioration. One of the most tragic indicators of the continent's chronic economic problems is the extremely high incidence of refugees. Although sub-Saharan Africa only has about 10% of the world's population, it now accounts for about 25% of the world's refugees, approximately 2.5 million persons. A high rate of infant mortality has also emerged as one of the effects of pandemic economic difficulties, and Africa's rate of infant mortality is now approximately double that of other developing regions. There is also increasing evidence that, among children who survive, chronic food deficits are beginning to take an additional toll in terms of such ratios as height or weight to age. Since life expectancy in developing countries is almost directly related to per capita income, it is not at all surprising that Africa's poorest countries also rank among the lowest in the world in this regard.

Unless there is a major reversal of current trends in the agricultural sector, there is little basis for optimism about Africa's economic future. Major economic indicators seem to point in the direction of a further worsening of the present situation. During the past decade, for example, African countries have accumulated an enormous debt burden, and this diminishes the prospects of economic recovery because debt servicing has now become an enormous drain on dwindling foreign exchange reserves. As late as 1974, the total outstanding debt of Africa's *low-income countries* was only about $7.5 billion and the debt service ratio for these countries (total debt servicing as a percentage of export earnings) was only about 7%. Within just a decade, the foreign debt of these countries increased nearly four-fold, to approximately $27 billion, and the debt service ratio had increased almost four and a half times, to almost 31.5%. Africa's debt service ratio is now substantially higher than that for Latin America and the Caribbean, generally regarded as the world's principal debtor area, where the debt service ratio in 1983 was approximately 26%. Indeed, if the debt burden is figured on the basis of the ratio of total debt to annual export earnings, Africa's low income countries have a debt burden nearly three times as great as that of Latin America and the Caribbean (World Bank, 1985). Since debt servicing competes with imported agricultural inputs for foreign exchange, the debt problem has a direct and negative bearing on the prospects of agricultural recovery.

The deepening character of Africa's agricultural crisis has occasioned an intense debate among policymakers and academics as to its causes and potential remedies. Broadly speaking, the participants in this debate can be divided into two schools of thought, internalists and externalists. Internalists tend to believe, on the whole, that the basic cause of agrarian breakdown in Africa can be found in the economic policies pursued by African governments since independence. They believe that economic recovery is effectively within the political jurisdiction of governments that are prepared to undertake serious efforts at policy reform. Externalists, on the other hand, tend to place their principal emphasis on adverse features of the international economic environment faced by countries that depend upon primary agricultural exports. There is a strong presupposition that unless the international economic environment can be made more hospitable, internal policy reforms will fall short of triggering economic recovery. Such broad categorization inevitably involves an unfortunate degree of oversimplification, as very few policy analysts or scholars fall unambiguously into either of these categories.

Today, there is much occasion to regret that these schools of thought were posed as adversarial interpretations, for they can as easily be thought of as complementary explanations, since each sheds valuable light on some of the complex causes of the present crisis. Indeed, it would be practically impossible to understand the agricultural performance of any African country without reference to the interaction between domestic agrarian policy and the character of the international economic environment. Thus, the purpose of this chapter is not, in any way, to suggest that internal factors provide either an exclusive or an exhaustive explanation of Africa's agricultural decline. Its intent is to suggest that the agricultural policies pursued by African governments have been of great importance in bringing about the widespread pattern of agricultural decline and, by extension that, until these policies are reformed, it is all but impossible to gauge the effects of the international economic environment.

Internal explanations

The theoretical point of departure for an internally focused analysis of Africa's agricultural crisis is the proposition that African governments have intervened in rural markets in ways that pose fundamental disincentives to agricultural production. The most influential scholarship elaborating this position has been that of Robert Bates and Elliot Berg (Bates, 1981; Berg, 1981). Bates presents a convincing argument that, since independence, African governments have adopted a set of economic policies that effectively reduce the economic incentives to agricultural producers by shifting the internal terms of trade against the countryside. The common denominator of these policies has been an attempt to use the agricultural sector as a source of economic resources to be dispensed elsewhere in the society. But the cumulative result of these policies has been a virtually pandemic pattern of agricultural stagnation as manifested in falling per capita food production and a steady loss in the continent's share of the world trade in exportable agricultural commodities.

The first question of importance in the internalist approach is precisely why such a large number of African governments have framed agricultural policies that have adverse effects on the most important sector of their national economies. The basic reasons are not difficult to identify. Agricultural policy in Africa is driven by three overriding imperatives: (*a*) the commitment to a radical expansion of public services; (*b*) the commitment to the promotion of an industrial sector; and (*c*) the need to respond to the political demands of powerful urban interest groups. With the advent of independence, African governments assumed a set of socioeconomic and political objectives that required vastly greater economic resources than had previously been available. All of these goals required a dramatic expansion of the financial resources of the state, and agriculture was clearly the only economic sector large enough to provide revenues for this expansion (Hart, 1982). During the generation that followed the end of colonial rule, agriculture came to be treated as an object of economic extraction, an almost infinitely elastic source of tax revenue for state expansion and finance capital for enormously expensive industrial projects.

Agricultural policy in independent Africa can be understood as an attempt to transfer economic resources from rural producers to the state, to be expended on such services as health, education and the expansion of rural infrastructure. Newly independent African governments have also considered themselves under a deep obligation to deal with urgent social problems such as urban unemployment. Though rarely articulating explicitly the principle of the state as employer of last resort, many African governments have in fact behaved as if this were an essential basis for ongoing political legitimacy. They have rapidly expanded their employment rolls to provide jobs in the public sector to persons who could not be absorbed elsewhere in the economic system (Abernethy, 1983). The result was a wholesale enlargement of the state bureaucracy and this expansion, like the proliferation of public services, required revenues that could only be provided by the agricultural sector.

A second broad imperative that underlay adverse governmental intervention in the agricultural sector had to do with the commitment to promote industrialization. Until the early 1970s, there was a strong consensus among development economists that the most expeditious means to achieve this goal was through the creation of industries based on the principle of import substitution. This strategy of industrial development seemed to offer a quick and efficient means not only of generating urban industrial employment but of conserving foreign exchange resources that were being spent to import light consumer goods. The only question for debate was where to obtain the capital with which to finance these new industries. The obvious answer was to obtain it from agriculture for, although some capital investment might be forthcoming from abroad, it was unlikely to be sufficient, and in any case many African political leaders harbored deep suspicions about multinational corporations and their investment activities in the Third World. Import substitution as an industrial strategy, then, required that agriculture be treated as a source of investment capital for the establishment of urban manufacturing, rather than as the object of policies designed to further its own development.

Of all the factors that have stimulated adverse governmental intervention in the agricultural sector, however, none is so important or powerful as the simple imperative of political survival. African governments are desperately concerned about the volatility of their urban constituencies. Keith Hart (1982, 10) has stated this point bluntly with respect to western Africa:

> The short-term preoccupation of West Africa's rulers is with the immediate danger of an unsatisfied urban mob. Long-term planning for the countryside is entirely incompatible with the siege mentality of politicians, soldiers and bureaucrats who are literally counting the days before they lose their power (and lives) in the face of growing anger . . . This anger means most in the major cities; it commands constant attention and the award of temporary palliatives, one after the other, all adding up to the relative impoverishment of farmers.

Otherwise highly diversified and potentially antagonistic urban interest groups find common cause in the implementation of agricultural policies that reduce the incomes of rural producers. Wage earners, for example, benefit from cheap food but, since food is a wage good, industrialists as well have an interest in keeping the price of foodstuffs as low as possible. Governments also satisfy their own interests by cheapening the price of food, since they can stabilize the wage scales for lower-paid governmental officials. Political leaders, industrialists and workers alike benefit when the export sector of agriculture can be compelled to yield up abundant supplies of cheap capital for urban investment. Thus, suppression of the agricultural sector is a policy that unites the total ensemble of urban interests.

Patterns of intervention

Price regulation

African governments have intervened in their agricultural sectors in a variety of ways. Of these, by far the most common is direct governmental regulation of producer prices. Bates (1981) presents a convincing argument that African governments, since independence, have employed their ability to establish monopolistic control of agricultural pricing to suppress the farmgate prices of agricultural commodities far below levels that would have prevailed if a free market in agricultural commodities had been allowed to operate. Surveying a range of producer prices for export crops, for example, Bates (1981, 29) concluded that 'in most instances, they (agricultural producers) received less than two-thirds of the potential sales realization, and in many cases they received less than one-half'.

The suppression of export crop prices was not begun by Africa's independent regimes. It was initiated by colonial governments that also viewed the export sector as a source of tax revenue and foreign exchange. Independent governments, under pressure from populations that had been politically mobilized during the nationalist era, and therefore especially needy of revenues with which to carry forward the projects that had been promised as the benefits of political freedom, pursued this policy with intensified vigor.

Bates (1981) found a similar pattern in the food-producing sector where newly independent African governments had also enlarged upon policies of price regulation that were initially set in place by colonial regimes. Here, too, it seemed that African governments, responding to the wide array of political pressures for cheaply priced food staples, had suppressed the prices of domestically produced foodstuffs far below the world market price levels of similar commodities. John de Wilde has argued that suppression of producer prices for domestic food crops is far more severe than that for export commodities where there has been some tendency to pass through world market prices. He has argued as follows:

Government intervention in the marketing and pricing of food

> products has probably been marked by the most serious
> deficiencies. In this field, governments have been caught up in
> a conflict between the political necessity they have evidently felt
> to keep food prices low for urban consumers and the need to
> stimulate domestic production of food. Reconciliation of this
> conflict has been virtually impossible. (*de Wilde, 1984, 118*)

de Wilde's observation that producer prices for food staples are determined
on the basis of the political influence of urban interests is shared
overwhelmingly by observers of African agriculture today. If correct, this
point has disturbing implications, for it suggests that it may be difficult to
identify or even create a social coalition that would facilitate reform of this
key policy.

African governments have frequently coupled the tendency toward price
suppression for staple foodstuffs with systems of uniform or pan-territorial
pricing. This has introduced further harmful distortions into the market
for food staples. It has, for example, added to the disincentive for food
producers who are located adjacent to urban markets. Since food staples
are relatively expensive to transport (their value per unit is low), uniform
national pricing in effect compels nearby producers to absorb the
transportation costs of farmers who are located at greater distances from
their markets. These farmers are thereby discouraged from growing food
staples, while those in more remote regions, whose transportation costs are
indirectly subsidized, have little or no incentive to shift to crops of higher
value per unit. Moreover, since government-set prices remain constant
throughout the year, farmers tend to concentrate their sales to government
purchasing agencies during the period immediately following the harvest.
(In a fixed price system, seasonal variations in supply have no effect on the
farmgate price.) This can easily result in severely overtaxing purchasing,
processing and storage facilities during the harvest period, while offering
farmers no incentive to hold back a portion of their production for the
off-season, when national supplies might be lower.

Since it is impossible to measure the cumulative impact of these
distortions, it would be premature to suggest that liberalization of Africa's
food marketing systems would result in food self-sufficiency. There is,
however, a compelling reason to believe that African countries could move
substantially closer to this goal than is currently the case. At the same time,
the concerns of those who fear that a freer food market would inevitably
lead to urban inflation seem ill-founded. It is at least equally conceivable
that a more efficient allocation of productive resources would result in a
greater supply of food staples at real price levels not measurably different

than those obtained today because, as statist systems of food distribution have become less and less effective, greater and greater proportions of Africa's urban populations have been compelled to obtain their food supplies on high-priced informal markets.

Currency overvaluation

A second government policy that has contributed to the present agricultural crisis is the widespread tendency toward currency overvaluation. (Francophone African countries which use the CFA franc as their national currency are, of course, a major exception to this generalization. This currency is directly convertible to the French franc.) The practice of setting official exchange rates at levels far higher than would prevail if there were free convertibility has set in motion a whole set of powerful disincentives to agricultural productivity. Perhaps its greatest impact has been to tilt the internal terms of trade within African countries massively in favor of the cities and against the countryside (Gulhati, Bose & Atukorala, 1985). Currency overvaluation tends to lower the cost of living of urban consumers by cheapening the prices of imported goods, including foodstuffs, while raising the prices of high-demand items in the countryside. It has thus contributed to major economic anomalies. Africa's upper and middle class elites have grown increasingly accustomed to a material lifestyle that includes a host of artificially cheapened imported goods that ranges from automobiles to expensive household appliances and luxury items. Since the foreign exchange utilized to finance these imports is typically generated by agricultural exports, overvaluation penalizes the rural producer by just about the same amount it rewards the urban consumer.

The impact of currency overvaluation on agricultural exports has been disastrous, since the producer price of export crops is normally a direct function of the conversion rate between foreign and domestic currencies; the greater the overvaluation, the fewer units of local currency per unit sold realized by the agricultural producer. This problem may help explain the severe cash squeeze experienced by many of Africa's export-oriented agricultural marketing boards. To the extent that their income in local currency is also a function of the conversion rate, they are penalized, as well, by a policy that overvalues that currency in relation to the foreign exchange their crops have generated. The general tendency, however, has been for the marketing parastatals to pass the systemic penalty on to the farmers by making either partial or late payments for their crops. African governments concerned about the political loyalties of parastatal personnel invariably make up for parastatal operating deficits through supplemental budgetary appropriations. The final result is that only the producers really suffer from the artificial exchange rate. Since their prices remain low and since governments rarely seem to feel that it would be as worthwhile to subsidize the growers as they do the parastatals, the net effect of overvaluation is an income transfer from rural farmers to the urban middle classes.

Overvaluation has also resulted in an indirect heightening of the price levels of consumer goods and vital necessities purchased in the countryside. As one of the building blocks in the industrial policy of import substitution, its intent has been to encourage the growth of urban industries by lowering their costs for capital goods, raw materials and spare parts. Despite the availability of artificially cheapened inputs, these industries have tended to be high-cost producers due to inefficient operation. To a very large degree, however, the goods produced in these industries are of special importance to rural consumers. The list includes such basic household items as kitchen utensils, cooking stoves and lamps, such important personal items as cotton cloth and soap, and such vital agricultural implements as hoes and plows. Since imported substitutes for these goods are not available or are available only at high prices due to protective tariffs, rural consumers are compelled to acquire the more expensive (and frequently lower quality) local items.

Although currency overvaluation theoretically lowers the cost of imported agricultural inputs such as fertilizers, equipment and pesticides, its practical effect has been entirely the opposite. The reason is not difficult to discern. When a local currency is overvalued, the most immediate effect is that foreign exchange becomes scarce due to the high demand in local currency for imported products. As a result, foreign currency itself must be rationed. In this process, as in so many others, the tendency of African governments has been to favor urban consumers at the expense of rural producers. Thus, although imported inputs are technically inexpensive, there is an extreme shortage of supply. This generates the need for a system of rationing and, when this occurs, the real prices paid by consumers are substantially higher due to intensive demand pressures. The rationing process also gives rise to serious distributional inequities, since large capital-intensive farms are in a stronger position to bid for vital inputs than smallholders or peasant producers.

Overvalued exchange rates have also had a serious impact as a disincentive on local food production. Since the costs of foreign foodstuffs are artificially cheapened along with those of other imported goods, it has become increasingly commonplace for western grains such as wheat, corn and rice to be less expensive in African markets than the same items, or equivalent staples, produced by local farmers. Sometimes the price structure of imported grains is lowered still further, as a result of the fact that they enter the country on concessional terms as food aid rather than on the basis of direct sales. Since it is all but impossible for donor agencies and African governments to prevent food that enters the country on assistance programs from entering local markets, food aid sometimes competes directly with locally produced foodstuffs. Occasionally, this competition has been actively encouraged by governments that are interested in holding urban food costs at the lowest possible levels (Jackson, 1982). The resulting disincentive effect on local producers has often been so great that it completely discourages local production for urban markets and launches a cycle of ever-increasing dependency on food imports.

When this process is under way, the difficulty of rebuilding local food-production systems is compounded because, once local food producers have withdrawn from markets dominated by artificially cheapened imports and have become accustomed to trading their goods in informal or illegal cross-border markets, it becomes extremely difficult to recapture their involvement in the official national economy. At the same time, Africa's urban consumers frequently become acclimated, in terms of diet as well as price, to inexpensive foreign grains. Once this has occurred, governments that are more concerned about short-term political stability than with the economic recovery of the peasant sector find it all but impossible to disengage from exchange rate practices that undermine the livelihood of their own rural populations.

Parastatal corporations

The principal means that African governments have used for regulating and administering their agricultural sectors has been the creation of a system of official agricultural marketing boards. As parastatal corporations, these marketing boards are typically given a legalized monopoly over the acquisition, processing and vending of stipulated agricultural commodities. If a crop has been given over to the authority of an official marketing agency, it is generally illegal for a producer to sell more than a specified amount to any other agency or for a purchaser, whether individual or commercial, to obtain a supply from any other agency. In many cases, the agricultural marketing boards are also entrusted with the responsibility for carrying out a number of critically important service functions such as credit provision, the distribution of inputs and the conduct of research on improved methods of crop husbandry. The marketing board system was originated during the colonial era to conduct the international vending of export crops, but official marketing agencies are as common today in the field of domestic food production. Since it has become increasingly commonplace for African governments to establish an official marketing agency for any crop of economic significance, many African countries have as many as a dozen of these parastatal corporations operating in the rural sector.

Contemporary observers of African agriculture are almost unanimously of the opinion that the performance of the agricultural marketing boards has been abysmal and that they are among the major reasons for the continent's agricultural decline. They seem almost universally to be characterized by waste, inefficiency, mismanagement, and corruption. The list of documented administrative shortcomings of many agricultural marketing boards is so extensive that it is sometimes surprising that they function at all. Standards of accountability, for example, are woefully inadequate or nonexistent, with the result that it is often impossible to determine how much revenue is being expended for a particular purpose over a given period of time. Bribery of parastatal personnel is commonplace and since parastatals have an official economic monopoly enforced by the police powers of the state, they have an almost limitless opportunity to extort illicit payments from farmers under their jurisdiction. Parastatals are sometimes viewed by political leaders as instruments of patronage and, as a result, they have been subjected to gross overstaffing. The selection and promotion of key parastatal administrators is sometimes based on political criteria that have little to do with administrative performance. Although parastatal salary scales and fringe benefits are often extremely generous, sometimes exceeding those available in the governmental civil service by a wide margin, there are virtually no incentives for efficient performance or penalties for maladministration.

The end result of these shortcomings is that Africa's agricultural parastatals have performed atrociously, not only in their basic function as crop-marketing agencies, but in their less important role as providers of agricultural services. Parastatal operating margins (the percentage of income from crop sales absorbed in internal operations) are often ruinously high, leaving little income with which to make payments to the producers. Indeed, there have been some cases in which the managerial expenses of agricultural parastatals actually exceeded the return they receive from market sales! As a result, they have sometimes fallen far behind in their payments to farmers, in some cases delaying payment by as much as 2 or 3 years; or they have made only partial payment, or made payment but only in worthless scrip that farmers find they cannot use in the marketplace. Administrative inefficiency and financial corruption have generated a deeply strained relationship between the agricultural parastatals and their producer clientele, a factor that, in itself, has led innumerable farmers to abandon official state-controlled markets for the greater freedom and flexibility of the informal marketplace. Perhaps more importantly, the compulsion to deal with inefficient and extortionary bureaucracies has operated as a fundamental disincentive to increased agricultural production.

The strategy of import substitution

The policy of industrialization through import substitution has also had disastrous effects on the agricultural sector. The ongoing capital requirements of the new industries have proven to be so great that agriculture has been virtually starved of the capital necessary for its own modernization. In retrospect, it is surprising that so little attention was paid to the amount of hard currency that would be required to launch a series of consumer goods industries and to keep them supplied with up-to-date capital equipment, spare parts and imported raw materials. It may be useful to recall, however, that agricultural commodities had enjoyed an unprecedented price boom during the mid-1950s, following the Korean War and that, as a result, African countries came to independence with foreign exchange reserves that appeared to be fully adequate. Of the many constraints on an industrial strategy that were initially considered to be important, the most important seemed to revolve around the lack of skilled personnel and the absence or relative weakness of the entrepreneurial class. Very few developmental experts considered the foreign exchange constraint to be a major obstacle to successful industrialization, perhaps, in part, because import substitution was considered to be a method of conserving, not expending, foreign exchange.

The agricultural marketing boards dealing with export crops were the principal source of hard currency for financing Africa's new industries. Almost universally, African governments, hungry for the financial wherewithal to launch an industrial revolution, turned to the marketing board reserves as their principal source of investment capital. The operative idea was that the difference between the prices that the marketing boards paid their farmer clientele and the prices they received on world markets would be used as venture capital for the new industrial sector. This decision converted the marketing boards from rurally oriented institutions, whose principal mandate was the economic vitality of export-oriented agriculture, to a wholly new function as the financial launching pad for urban industrial development.

The siphoning off of marketing board reserves can be assigned a great deal of the blame for the steady deterioration of Africa's position as a leading exporter of agricultural commodities. As the capital needs of urban industries grew, producer prices for export crops had to be reduced to a lower and lower percentage of the world market price so as to replenish the marketing board reserves. The capital requirements of urban industries also reduced the capacity of the marketing boards to conduct their other assigned functions such as research and the provision of services. Within a few years, the marketing boards were converted from a position as ally of the rural producer to that of adversary, exploiting the productivity of export farmers but delivering little, if anything, in return.

Africa's attempt to industrialize through import substitution has now been generally acknowledged as a costly experiment that has failed dismally in the majority of cases. Michael Roemer (1982, 132) is expressing a fairly widespread sentiment in the following criticism:

> Economists have universally condemned this strategy as typically practiced. Yet, although it is losing favor everywhere, it is still widely in use in Africa. The fundamental problem with import-substitution strategies in Africa is that they focus development efforts on industrialization, although it is agriculture that remains the base of the economy and that employs the great majority of workers.

Import-substituting industries in Africa have represented a continuing burden on the earnings from agricultural exports and, to this degree, have prevented a reinvestment of capital in the agricultural sector, gradually depleting its capital base and diminishing its capacity to respond to changing conditions of demand in the world market. Lack of flexibility to respond to short-term market niches has sometimes been cited as a significant reason why Africa has fallen so far behind other developing areas in maintaining its share of the world's agricultural trade.

Import-substituting industries have absorbed capital that might have been more effectively used to finance the development of industries for processing agricultural exports, thereby enhancing a country's foreign exchange earning capacity rather than undermining it. This pattern of industrial development has also been an important contributing factor in shifting the internal terms of trade within African countries against the rural sector. Because of their lack of competitive efficiency, import-substitution industries have had to be given a high level of protection from foreign competition. Thus, the goods that they produce are almost invariably much higher priced than those available from abroad. Since consumer goods have now become an integral part of Africa's rural lifestyle, the result of this protectionism has been to force rural consumers to pay higher prices for the goods they purchase at precisely the time they are being offered lower and lower prices for the goods they produce. As the effective incentive for production has fallen, so too has the level of marketed production. In the most extreme cases, export farmers have turned away from export crops that are subject to government controls to the cultivation of unregulated commodities. Since these are typically food items for local consumption, the impact on export volumes, and hence on export earnings, has been extremely serious.

Conclusion: the need for policy reform

Taken cumulatively, the patterns of policy intervention pursued by African governments toward their rural sector have resulted in a shift of the internal terms of trade adverse to the economic well-being of the rural sector. Driven by powerful coalitions of urban-based interest groups, African political leaders have adopted a set of economic measures that cater to the economic needs of city dwellers at the expense of the steady impoverishment of rural social classes, especially smallholder farmers. Economic policies based on short-term political expediency can now be shown to have disastrous long-term effects. In contemporary Africa, both rural starvation and rising levels of urban unemployment are the outcome of a set of agricultural policies designed to subsidize the cost of living of urban consumers at the expense of rural producers. Today, observers at virtually all points on the political spectrum strongly agree that the major starting point for agricultural recovery must necessarily be the implementation of a set of measures that will shift the internal terms of trade back toward the rural sector and improve real economic incentives for agricultural producers.

Price reforms

The first step should be a substantial, indeed, radical improvement in producer prices. Unless farmers' incomes are dramatically increased, the current trend toward deteriorating per capita production in both the food and export crop sectors will only continue. An improvement in the price structure for basic food crops is especially essential, since the decline in food production is directly related to famine conditions of deepening severity; and these, in turn, have created a need for food imports so great as to crowd out the importation of goods vitally necessary for the industrial sector. The impact of increasing food prices on political stability must be addressed squarely. African governments that view this policy reform as politically suicidal are extremely unlikely to undertake it.

The evidence that rising food prices lead directly to urban unrest and, indeed, to critical problems of regime stability is overwhelming. The list of African cities where heightened food costs have been directly related to recent major episodes of political instability now includes such disparate locations as Khartoum, Cairo, Tunis, Rabat, Kampala, Nairobi and Monrovia. The relationship between urban inflation and political instability is so direct that it is not surprising that so few African governments have contemplated serious reforms in their food pricing systems, or, for that matter, in other areas of agricultural policy (Singh, 1983; World Bank, 1983). Governments that place a value on their own political survival have found it far easier to cite factors beyond their control as the source of the problem and to use the famine conditions that result from poor policy as a source of leverage on the donor community for food assistance. While this strategy may help assure day-to-day political survival, it is valueless for long-term economic recovery.

The critical question, then, is how to induce reforms in the pricing system that do not fundamentally threaten the political security of governing elites. There is no simple answer to this question and those, such as Africa's more monetarist donors, who urge the full and immediate adoption of a free-market pricing system simply avoid the political ramifications of this remedy and, therefore, the issue of its political feasibility. It would be far more realistic to develop a course of action that, by combining partial and gradual solutions, achieved an element of political practicality. The essential presupposition underpinning a gradualist strategy is the recognition that today's agricultural crisis is the product of many years of questionable agricultural policies and, precisely for this reason, it is not amenable to easy remedies. The proponents of producer price increases must recognize that the problem cannot be addressed on a piecemeal basis, because a system of agricultural producer prices is a fragile and delicately balanced mosaic of individual crop prices and abrupt changes in any one of these could easily have unintended and unwanted consequences for others.

A workable strategy of pricing reform would probably need to have a number of components. Pricing changes would need to be gradual, not only to avoid an inflationary shock in the cities but also to provide opportunity to assess the effect of change in any one sector of the agricultural economy on the productivity and vitality of other sectors. Massive increases in producer prices for food staples that simply lead to the wholesale substitution of food crops for export crops are not really a solution to the problem. During a transitional period of pricing reform, the donor community would have a critically important role to play. To help hold down the rate of urban inflation, donors would need to be prepared to continue to supply food assistance, much of it on concessional terms. The precise level of food assistance would need to be carefully calibrated so that food aid would not constitute a disincentive to local production. Food aid could be an important political condition for local pricing reform, but it would also be essential to reduce the level of food assistance as increased producer prices triggered an increase in the level of marketed domestic production.

An additional ingredient in a politically workable program of price reform would be the willingness of the donor community to provide capital assistance for the industrial sector. Industrialization by import substitution may have been a flawed strategy from its inception, but these industries do offer an important vehicle for quickly regenerating urban employment. This is not only economically essential, but, insofar as an increase in employment provides an additional basis for political stability, it is a vital part of the broad assurance to African governments that willingness to engage in policy reform need not be self-endangering. Most importantly, the regeneration of the consumer-goods sector is integral to the success of pricing reform. Unless a supply of consumer goods to the countryside can be re-established, price increases are meaningless. In some African countries, the process of economic decline has gone so far that essential consumer items such as soap, cloth, sugar, radio batteries, bicycle tires and cooking oil have simply disappeared from the countryside. Where this has occurred, farmers have no incentive to respond to price increases, because an improvement in cash income does not really produce an improvement in living conditions.

Currency devaluation

Currency devaluation is an additional prerequisite of agricultural recovery. No other single policy so conspicuously reflects the urban bias of African governments or so drastically tilts the internal terms of trade against rural producers as does the imposition of an overvalued exchange rate. By facilitating the tendency toward conspicuous consumption in the cities and by draining the rural sector of foreign exchange to finance the needs of import-substitution industries, currency overvaluation deprives agriculture of badly needed capital. It is causally related to agricultural decline as an integral feature of pricing systems that suppress production by lowering the real returns to agricultural producers. This is especially evident in the case of export crops which are, in effect, 'taxed' by the amount of the overvaluation, but it is also operative in the food crop sector where food prices are typically set sufficiently low in order to minimize any possible tendency to substitute food crops for cash crops in the rural areas.

The precise importance of devaluation is sometimes missed by critics who argue that it is not likely to stimulate exports by reducing their prices on world markets. Those who hold this point of view point out, quite correctly, that the world market prices for agricultural commodities are set in hard currencies, typically the US dollar, and that devaluation does not have any effect on these prices. The purpose of devaluation, however, is not to lower export prices, but to facilitate a substantial increase in the domestic producer price (this is paid in local currency) and, in this way, help to stimulate marketed production. Since one of the major symptoms of Africa's agricultural decline has been a falling share of world markets for key commodities, devaluation provides a direct means of re-establishing its former market position.

Like other strategies that seek to address the problem of agricultural decline by shifting the internal terms of trade back toward the countryside, currency devaluation poses a serious political problem because of its potential for inflationary impact on the urban population. The critical question with respect to this reform, then, as with respect to reform of producer prices, is how to achieve an economic goal while minimizing the political strain it imposes. The answer is not radically different. Currency devaluation is not an all-or-nothing proposition; it can be achieved in a sequence of phased steps so as to reduce the impact at any given moment. Devaluation can also be accompanied by other measures such as wage increases or tax reduction that are intended to yield a partial offset of its effects especially on poorer segments of the urban population.

If donor nations cooperate by making financial assistance available during the difficult period of transition to a realistic exchange rate, the inevitable increase of urban discontent should not be so great as to destabilize governments intent on the reform process. Much may depend, ultimately, on the character of the specific government in question: those that have a democratic base and that seek to distribute the burdens of reform equitably may have a greater chance of survival than those that seek to maintain the privileges of the few while lowering the standard of living of the majority of the population.

Reform of the parastatal sector

The third broad area where policy reform is essential to economic recovery has to do with the continent's ubiquitous agricultural parastatals. It is extremely doubtful that a process of agricultural recovery can be launched as long as state corporations continue to hold legalized monopolies over the purchasing, processing and vending of major commodities. These corporations have few, if any, incentives for efficiency or economy of operation; they have become the subject of gross political abuse, including rampant corruption and nepotism; perhaps more importantly, they have generated an atmosphere of cynicism and mistrust among their producer clientele. Under these conditions, it would be tempting to conclude that the most fruitful course of action would be to abolish the parastatals altogether and, indeed, parastatal performance in Africa has been so dismal that genuinely moderate observers have sometimes been driven to this conclusion along with ardent proponents of the free market.

This point of view, like other extreme remedies, does not fully address the political and economic complexities of the present situation. Because of their substantial resources, large numbers of personnel, and organizational linkages throughout society, parastatal corporations wield enormous political power. African political elites typically regard the parastatal sector as a key source of political support and are understandably reluctant to contemplate a policy change that might alienate this powerful constituency. Moreover, certain of the economic functions assigned to the parastatals remain of great importance, even if the parastatals, as currently constituted, do not perform them very well. The central question, then, is what sort of policy reform might make it possible to reform the parastatal system in order to achieve some of the efficiency and economy of operation of the private sector while, at the same time, maintaining a mechanism to provide important economic services that are not likely to be delivered by the private sector.

Defenders of the parastatal system point out that these organizations are given a virtually impossible job to perform and that they are required to operate within a framework beyond their jurisdiction. Agricultural parastatals, for example, are expected to stabilize the financial return to the farmers (a goal that almost no agricultural system in the world has been able to achieve), but they are given almost no control over the pricing system, since the prices at which they must buy and sell the commodities for which they are responsible are generally set at the cabinet or ministerial level. Parastatal corporations are also characteristically required to purchase whatever supplies of a given crop are offered to them rather than to operate on the basis of their own judgment about what the market can absorb at a given time. Parastatal supporters point out, as well, that the operating costs of these organizations are entirely out of their control, since decisions about staffing levels and salary scales are also made externally by members of the cabinet or special cabinet committees. The managers of export parastatals operate under a particularly severe disability, because the prices they can pay to growers are a function of officially set exchange rates and, where currency overvaluation is a serious problem, they are unable to avoid the downward distortion of producer prices that this causes.

The ultimate defense of the parastatal system rests on an implicit comparison with the performance of the private sector. While acknowledging that corruption and inefficient performance are extremely serious matters, supporters of agricultural parastatals argue that a system of state control ultimately provides far more benefits for peasant farmers than would a free market. There is an intense conviction that private merchants, who might in many cases be racially differentiated from their farmer-producers, would behave far more exploitatively than the parastatals as regards pricing and that the private sector would fail altogether in terms of such vital goals as price stabilization or the conduct of such important functions as crop research.

The debate over the appropriate role, if any, for agricultural parastatals is as intense as any in the field of African development today. Numerous critics of the parastatal system tend to believe that its problems are inherent in the very concept of official monopoly and argue strongly for thorough-going privatization of the agricultural sector as the only solution to today's crisis. Others tend to believe that a more workable and politically feasible solution lies in the search for a formula that will combine the best features of competitive private trading with retention of parastatals to guarantee the carrying out of certain functions that only state institutions are well equipped to provide. Those who propose a more mixed system tend to argue that parastatals should be buyers and sellers of last resort, establishing a price band within which private traders could operate freely. They also tend to believe that only state institutions have the capacity to carry out such important functions as (a) monitoring the behavior of the private sector; (b) the creation of a strategic food reserve; (c) the dissemination of food aid to especially needy regions; and (d) the conduct of agricultural research and dissemination of its results. It is a measure of the failure of the parastatal system that virtually no one defends its retention in its present form.

Reform of the industrial sector

It is not in Africa's long-term advantage to retain an industrial strategy based on import substitution. Industries based on this principle seem inherently prone to inefficient and high-cost operation, for industrial managers who can depend upon protected markets buffered by high tariff barriers and outright import restrictions have neither the incentive nor the impetus to improve their industrial processes. Rather, they tend to focus their efforts on maintaining the goodwill of the politicians in whose power it is to maintain the protectionism that guarantees profitability. Moreover, as a result of protectionism, industries based on import substitution compete unfairly for labor and other inputs with industries that are export-oriented and that might make an economic contribution by adding to a country's foreign exchange earnings.

African governments inclined to withdraw their support from heavily protected industries face a difficult conundrum. In the midst of the current crisis, these industries do offer an excellent prospect for increasing urban employment and for improving the availability of consumer goods. In addition, in the absence of a viable alternative, governments that withdraw their support from import-substituting industries run the immediate risk of having no industrial sector at all. On the other hand, governments that continue their support for import substitution will continue to find themselves drained of foreign exchange, burdened with a high-cost, low-productivity industrial sector, inherently prone to corruption and unable to compete in foreign markets and, therefore, chronically short of the financial reserves necessary to replenish the agricultural realm. The most difficult question about policy reform poses itself once again: how to disengage from a failed policy of import substitution in such a way that the disengagement itself does not worsen the problem of economic malaise.

The only workable answer to this question is that the process of economic disengagement from import substitution must be characterized by the same prudent gradualism that marks the process of reform in other policy areas. Industries that have enjoyed a long period of protection cannot be expected to adjust overnight to economic competition from the world's most efficient industrial exporters. A certain proportion of the protected industries may never be able to compete on the basis of their productive efficiency. The process of ending the protectionist policies that encourage infant industries never to mature is one of endless difficult choices: how quickly to end tariff protection and where; how quickly to allow foreign imports to invade domestic markets and precisely which ones; and whether there might be certain protected industries that, despite their economic inefficiencies, merit continued protection for other than purely economic reasons. The movement away from import substitution is one that calls for continuous and meticulous empirical research.

The decision to end import substitution could be taken to imply acceptance of the view that there is an alternative, export-led growth strategy available, and that this strategy has greater potential for improving the material conditions of the African peoples. This is an issue that merits full exploration before a final rupture with the strategy of import substitution is implemented (Roemer, 1982). Even World Bank economists have begun to cast doubt on the feasibility of the export-led growth model. In recent years, the prices of primary agricultural commodities have declined sharply in world markets and, given the intense competition among primary agricultural exporters, as well as the relatively slow growth in world demand for these products, it is unrealistic to expect commodity prices to rebound in the foreseeable future (Singh, 1983; World Bank, 1983). Nor, as another World Bank study suggests, is it feasible to turn to the processing of primary agricultural commodities as a means of attaining an improved position in the world economy, since in an effort to maintain their own industrial base, the world's industrial nations have erected their own systems of protectionism to discourage the import of finished goods (Nelson, 1981). As a result of neoprotectionism in the industrial world, African nations that begin to process primary agricultural commodities, such as cotton and rubber, could well find their world market position even worse than it was before.

The selection and implementation of a strategy of policy reform is, therefore, an extraordinarily difficult and complex problem. Not even the most zealous advocates of policy reform would suggest that these remedies alone can elicit a greater supply response from Africa's peasant farmers. Attention must be paid, as well, to improving the physical infrastructure, to removing the innumerable petty bureaucratic restraints on production, such as bans on the interdistrict movement of goods and to providing greater physical security in the countryside. One of the greatest constraints on fuller farmer participation in the marketplace is the sheer physical difficulty of moving goods from the farm to the nearest market or transportation center; another is the danger of robbery by criminal gangs or extortion by corrupt local officials.

For the time being, the most that can be said is that thorough-going reform of the policies that most affect agriculture is an absolute precondition for the revival of overall economic vitality. If such reform is undertaken successfully, Africa may well be in a strong position to recapture its former share of world market trade in key agricultural commodities and to regenerate its badly faltering systems of food production. Moreover, unless agricultural policy is successfully reformed, no industrial policy is likely to succeed, because there will simply be no resources with which to implement one. Reform of agricultural policy may not insure a more prosperous and self-sufficient future, but continuation of present policies rules out that possibility.

References

Abernethy, D. B. (1983). Bureaucratic growth and economic decline in sub-Saharan Africa: a drama in twelve acts. Berkeley, CA: African Studies Association.

Bates, R. (1981). *Markets and States in Tropical Africa: The Political Basis of Agricultural Policies*. Los Angeles: University of California Press.

Berg, E. (1981). *Accelerated Development in Sub-Saharan Africa: An Agenda for Action*. Washington, DC: World Bank.

de Wilde, J. C. (1984). *Agriculture, Marketing and Pricing in Sub-Saharan Africa*. Los Angeles: African Studies Center and Crossroads Press, University of California.

Food and Agriculture Organization of the United Nations (1985). *Food Situation in African Countries Affected by Emergencies: A Special Report*, pp. 2–5. Rome: FAO.

Gulhati, R., Bose, S. & Atukorala, V. (1985). *Exchange Rate Policies in Eastern and Southern Africa, 1965–1983*. World Bank Staff Working Paper No. 720. Washington, DC: World Bank.

Hart, K. (1982). The state in agricultural development. In *The Political Economy of West African Agriculture*, pp. 83–109. Cambridge University Press.

Jackson, T. (1982). Project food aid as competition with local food production. In *Against the Grain: The Dilemma of Project Food Aid*, pp. 85–93. Oxford: Oxfam.

Nelson, D. R. (1981). *The Political Structure of the New Protectionism*. World Bank Staff Working Paper No. 471. Washington, DC: World Bank.

Roemer, M. (1982). Economic development in Africa: performance since independence and a strategy for the future. *Daedalus*, Spring, 132.

Singh, S. (1983). *Sub-Saharan Agriculture: Synthesis and Trade Prospects*. World Bank Staff Working Paper No. 608, pp. 14–20. Washington, DC: World Bank.

US Department of Agriculture (1981). *Food Problems and Prospects in Sub-Saharan Africa: The Decade of the 1980s*. Foreign Agricultural Research Report No. 166. Washington, DC: Economic Research Service.

World Bank (1983). *The Outlook for Primary Commodities*. World Bank Staff Commodity Working Paper No. 9. Washington, DC: World Bank.

World Bank (1984). *Toward Sustained Development in Sub-Saharan Africa: A Joint Program of Action*, p. 10. Washington, DC: World Bank.

World Bank (1985). *World Debt Tables: External Debt of Developing Countries*, pp. 6–7 and 158–9. Washington, DC: World Bank.

6

Internal factors that generate famine
ROBERT J. CUMMINGS
African Studies and Research Program, Howard University

Introduction

The purpose of this chapter is to provide a critique of the importance of *internal* factors that generate famine in African nations. It is understood that internal and external factors are not isolated from one another, that there is not a typical 'African case', and that the colonial and independent periods of African history are inseparable for understanding the processes by which contemporary Africa arrived at such a significant decline in its food-producing capacity. In fact, Africa has not always been unable to produce sufficient foodstuffs for itself, although the present food crisis has been in the making for a longer period than the past few decades. In an historical context this has been a more recent phenomenon in no small part because of an extraordinary set of forces that have served to change both African social and ecological systems.[1]

A focus on internal factors is necessary because it can guard against the tendency to impose only external solutions. Only by a consideration of internal and external factors might a proper complementary role be developed for foreign assistance. As Ugandan social scientist, Mamdani, noted, 'Only that [foreign] relief is worthwhile which undermines itself in the long run; which restores the initiative of the victims and does not strangle it; which sees victims not simply as objects to be helped, but as subjects potentially capable of transforming their disaster-prone situation' (1985, 96; see also Lappé, Collins & Kinley, 1980; Lofchie & Commins, 1984).

Internally, African leadership must finally develop the *political will* and *imagination* necessary 'to arrest the continent-wide ecological deterioration before it leads to further declines in income and a far greater loss of life from starvation' (Brown & Wolf, 1985, 65). Concurrently, African governments and the international community can then move to 'adopt a development strategy based on environmental rather than narrow economic goals, one that restores and preserves natural support systems – forests, grasslands, soils, and the hydrological regime – rather than meeting a specified rate of return on investment in a particular project' (Brown & Wolf, 1985, 66).

Internal factors that generate famine can be categorized (not in order of priority) as follows: (*a*) the insufficient development of human resources; (*b*) the continued maintenance of inherited colonial institutions and structures; (*c*) the existence of too many small economies; (*d*) the negative trends in food production; (*e*) poverty; (*f*) high population growth rates; (*g*) increased dependence on food imports; and (*h*) the violation of the delicate balance between Africa's environmental realities and the dire necessity of agricultural development. With respect to the last factor, soil erosion, deforestation, desertification, overpopulation, overgrazing, and poor planning and management strategies are the result of the human activities that are mismatched with environmental realities. Yet, we know that the responses (or lack thereof) of Africans to these existing as well as impending social disasters signify neither malice nor ignorance on their part. Rather, they illustrate the depth of linkages between the internal and external factors that affect African development. No attempt is made to discuss each of these variables separately, because the first six are identified as economic questions. Nor do I discuss as internal factors herein military coups, political corruption and graft.

History

The colonial period for most of Africa created an artificial pattern of growth that did little to develop a foundation for African economies. The parasitic nature of colonialism caused it to subsume so much of the continent's indigenous socioeconomic structures and resources (providing self-sufficiency) that the eventual departure of colonial powers left only a shell of the original socioeconomic structures. This fact has become appreciated only recently. Evidence shows, for example, a growth in food production rate in Africa of less than 2% per year between 1960 and 1980 but a decline in per capita food production during the same period (US Department of Agriculture, 1981). The population growth rate increased from 2.1% in 1955–56 to 2.7% by the late 1970s. These figures are instructive because they introduce the rationality of and the choices available to the small peasant landholding farmers and the complexity of the social and economic relations that entraps them.

Africa was generally integrated into the colonial (imperialist) economy by being transformed into either (*a*) a cheap labor reservoir or (*b*) a cheap raw materials reserve. As Mamdani points out, the former case 'was the migrant labour systems where the wife remained a peasant producing food in the village but the husband migrated as a worker to a plantation. He was employed only part-time; the rest of the year, he returned to the village and lived off the food cultivated by the wife'. The latter case simply 'collapse[s] the distance between husband and wife, with the wife still producing food and the husband producing an export crop' (1985, 93).

The entire colonial system of cheap raw material production required that (*a*) labor met a considerable share of its own cost of reproduction (food costs), and (*b*) labor remained the primary input in production, or more specifically, the technological input remained low. This colonial approach has a serious exploitative impact on peasants, the major producers of African wealth.

While the first type of peasant exploitation, (*a*), is relatively indirect, as a result of unequal market relations which compel peasants to sell cheaply and buy expensively, the second, (*b*), is direct, as a result of forced labor, forced cash-crop production and, in the case of the church or the party, forced cash contributions. The single most important result of this dual exploitation of the African population (unequal market relations and direct force) is that peasants operate with a permanent handicap, because their surplus produce is continuously siphoned off. Walter Rodney (1972) was correct, when he observed that the African peasant entered colonialism with a hoe and came out of it with a hoe. It is clear now, however, that the hoe carried into colonialism was locally produced, while the one coming out was imported. Thus, independent African peasants have come to a point where they have no control (or choice) over their implements of labor, which results in their attempts to impose control over the remaining factors of land and labor available to them.

The peasants are 'guilty' of increasing cash-crop activities, further reducing the production of food crops, overworking the land with subsistence farming, and having many children to maximize the family's labor output. Yet, it is fair to note that these African actions, though destructive, are economically rational from the viewpoint of the African peasant. In the case of family size, these peasants are not poor because they have large families. Rather, they have large families because they are poor. The nature of their condition allows only contradictory solutions: short-term responses (to long-range problems) that result in the intensification of such problems as soil erosion, overpopulation, declining food productivity, and dependency.

Colonial approaches to African growth patterns are directly related to such factors as education, urbanization, income levels, social services, health facilities, diseases, poverty levels and the production (and extraction) of resources for urban markets. Plans for Africa were made within the framework of colonialism. They rearranged the 'traditional' economic activities within the colony to fit *new* requirements, requirements that existed in countries outside Africa. Paul Bairoch (1975, 19) notes that 'Colonialization usually endowed agriculture with a dualistic character: on the one hand, there were export crops generally, but not invariably, dominated by a plantation system . . . On the other hand, there were subsistence crops, often excluded from the best land'.

Today, a number of researchers argue that the primary causes of underdevelopment in Africa are natural factors including extreme climatic variability and variable soil quality. Certain areas are said to have been devoid of resources sufficient to enhance development. In those areas where resources are available, however, the argument shifts from one focused on 'resources' to one focused on the 'backwardness' of the local populations who are somehow unable naturally to appreciate, manage, or use local resources efficiently. Nii Plange (1979) argues vehemently against such 'naturalistic fallacies', believing that the local historical structures and patterns of human life activities that are said to represent underdevelopment are in fact *inauthentic*; that is, they did not result from direct interaction of the local population with its environment. He argues that they are the effects of the expansion of European capitalism on local communities, noting that

> Within the context of this 'colonial' development there occurred the domination over and distortion of precapitalism modes of production. The economic logic behind this was to rearrange the existing patterns of economic activity with a view to extending areas of capital investment, develop cheap sources of raw materials and labour, and finally to establish captive markets: in sum, the creation of a colonized economic system. To accept these conditions as 'natural' is to be ahistorical, theoretically myopic, empirically erroneous, and furthermore, to demonstrate lack of understanding of the dialectics of man and nature and the consequences of this for the development of society. (*Plange, 1979, 4*)

This economic interpretation is reinforced by Mamdani, who observed that:

> The Karamoja famine can't be understood without an historical analysis. Its starting point must be to understand that the

Karamojong people lost roughly 20% of their grazing land, in phases, through either the redrawing of administrative boundaries in the 1920s or creation of National Parks like Kepepo later. It is this fact which called forth a *change in pastoral practices*. No longer could dry grasslands be rested up to the annual burning. Not only was all grazing land used throughout the years without any annual burning, forests were progressively cut down to increase the grazing area. The British saw this as a problem of 'overgrazing', not one of land confiscation! So, having first grabbed some land, their 'solution' was to grab some cattle now!! This they did indirectly by introducing simultaneously a tax with a government monopoly of buying, so that the buying price of cattle could be adjusted to the tax assessed . . . the end *result was a change in the whole ecology of Karamoja* [emphasis added]. (*Mamdani, 1985, 95; see also Hedlund, 1979*)

Rodney and other underdevelopment theorists have forced the debate regarding the adverse impact of colonialism on African development. Eicher (1982, 157), for example, notes how little attention colonial administrative authorities paid to 'investments in human capital, research on food crops, and strengthening of internal market linkages'. He continues his comments, noting that

. . . colonial governments gave little attention to the training of agricultural scientists and managers. By the time of independence in the early 1960s, there was only one college of agriculture in French-speaking tropical Africa. Between 1952 and 1963, only four university graduates in agriculture were trained in francophone Africa and 150 in English-speaking Africa. By 1964, there was a total of three African scientists working in the research stations in the East African countries of Kenya, Uganda, and Tanzania. (*Eicher, 1982, 157*).

Political power at independence was transferred to African leaders who were generally perceived to be the potential indigenous middle class necessary to pursue (and implement) neo-colonial strategies. These leaders have tended to maintain the inherited transportation and communication systems, as well as the inherited export-oriented models of trade and exchange, while assigning relatively minimal government investments to new institutional arrangements designed to meet their respective national needs, for example, for the purpose of increasing agricultural (including food) production. Eicher (1982, 163) writes that

Throughout much of the post-independence period, most states

have viewed agriculture as a backward and low-priority sector,
have perpetuated colonial policies of pumping the economic
surplus out of agriculture, and have failed to give priority to
achieving a reliable food surplus as a prerequisite for basic
national, social and economic goals.

Thus, food and agricultural policy has been left to the purview of relatively
low-level managers, as suggested by the dearth of agricultural ministerial
appointments throughout the continent in spite of the fact that more than
70% of the entire continent's work force is rural and agricultural.

In the few decades since political independence, African leaders have
debated the following five key questions on food and agriculture (for more
detailed comments on these debates, see Eicher, 1982):

1) What are the appropriate priorities for industry and for agriculture in
   development plans and budget allocations?
2) What is the relevance to Africa of western economic models, as opposed
   to the 'political economy' (primarily dependency and class structure)
   and the so-called radical models of development?
3) To what degree should pricing and taxation be used to achieve
   agricultural and food policy objectives?
4) Should Africa pursue an agrarian socialist or an agrarian capitalist
   strategy of development?
5) How might (could) the Green Revolution influence (increase) the
   availability of choices to African farmers?

These debates demonstrate the difficulty facing African countries as they
entered the world of independent states in the 1960s, a time when
industrialization, intense international competition, and the successful
modern state represented the reality of the times. These newly independent
states were caught up in the currents of a rapidly changing world and were
forced either to participate or to be left behind. Too many factors made
these leaders reluctant, often unable, to employ as development models the
histories of nation-building of those countries that used an agricultural base
and self-sufficiency in food production as the foundation of a viable modern
state.

These debates also demonstrate the complexity of Africa's problems.
However, it is the 'failure of most African states to develop an effective set
of agricultural policies to deal with the technical, structural, institutional
and human resource constraints' (Eicher, 1982, 163) that is most central to
Africa's current state of development. This internal factor (i.e.
developmental position) generates famine. The refusal or fear (i.e. lack of
political will or imagination) of African leaders to restructure their

agricultural institutions perpetuates the present state of affairs. However, this in no way negates the fact that a significant part of the failure of African development falls squarely on the colonial legacy as well as on foreign economic advisers in Africa and in donor countries.

Yet, after three decades of political independence, African leaders must ulimately look to themselves for having retained the structures of their own entrapment. They can no longer afford, economically or politically, to allow themselves to remain wrapped in the cloak of colonial capitalist exploitation, as it will further delay their ability to meet the challenges of development. It appears that they have finally come to this realization.

Significance of internal factors

The Lagos Plan of Action can be used as a point of departure to critique the importance of the internal factors that generate famine in Africa. Internal factors are important because they have served (*a*) to forge the relationship between emergency assistance and development; (*b*) to clarify the problem of 'welfarism' as a development tool; and (*c*) to identify, on the one hand, the inability of African leadership to take the necessary initiatives to confront questions of internal economic development and, on the other hand, the frequent unwillingness of that leadership to demonstrate the political will through continental cooperation necessary to transform inherited institutions which serve to maintain those internal factors most supportive of sociopolitical and economic growth, but *without development*.

Today, there is a growing acceptance in many quarters of the serious impact that poor planning and policy implementation have had in giving famine continued access to the continent. There is also the fact that the reality of the past two decades of stagnation in output has forced a desire on the part of African leadership to provide an appropriate response to the continuing crisis associated with the deep, and ever-widening, economic stagnation.

The most meaningful response to date has come from Africa by way of the *Lagos Plan of Action for the Economic Development of Africa, 1980–2000* (Organization of African Unity, 1981; see also UN Economic Commission for Africa, 1981*a*, *b*, 1982*a*, *b*, 1983; Browne & Cummings, 1985). The Lagos Plan of Action (LPA) constitutes the first comprehensive continent-wide effort to articulate the desired long-term development objectives for Africa. The LPA, however, is concerned almost exclusively with questions of economic development for Africa. It does not include efforts to link the economic strategy with an environmental one, although present research data are pointing dramatically in the direction of required ecological

considerations and activities to rehabilitate water resources and depleted
soils among other resources in order to improve food supplies, to reduce
population growth, and to reestablish productive economies throughout the
continent. (The LPA does, however, note that 'restoration also deserves
special attention'.)

Briefly, the LPA envisions a new and more balanced path for Africa: the
path of collective, self-reliant development. The document, however, has
notable shortcomings. It states, for example, general rather than specific
goals, while failing to set forth quantifiable industrial volumes or monetary
terms, specific strategies or schedules. It does not set solutions (at this
juncture) to African social problems. The path proposed by the LPA,
however, does veer from the one currently pursued by most African
leadership. It calls for an Africa which (*a*) uses its extensive resource base
primarily for its own development rather than contributing to the
development of other (often industrial) societies as a result of export policies;
(*b*) expands its industrial base primarily for home consumption and only
secondarily for export; (*c*) relies primarily on its own technical skills and
not on those of foreigners; and (*d*) develops an industrial base and
consumption patterns suitable to African needs and customs rather than to
the blind adoption of models from abroad.

Since this type of development requires minimum size markets, the LPA
seeks to form effective African subregional groupings. Once formed, these
nations can cooperate on training, trade, technology, monetary issues,
energy, communications, environmental protection, and a plethora of other
matters. Subregional cooperation will gradually expand its continental
efforts, eventually leading to an African Common Market and, by the turn
of the century, to an Economic Community of Africa. This is the vision,
as articulated in the Lagos Plan of Action. The LPA itself evolved out of
the menacing impacts on African nations of internal factors.

The evidence today is replete with data demonstrating that certain internal
factors, such as food deficits, agricultural pricing policies, budgetary
imbalance, overpopulation and ecological imbalance, together with inherited
colonial developmental planning models, have served to generate famine –
one of the most obvious consequences of the breakdown between the people
and their natural support systems. The primary importance of these internal
factors to those concerned about Africa's (and the world's) survival is their
effect on our understanding of the great need that now exists to construct
a new set of relationships between the world's developed northern tier
countries and its developing southern tier countries.

Most African countries could be characterized in the 1960s by respectable economic growth rates, viable trade balances, and improved education and health programs. The first decade of African independence, however, was followed by the 1970s, characterized by high energy prices and the Sahelian and Ethiopian droughts, events that forced economic stagnation, balance of payments deficits, and an ever-increasing debt burden on already faltering national economies. This situation has changed little in the 1980s.

Economic, ecological, and social conditions have deteriorated to such an extent in the present decade that the Worldwatch Institute in its *State of the World – 1985 Report* has declared that 'Without a sharply expanded effort in both family planning and farming, much of Africa will drop back into the first stage of demographic transition' (Brown *et al.*, 1985, 21; see also Brown & Wolf, 1985). In short, the food–population balance in Africa is in a precarious, worsening state. Although roughly 80% of the people in the tropical and semi-tropical regions of the world are actively farming, the food needs of these populations are not being met (LeMelle, 1984). Malnutrition and related diseases are prevalent and starvation periodically ravages entire populations in many of the severely drought-stricken areas of Africa. The inability to meet basic food needs has now reached crisis proportions in some countries, where extended periods of undernourishment and malnutrition have become chronic. The working populations, as a direct result, have become more vulnerable to disease and less capable of sustained work with a resulting reduction in farm productivity.

The goals of modernizing agriculture and of establishing a complementary industrial movement in these independent states have been frustrated, with few exceptions, as a result of internal political and economic failures as well as of external shocks. The high expectations for a Green Revolution in Africa notwithstanding, Africa is moving rapidly toward a third decade of declining food production and increasing population growth.

While meaningful differences can be identified between the performance of those developing countries that chose a market-oriented economy and those that opted for a centrally planned one, the overall results of both of these approaches to African development have fallen short of even modest expectations. This admission underscores the deficiencies in our knowledge and understanding of the process of agricultural modernization in Africa and elsewhere.

Yet, 'experts' from industrialized countries have been advising with great confidence the developing countries for at least the past three decades. In fact, an improved understanding of the relationship in transitional and traditional societies of farm size, land ownership, rural migration, the role of peasant women farmers, sharecropping, farm mechanization, rural markets and farm income, to peasant productivity and to social barriers to the adoption of multiple-cropping systems or innovative agricultural practices could be of direct importance for the start of successful agricultural development in Africa in the next few decades.

The Lagos Plan has identified self-sufficiency in food as the first priority in the articulation of its goals. Food-production problems, highlighted by famine, are central to the present difficulties within many African countries. Three principal factors contributing to this problem are (*a*) population pressures, (*b*) production constraints in traditional societies, and (*c*) certain policy issues affecting food production (LeMelle, Keynote address to Inter-University Center for European Studies, Montreal, May 1984; Lofchie & Commins, 1984).

Overpopulation is at the center of the food problem confronting not only Africa but other underdeveloped areas as well. If the present trends continue, an additional billion people or so will expand the population of these countries alone to almost five billion by the turn of the century.

> Viewed in another way, by the 1990s the annual increase of some 95 million more people would result in the creation of a new population group the size of India's every seven years. As we have seen over the past thirty years, in countries of low rates of economic expansion and widespread poverty, the tendency of the population growth rate is to increase rather than decrease. (*LeMelle, 1984, 6*)

Confronted by harsh and inequitable social and economic conditions, a family's desire for more children is heightened by the reality of high infant mortality, the need for labor, and the prospect for security during old age. Generally, children (particularly girls) in their most nutritionally vulnerable years are the most severely affected in times of food scarcity. Thus, population pressures limit the availability of food per capita as well as contribute to malnutrition. Yet, the countries clearly identified as food deficit countries with high population growth rates are 'heading, according to the 1982 Report of the International Food Policy Research Institute and other studies, for increasing shortfalls in basic food crop production by the end of the century' (LeMelle, 1984, 7).

It is estimated that by the year 2000 the countries with the majority of the world's population will face an annual deficit in food staples of about 226 million metric tons, three times their 1977 deficit. With population growth increases outstripping the ability to produce and supply the necessary foods for mere subsistence, malnutrition would affect, by the year 2000,

1.3 billion of approximately 4.8 billion people living in underdeveloped countries. Without expanding their incomes, underdeveloped food-deficit countries would be unable to procure needed foodstuffs from the relatively more expensive food surpluses in the developed countries. By the year 2000 it is also estimated that Africa's export earnings will be able to cover only one-half the cost of its import requirements.

Although some developing countries with large populations such as Brazil, Mexico, and Indonesia have managed in recent years to decrease their rates of population growth, without continued high economic expansion, there is certainly no guarantee that such trends will persist for a long time. Here, the Republic of Kenya is forcefully instructive.

During the first decade of independence (1963–74), Kenya underwent impressive economic growth in the range of 6.5% per year. The government adopted a national family planning policy program and was perceived to be on the way to achieving stability in its population growth rate. Then, the oil crisis shocked the government's fiscal plans and the Kenyan economy suffered serious internal dislocations. By 1980 the government was shocked again to learn that its population growth rate was the highest in the world at 4% per year. Moreover, for the first time, it faced a national food shortage.

On the one hand, successful family planning programs in several developing countries have demonstrated that measures can be adopted to restrain burgeoning population growth. On the other hand, we have too little knowledge about the impact of such policies on the fast-changing but basically traditional societies (Cummings, 1986) to be sanguine about the timely reduction of population pressures upon food production in these countries during the next three decades. Whether population growth rates in Africa, or elsewhere, will decline permitting food production and general economic activity to expand, or whether economic activity will increase, permitting population growth rates to decline, is only one of the serious dilemmas troubling underdeveloped nations, as they face the future.

Food shortages in the underdeveloped nations are also a matter of low productivity associated with cultural practices of subsistent peasant farmers who are only slowly adjusting their behavior and adopting methods consistent with the demands of modern surplus-producing agricultural systems. While 67% of the world's population resides in underdeveloped countries, peasant farming accounts for only 38% of the world's agricultural production. According to the United Nations Food and Agriculture Organization, only one-third of the population in the underdeveloped world (excluding the People's Republic of China) lived in 33 countries whose agricultural production growth rates averaged over 3% per year in the 1970s.

The majority of these countries, parrticularly the low-income states, did not reach production growth rates anywhere near that level.

Production performance, particularly in Africa in recent years, has been falling below the level of population increase. Between 1976 and 1980, for example, Africa's agricultural production growth rates dropped from an annual rate of 2.8% to 1.4%. While the devastating Sahelian and Ethiopian droughts of the 1970s and the droughts of the 1980s in western, northwestern, and southern Africa have compounded production problems on the continent, these conditions, it is fair to say, are only part of sub-Saharan Africa's production problems.

Much of the land in underdeveloped countries has been used in highly extractive ways for centuries. What is left is soil depleted of nutrients. Even though new high-yield varieties of food grains, root and tuber crops are available to underdeveloped countries, the adoption of such crops has been slow. Where adopted, their yields have been only one-third to one-half of projected yields. No doubt, one of the principal constraints on the adoption of high-yield varieties of food staples is the cost of specially prepared seeds, expensive fertilizers, pesticides, power implements and irrigation systems. In the poorest of the underdeveloped countries in Africa (as elsewhere), peasant farmers do not have the resources to pay for the high cost of increased productivity.

While the high cost of technical innovation is an obvious barrier to increased productivity among low-income farmers in the developing countries, the political, social and economic costs of adopting new cultural practices and working in a competitive non-subsistent production environment pose equally limiting constraints on the achievement of higher productivity. In fact, the latter costs represent the principal obstacle to increased productivity. The change required of peasant farmers to adopt new agricultural technology will occur, only if they have confidence in the integrity and commitment of the national political leadership to stand behind the new system. Farmers must also see clear economic advantages in higher productivity. There must be tangible incentives for dropping traditional production technologies and adopting new ones. Increased crop production without a fair price, reasonably secure markets, adequate storage facilities, reliable transportation systems, and expected social benefits is an unreasonable objective for which there would be no real motivation. The failure to realize social benefits, such as the extension of basic health facilities and educational opportunities, also frustrates the incentives among the peasant population to change from subsistence to intensive agricultural production.

Among other social barriers to increased agricultural productivity in many developing countries is the generalized practice of discrimination against women in traditional societies. In most countries women do more than half the field work: planting, cultivating, harvesting, post-harvest processing and marketing. Women often manage rural markets and control a large portion of the national agricultural distribution and transportation system. Despite this dominant role, women are often ignored by agricultural extension and training programs and are invariably discriminated against in land-holding arrangements. It is difficult to imagine how significant increases in food production will be achieved, if such formidable barriers are not removed.

The Green Revolution has demonstrated that new technology of high-yield food varieties, ongoing research and extension, appropriate mechanization, proper incentives and new cultural practices can increase productivity in controlled and closely managed sectors in traditional societies. We have also learned that the technical problems of production may be less difficult to overcome than the political, economic and social barriers to expanded food production.

A third issue that must be considered in examining the capacity of the underdeveloped countries to achieve relative stability in food production is the impact of government policy on agricultural production. While the role of government in providing a suitable economic environment and appropriate incentives for farmers seems self-evident, in practice many governments in developing countries have demonstrated a peculiar ineptness in developing viable agricultural development policy. No doubt economic biases along the spectrum, from a strongly market-oriented option as practiced in the Ivory Coast to a tightly guided (and planned) approach in Tanzania, contribute in an important way to ultimate policy-making in this area. However, the experience of the past 30 years suggests that developing countries in Africa are in no way capable of adopting doctrinaire positions on such food-production-related matters as producer price policies, input policies, marketing strategies and trade and investment policies. Consistent, but flexible, policies on food pricing and procurement are called for in the face of serious structural weaknesses at every level in the food sector in most African countries. Pricing policies meant to encourage food production can easily be undermined by import policies that are often in conflict with those pricing policies. Input policies covering the costs of land, labor and capital must deal with such varied situations and broad uncertainties in underdeveloped African countries that only a stable but pragmatic approach to policy-making can insure realistic decision-making. The same holds true for basic government decisions regarding investment in food production or in revenue-earning export-crop development. The contradictions facing these countries in such matters are such that only a fresh start at determining new balances in food development policies, as in most other areas of economic activity, can hold the promise of achieving progressive development.

Let us not leave the question of internal policy failures, however, without observing the *national capacity* to engage, for example, in infrastructural reforms and changes and drought-mitigation programs. What is the capacity of a country to construct dams and reservoirs in order to meet fundamental needs for electricity and irrigation projects? How capable is a country of purchasing machinery for deep-drilling and rehabilitating harbors and ports, or of replenishing and protecting its forests and other natural resources?

Internal factors that have generated famine are important because they forced us, by way of the starkness of their adverse impacts on societies, to reconcile the assumptions of the past with the realities of the future. Assumptions and expectations aside, the 1970s demonstrated the inability of, for example, local research institutes to achieve stable development and scientific maturity. International institutes simply failed to provide the expected broad positive impact on national research centers and production programs.

> Moreover, while the international institutes have made impressive gains in increasing yields of rice, maize, wheat, chickpeas, cassava, potatoes, yams, animal production and water resource management, they have achieved considerably less in bettering our understanding of the socioeconomic side of the rural transformation process. While we know much more today through the work of the institutes on how to respond to the needs of the already modernizing traditional farmer, with a sufficient amount of productive land, funds for new inputs, irrigation, and other needed resources, we have made only slight progress in understanding the vast majority of resource-poor farmers in the developing countries who remain unmotivated to eagerly adopt the new high yielding technology package in food production. (*LeMelle, 1984, 16–17*).

Tunisia successfully reduced its national population growth rate but was unable to attain self-sufficiency in primary food production. Kenya's ability to employ the available high-yield food production technology while making modest institutional reforms resulted in the temporary achievement of self-sufficiency in food production. Kenya's success, however, appeared to be short-lived as its population growth rate of approximately 4% per year outstripped its capacity to meet the basic food needs of its expanded population. The cases of Tunisia and Kenya clearly illustrate that the complexity of technology pales before the difficulty of implementing social and institutional changes that must accompany any enduring acceptance of new agricultural technology.

There is evidence now that African governments are exhibiting considerably greater awareness of policy and institutional weaknesses with respect to the conception and implementation of national development programs. Policy decisions have been taken in some cases to improve agricultural prices and to improve the efficiency with which input and output markets operate. Many of these governmental expenditure programs have begun to reflect both the general shortages of specific resources and the need to improve the efficiency with which public resources are used (World Bank, 1981, 1983).

These cases represent only modest beginnings. More importantly, however, they represent the menacing effect of internal socioeconomic and environmental factors on efforts to mitigate the impacts of drought and to deny famine a future in Africa's development prospects. The Lagos Plan of Action, an internal African program plan, has served to initiate the articulation and definition of strategies to be adopted by Africans themselves as they seek answers to such questions as: what kind of development does Africa need? what kind of Africans do they wish to mold?

Notes

1. *Social system* is the mode of production and its associated superstructure, specifically, property rights, division of labor and kinship patterns; and *ecosystem* is the interrelationship between society and environment.

References

Bairoch, P. (1975). *Economic Development of the Third World Since 1900*. London: Methuen.

Brown, L. R., Chandler, W., Flavin, C., Postel, S., Starke, L. & Wolf, E. (1985). *The State of the World – 1985*. New York: W. W. Norton & Co.

Brown, L. R., & Wolf, E. C. (1985). *Reversing Africa's Decline*, World Watch Paper 65. Washington: World Watch Institute.

Browne, R. S. & Cummings, R. J. (1985). *The Lagos Plan of Action vs The Berg Report: Contemporary Issues in African Economic Development*. Lawrenceville, VA: Brunswick Publishing.

Cummings, R. J. (1986). Migration and national development: the Kenyan case. In *Migration and National Development*, ed. Beverly Lindsey. pp. 148–69. Middleton, PA: Pennsylvania State University Press.

Eicher, C. (1982). Facing up to Africa's food crisis, *Foreign Affairs*, **61**, 151–74.

Hedlund, H. (1979). Contradictions in the peripheralization of a pastoral society: the Maasai. *Review of African Political Economy*, **15/16**, 15–34.

Lappé, F. M., Collins, J. & Kinley, D. (1980). *Aid as Obstacle*. San Francisco: Institute for Food and Development Policy.

Lofchie, M. F. & Commins, S. K. (1984). *Food Deficits and Agricultural Policies in Sub-Saharan Africa*. The Hunger Project Papers, 2. San Francisco: The Hunger Project.

Mamdani, M. (1985). Disaster prevention: defining the problem. *Review of African Political Economy*, **33**, 92–6.

Organization of African Unity (1981). *Lagos Plan of Action for the Economic Development of Africa, 1980–2000*. Geneva: Institute for Labor Studies.

Plange, N. K. (1979). Underdevelopment in northern Ghana: Natural causes or colonial capitalism? *Review of African Political Economy*, **15/16**, 15–34.

Rodney, W. (1972). *How Europe Underdeveloped Africa*. Dar-es-Salaam: Tanzania Publishing House.

UN Economic Commission for Africa (1981a). *Implementation of the Lagos Plan of Action – Some Proposals and Recommendation for the Guidance of Member States*. Document No. E/CN. 14/TPCW. II/18. Addis Ababa: UNECA.

UN Economic Commission for Africa (1981b). *Report of the ECA/UNEP Seminar on Alternative Patterns of Development and Lifestyles for the African Region*. ECA Document No. E/CN. 14/1981b. Add 1. Addis Ababa: UNECA.

UN Economic Commission for Africa (1982a). *Critical Analysis of the Country Presentations of Africa's Least Developed Countries in the Light of the Lagos Plan of Action and the Final Act of Lagos*. ECA Document No. ST/ECA/PSD. 2/31. Addis Ababa: UNECA.

UN Economic Commission for Africa (1982b). *Progress Report of the Secretary General of the Organization of African Unity and the Executive Secretary of the United Nations Economic Commission for Africa on the Implementation of the Lagos Plan of Action and the Final Act of Lagos*. ECA Document No. E/ECA/CN. 9/1. Addis Ababa: UNECA.

UN Economic Commission for Africa (1983). *ECA and Africa's Development, 1983–2008: A Preliminary Perspective Study*. Document No. E/ECA/CM. 9/23. Addis Ababa: UNECA.

US Department of Agriculture (1981). *Food Problems and Prospects in Sub-Saharan Africa*. Washington, DC: Government Printing Office.

World Bank (1981). *Accelerated Development in Sub-Saharan Africa*. Washington, DC: World Bank.

World Bank (1983). *Sub-Saharan Africa: Progress Report on Development Prospects and Programs*. Washington, DC: World Bank.

7

Towards a political economy of the African crisis: diplomacy, debates and dialectics

TIMOTHY M. SHAW

Political Science & African Studies, Dalhousie University

We are the leaders of our nations. It is our responsibility to lead our people in making the right response to these disasters . . . It is our responsibility to see that, at least in the long term, our continent and our nations become less vulnerable to such external events. In particular, we know that Africa is constantly subject to droughts . . .

. . . the famine relief food supplies which are now, albeit too little and too late, being sent in to Africa are very welcome . . . But this is not the real answer to the problems which face Africa . . .

In the face of this response to world problems on the part of the economically powerful nations, we in Africa have no choice to increase our own endeavours to be self-reliant, both on a regional and a national basis.

President Nyerere (Address to OAU Summit Meeting, Addis Ababa, November 1984)

The current continent-wide African crisis was neither unexpected nor unavoidable. Rather, it is the latest and most dramatic evidence of Africa's developmental demise. While the incidence of recurrent drought could not be forecast with great accuracy, the vulnerability of most African political economies and peoples to it is undeniable. In addition, this vulnerability is exacerbated by constraints on, and contradictions within, Africa's post-neocolonial regimes. The vulnerability has multiple aspects, notably inadequate production, inefficient distribution and insufficient storage. Its roots lie in Africa's inheritance of impoverished, extroverted and underdeveloped economies compounded by adverse post-independence changes in the global division of labor. Africa is more marginal now than ever. It contributes a declining proportion of the world product; hence, its vulnerability to drought as well as to dependence and dominance. Drought results in famine only when food-security measures fail. Emergency provisions and strategic reserves are negligible in Africa, a continent preoccupied with survival.

Africa's present problem may be characterized as a 'developmental' drought. Climatic variability is so devastating, resulting in famine, because the continent has never been able to develop itself over an extended period of time. Thus, underlying the vulnerability of so many African countries and classes is the condition and process of underdevelopment. Yet, Africa has never been a passive actor, despite its disadvantages. Its collective and official (as well as communal and unofficial) responses to underdevelopment and drought need to be taken into account in any evaluation of its prospects. These, in turn, should inform consideration of alternative development strategies and development scenarios. There is a concern regarding the short- versus long-term responses both within and outside the continent and, more specifically, the degree to which emergency aid and structural change are (or are not) compatible.

Departures: alternative emphases and explanations of the crisis

The essential causes of Africa's vulnerability to drought are two-fold. The first is *external*; Africa's marginality is a function of the new international division of labor in which the products of an export-oriented continent are declining in value relative to the prices of imported capital and consumer goods, energy, money and services – the demise of the neocolonial nexus. The second cause is *internal*; its vulnerability is exacerbated by urban-oriented assumptions, inclinations and prescriptions. Yet, these were the correlates of survival for neocolonial regimes, as indicated by the famine report from the Independent Commission on International Humanitarian Issues (ICIHI):

> Intertwined with the crisis of the rural economy in Africa is a crisis in government itself. One reflection of this is the urban bias in policy-making and the absence of rural participation. Governments have been preoccupied with the needs of the new, and economically under-productive, cities. Indeed, catering to the aspirations of the urbanites has sometimes been a prerequisite for political survival. The price has been paid in the countryside.
> (*ICIHI, 1985, 33–4*)

Thus the occurrence of famine, in part a reflection of the inability of regimes and peoples to cope with drought, is a dramatic indicator of Africa's vulnerability as well as its marginality; that is, the orientation and impoverishment of post-neocolonial regimes in the emerging post-industrial world order. This suggests a reason for the apparent inability of either internal or external authorities to recognize the signs of impending disaster, as both were caught in their webs of past assumptions, positions and precedents. Until both international and domestic conditions change, Africa will remain vulnerable to drought. Given the continent's arid and semiarid zones, droughts are inevitable; yet resultant famines are not so. However, the necessary restructuring of internal and external institutions to mitigate the impacts of drought will be slow and painful. Famine as a sufficient shock, like war, can generate the necessary willingness and context for substantial reordering, e.g. from urban to rural bias and from extroversion to introversion (Luke & Shaw, 1984).

The notion of vulnerability – a structural condition – diverts attention away from short-term responses to longer-term situations (Shaw, 1985*b*, *d*, *e*). A broad historical context – the future as well as the past – is essential for any explanation of causes, expectations, or resolutions (North-South Institute, 1985). It is important to consider why, over time, crises have become more serious and whether, again over time, they can be resolved. To be sure, short-term amelioratives are imperative to preserve life, of cattle and trees as well as of people and regimes, but these should not be at the expense of longer-term restructuring (Jackson, 1985). Such a notion already implies a particular perspective on causality: that it is not just lack of rain. This is just one of several distinctions not sustained in the literature to date.

Finally, neither drought nor famine is a simple concept. Different definitions of them are revealing of alternative formulations (Glantz, 1976). Drought is not just lack of rain and food. Famine 'is more than people dying from starvation. It is an acute breakdown of society that brings turmoil that cannot be ignored' (ICIHI, 1985, 25). Africa has in recent decades, even centuries, not been an easy environment. Underdevelopment has meant that life for most Africans has been nasty, brutish and short. However, famine is different from continual or gradual decline; it is a general crisis. As the ICIHI report proposes, 'What distinguishes famine from starvation and run-of-the-mill food shortages, however severe, is that famine is political' (ICIHI, 1985, 26). A variety of individual and collective responses to famine result; from human migration to debt renegotiations and from food distribution centers to agricultural policy redirection.

It is important to distinguish famine-caused social changes from those that have other causes and to recognize that many famine-induced changes are confined to particular localities and classes and, therefore, do not register at the national level. However, the famines that have affected Botswana, Burkina Faso, Chad, Ethiopia, Ghana, Niger, Tanzania, among others, have resulted in profound and sustained social change, the implications of which are still being identified (Bush, 1985).

The range of implications is in part a function of the paradigm employed; if drought is treated as a short-term, cyclical and restricted problem, then explanation and prescription are similarly limited, e.g. weather and technology, respectively. By contrast, if famine is conceived in longer-term, structural and comprehensive terms, then explanation and prescription are equally broad, e.g. incorporation and self-reliance, respectively (ODI, 1984; North–South Institute, 1985). These two formulations not only may be distinctive, they may also be incompatible, in the sense that impassioned short-term reactions tend to exclude considered long-term responses. In addition, both perspectives, whether short- or long-term, tend to exclude from consideration any treatment of Africans' responses or Africa's position, respectively.

Debates: alternative theories and treatments of the crisis

Africa's experience with drought in the 1980s has been widespread and unsettling (ADB/ECA, 1985). When combined with other developmental difficulties, it has exacerbated an already unpromising and problematic situation (World Bank, 1981, 1983, 1984). Even if one eschews a radical or structural conceptualization, the African drought embodies several salient strands, as indicated in the formation of the Director-General of the FAO, Edward Saouma: political, economic, environmental and technical (FAO, unpublished). However, the ranking of these elements is crucial and controversial, with theoretical and practical implications. Moreover, there tends to be a correlation between perspective and position; non-African and bourgeois analyses tend to adopt short-term, specific approaches, whereas African and proletarian analyses tend to espouse long-term, broad formulations. The latter also tend to recognize and respect a range of African responses both popular and collective, whereas the former tend to assume that only non-African reactions are relevant.

The first section of this chapter treats distinct perspectives on vulnerability and marginality informed by alternative sets of assumptions and prescriptions. Like the continuing debate about Africa's development strategies, there are discernibly divergent positions over the issues of famine and responses to them.

If the extra-African emphasis is on ecology, then that within Africa is on economy. The former emerges out of the modernization and development literatures – with the right mix of policy, technology and money, anything is possible – whereas the latter is more compatible with the *dependencia* and underdevelopment perspectives – no matter how hard Africa tries, its best efforts are likely to be frustrated by international structures of inequality. For the former, drought is just another difficulty to be overcome, whereas for the latter it is further evidence of the impossibility of sustained change within inherited parameters. Symptomatic of these two divergent positions are the report of Canada's Emergency Coordinator for the African Famine

(CIDA, 1985) and the recent statements of President Julius Nyerere (1985*a*, *b*), respectively.

In early 1985, David MacDonald, on behalf of the Canadian government, argued that solutions lay in emergency measures from medical teams to media centers and from chartered ships to student assistance (CIDA, 1985). Neither longer-term antecedents nor consequences were identified (Shaw, 1985*d*, *e*). The emphasis was on people rather than structure. Yet, even in terms of narrow bureaucratic considerations, hard choices cannot be avoided. Assuming the Canadian International Development Agency's (CIDA) budget remains constant, any short-term 'band-aid' is at the cost of long-term assistance (Shaw, 1985*c*).

Characteristic of such an uncritical and ahistorical approach to African agriculture and drought is an ODC *Policy Focus* (Lancaster, 1985). Rather than treat underlying structural causes related to commodity exports, the author identifies a set of superstructural issues. From a political economy perspective, these would be symptoms rather than causes. Lancaster's (1985, 6–7) World-Bank-style analysis of Africa's food problems cites the following as 'barriers to progress':

1) *agricultural policies* which prevent agricultural expansion, by overtaxing agricultural products, particularly exports, and offering farmers too-low prices for products;
2) *inappropriate agricultural techniques* which are labor-intensive and low-yielding; and
3) *inadequate agricultural support institutions*, particularly for women, the major producers of domestic food.

Rather than situating these factors in the context of colonial-type commodity economies, Lancaster proposes a litany of responses, from higher producer prices to new techniques and better research and seeds. Yet, until the post-colonial political economies which perpetuated such inherited extroverted structures are transformed – either by internal change (less likely) or external pressure (more likely) – the root causes of the crisis will not be tackled. To be sure, policy, pricing and technological factors also need to be treated. Yet from a radical perspective, this is to deal with symptom, not cause. Critical analysis is neither as immediate as instant relief, nor is it acceptable to various establishments. Even the supposedly Marxist government of Ethiopia would rather welcome European pop stars than engage in self-criticism; its espousal of the Western media is remarkable, given its antifeudal revolution (Hancock, 1985).

Almost as dangerous or diversionary as the short-term emergency response is a preoccupation with long-term ecological degradation.

Extending a perspective popularized by the Club of Rome, the 'ecological fallacy' attributes Africa's problems to a combination of environmental mismanagement and external orientation. Earthscan (Wijkman & Timberlake, 1984; Timberlake, 1985), brings Africa's past and present together, arguing that 'the real evil of this overemphasis on cash crops has been not in economic but in environmental terms'. Thus, Africa's outward orientation and emphasis on commodity exportation has caused not only food importation but also 'environmental bankruptcy'. Earthscan blames both indigenous regimes and international pressures for this legacy: 'The seeds of this environmental bankruptcy have been sown by government policies and watered by three decades of misdirected foreign aid' (J. Tinker, unpublished). Yet, an inheritance of colonial exchange and neocolonial structures prevented radical revision of development directions. Post-colonial political economies were hardly sustainable, yet the new class hardly wanted to admit or transcend this.

A dramatic illustration of the very real dangers of radical analysis in Africa of the issue of drought is provided by the treatment of Mahmood Mamdani, a radical Ugandan political scientist. Mamdani's lecture at a Uganda Red Cross Conference on 'Disaster Prevention' (March 1985) focused on the structural causes of famine in a colonial economy (like Uganda) in which peasant exploitation generates inequalities leading to peasant marginalization and vulnerability so that implementation of better agricultural techniques or inputs is impossible.

> Whereas this first type of exploitation of the peasant is indirect, as a result of dull compulsion of economic forces, the second type of exploitation is direct, through the use of force by the state or state-connected organizations like the party or the church.
> (*Review of African Political Economy, 1985, 94*)

Incremental impoverishment leads to contradictory short-term responses, e.g. poorer crops, methods and results. Mamdani situates disaster-proneness in the context of exploiting and contracting colonial-type political economies and calls for informed, appropriate and radical responses, warning against 'utopian thinking', aid dependence and further land alienation: ' . . . any strategy that claims to be a solution must seek to revive the creativity and the initiative of the people. Central to this must be to educate people about those relations which make them disaster-prone' (*Review of African Political Economy*, 1985, 96). For this exercise in radical analysis, Mamdani had his Ugandan citizenship abrogated, a controversial and arbitrary move which only served to confirm his warnings about the vicissitudes of state power (*The People*, 1985).

While Mamdani's fate constitute a dramatic human indicator of the dangers of critical scholarship, there are other less obvious pressures tending toward conformity. Most of the aid establishment prefers shorter-term palliatives than longer-term reforms, and the immediacy of the African crisis has encouraged such tendencies. To focus on structural constraints when people are dying smacks of a lack of compassion. On the other hand, the Ethiopian and Sudanese famines have been veritable windfalls for some aid organizations who specialize in fund-raising rather than in development projects; why not exaggerate the poignant and downplay the context (Hancock, 1985)? Yet, disillusioned donor constituencies are dangerous, by discouraging continued efforts. While the salience of aid, if not redistribution, has been revived by the African famine, it may disappear once again, if benefits are not delivered and famines recur. Because of the intensity of the recent drought, aid, not a New International Economic Order (NIEO), has once again become 'high politics'.

Diplomacy: alternative levels and institutions responding to the crisis

The images of, and responses to, the Ethiopian famine impacted upon a world already debating the continent's crisis (Luke & Shaw, 1984). Such images heightened awareness but diluted the discourse, as emergency food aid replaced longer-term structural change as the focus. Like the populist appeal of the basic human needs preoccupation, 'Band Aid' diverted attention from difficult historical questions. The global conjuncture of the mid-1970s had already produced an African reaction; the late 1970s OAU/ ECA Monrovia Symposium and the IBRD African governors' call on the International Bank for Reconstruction and Development (IBRD) (Shaw, 1983*a,b*). The divergent interests reflected in the preliminary debates at these two meetings were represented, respectively, in the subsequent Lagos Plan of Action (OAU, 1981) and the Agenda for Action (also known as the Berg Report) (World Bank, 1981). Thus, these proposals and their receptions were already rehearsed, as the need for them increased, because of the intensification of development difficulties and the decline into full-blown economic crisis. The impoverishment about which the Brandt Commission (1981, 1983) cautioned became all too apparent in droughts, food riots, *coups d'état* and increased antagonisms. Yet the causes, unlike the catalysts, were not just high prices for oil, food and money or declining outputs of food, manufactures and services. Rather, the fundamental factor in Africa's post-1970 decline was its increasing marginalization in the global economy. For

Africa the post-Bretton Woods order meant the demise of the familiar neocolonial nexus. The continent lost its *raison d'être* in the post-industrial world of microchip and laser technologies.

Although Africa's marginality in the emerging international division of labor also increased its vulnerability (by contrast to the immediate post-colonial period; C. Young, unpublished) it at least had an indigenous response in place: the *Lagos Plan* for the post-neocolonial era (Adedeji & Shaw, 1985). Africa's collective, consensual espousal of self-reliance and self-sufficiency was intended as a reaction to the global shocks of the mid-1970s, but it has been elevated into a timely alternative to extra-continental proposals (Shaw & Aluko, 1985). The Lagos Plan/Berg Report dispute has been succeeded by a Lagos Plan/conditionality debate in which self-reliance is pitted against structural adjustment, food aid and market forces (Shaw, 1985*a*). Without the 1980 African summit and scenario, the mid-1985 Addis Ababa economic conference could not even have been considered (Shaw, 1985*c*), for it was constructed upon the foundation provided by the sequence of the Monrovia Symposium, *Lagos Plan* and the ECA silver jubilee prospective study (ECA, 1983).

In contrast to the World Bank (1981; 1983; 1984) studies, UN emergency operations and conferences, and assorted national responses (all of which emphasize structural adjustment, donor coordination and aid dialogue), Africa has emphasized a pair of alternative related directions: continental self-reliance and international restructuring (Adedeji, 1985; Adedeji & Shaw, 1985). Africa can advance the former proposal at national, regional and continental levels, whereas it can only advocate the latter in global fora. In addition, the former (continental self-reliance) constitutes a response to disappointing rates of development over the last decade, while the latter (international restructuring) consists of a unilateral attempt to revive the moribund NIEO debate. While non-African agencies seek to advance a limited dialogue over aid, Africa demands a renewed global dialogue over North–South inequalities (North–South Institute, 1985, 9–12). Yet, without international restructuring, the continental self-reliance is pointless, making future emergencies inevitable. There is a relationship between famine and the future, as President Nyerere (1985*b*, 4) delicately noted:

> The people of Europe and America respond with great generosity to knowledge of actual starvation arising from famine. But a less internationalist attitude is shown by their governments when it comes to helping Africa to develop self-sustaining economies which might prevent future famine. Instead, Overseas Development Assistance is cut in real terms, and there is an

increasing tendency to use aid for political purposes. Then, when poor debtor countries have difficulty in making due payment, their creditors refer them at once to the IMF, whose conditionality is always heavily deflationary, often highly political, and most usually based on the idea of immediately exporting more and importing less.

Because the proposed remedy is intended at best to treat symptoms rather than causes, it may exacerbate rather than relieve the fundamental condition of underdevelopment and vulnerabilty, the very factors that the *Lagos Plan* intends to treat unilaterally if extracontinental assistance and restructuring are not forthcoming on a long-term basis.

Symptomatic of the imperative of theoretically as well as empirically informed responses to the African crisis is the wide range of often incompatible proposed solutions, from airlifts to population control, and from pop concerts to long-range weather forecasting. Unless we puzzle over causality, we cannot reach a political or pragmatic consensus over response. Typical of misplaced benevolence is the fact that famine is quite predictable in terms of its spatial and social incidence. As Timberlake (1985, 22) points out, 'Drought is not like most other disasters: it can be seen coming, slowly, from a long way off. Drought, more than any other disaster, chooses its victims . . . the wealthy are *never* killed by drought'.

Given Africa's place at the global periphery, its vulnerability is largely a function of external forces. However, the internal impact of such underdevelopment is the responsibility of national regimes and bourgeoisies. It is this joint recognition that lies at the heart of new attempts to reorient and revive Africa's lagging economies. As Timberlake (1985, 7) recognizes, 'The tragedy has had one positive effect: it has started a painful reappraisal among those responsible for Africa's 'development', and brought a new willingness to admit mistakes'.

Dialectics: compatibilities and contradictions in praxis

The difficulties (and dilemmas) of responding to the African crisis should not be dismissed as academic piety or procrastination. Rather, they might provide pause for thought. If there is one lesson in the current condition, it is that there has been too much, very inconsistent, quite changeable, and often gratuitous advice from various sources, primarily from the 'development set' of the Western world. Nowhere is the 'science' of economics revealed to be so threadbare as in the spreading deserts and shanties of Africa. Yet, despite the shock which Africa administers to our senses and our theories, assorted economic and environmental quick fixes are still promised. Indeed, the quickest fix of them all is (one-way) 'conditionality', that is, fiscal arrangements rather than structural ones, when clearly well-established assumptions, prescriptions and expectations are at fault.

The Ethiopian drought and famine and other such cases in Africa have generated their own set of explanations, as well as projections. In order to understand and differentiate these responses, it is important to situate them in historical and contextual perspective. This implies rejecting a simplistic notion that the only or primary cause of famine is the lack of rain. Rather, it means accepting a longer-term perspective on both genesis of and response to famine. Yet the origins of famine and its elimination remain controversial. As already noted, the primary tension is between economic and ecological approaches, although with respect to economic approaches, distinct political, policy, and social factors have been identified.

In addition to going beyond long- versus short-term distinctions, it is also important to transcend any internal versus external apportionment of blame, primarily because vulnerability is a function of multiple forces, each of which is the result of both inter- and intranational interests and pressures. The cautious note of the ICIHI (1985, 24–5) report should be heeded:

> At each stage, from its genesis in rural poverty and food production failures through to the reduction of communities to destitution and starvation, famine is avoidable. More than that, its causes are much more complex than just bad luck with the weather. The simple assumption that, if the rains fail, as they have in recent years in much of Africa, less food will be grown and people will inevitably starve, may be a comfortable abdication of any human responsibility for what has happened. But it is a misleading simplification.

To be sure, the ecological perspective does include a more sophisticated variety which incorporates some factors of political economy such as the social bases of desertification. Typical of this genre is the Earthscan/ International Red Cross analysis (Wijkman & Timberlake, 1984) *Natural Disasters: Acts of God or Acts of Man*. However, the notion of 'ecological bankruptcy' (Timberlake, 1985) does not incorporate the political economy that produces such bankruptcy, e.g. the economic, social and environmental conditions of deforestation, soil erosion, inappropriate grains and technologies, inadequate water supply and distribution and so forth. It identifies and treats effects rather than causes: desertification rather than commodity production, deforestation rather than energy needs, soil erosion rather than agricultural priorities, practices and institutions.

Different theoretical perspectives treat ecological variables as either dependent, independent or intervening. 'Ecological determinists' treat environment as cause, whereas political economists treat environment as effect. In between are a variety of 'radical environmentalists' who consider environment to be integrally related to economic forces, i.e. an ecological bankruptcy perspective with a developmental bankruptcy perspective clearly determines policy response – ecological, economic, or mixed treatments, respectively. There is a danger that environmental determinists will be persuasive in the short term and, as a result, concern for economic issues will be postponed to the longer term. Conversely, any sustainable developmental response must incorporate an ecological dimension.

In addition to the issues of priority and time scale, there are by-products to be considered such as 'environmental refugees' (Timberlake, 1985, 10) and military responses. Environmental refugees may be expected to multiply, as populations increase while land resources dwindle. Likewise, determined military or other authoritarian reactions are likely to spread as mistaken forms of 'decisiveness', which serve to repress, not remove, both symptoms and causes of drought and underdevelopment (*Review of African Political Economy*, 1985).

Yet in addition to appropriate longer-term responses being required, there is also a need for an informed historical awareness. Drought is not new to Africa; neither is famine. The real contemporary issue is whether drought automatically produces famine. Are there not sufficient reserve systems, technological inputs, development theories and emergency responses to counteract the alleged inevitability of famine? As Earthscan suggests, 'The traditional agricultural and environmental safety-valve had been destroyed by development' (J. Tinker, unpublished) – or at least by mal- or underdevelopment. Traditional methods of drought control such as food storage, land reclamation and collective migration cannot be readily revived as the economic, political, social and ecological context has changed so dramatically. They can, however, at the least be recalled and reconsidered in case some appropriate technological and theoretical lessons can be learned.

An example of the environmental-determinist school is *Africa in Crisis* (Timberlake, 1985), a comprehensive popular analysis that eschews either a hierarchy of factors or a ranking of causes. Instead, a 'populist' stance is proposed – peasants know best so that insufficiencies or inequalities of resources (land, water, fuels, animals, seeds) are blamed on a combination of indigenous politicians and exogenous policies. Yet, although history, economy, polity and ecology are treated, there is no framework for analysis and praxis other than a faith in the grassroots. Capital and class are never mentioned; neither are accumulation and exploitation. Non-materialist assumptions are pervasive:

> Environmental bankruptcy has not been caused by the stupid African peasant, as so many . . . governments and aid agencies and their experts like to think. On the contrary, it is the African peasant who best understands how and why he or she has been forced to damage the environment on which they depend, and it is he or she who is the key to rebuilding the continent. To ground Africa's future in an environmental reality which is maintainable, to produce development that is sustainable, will require a great deal of common sense . . . (*Timberlake, 1985, 224*)

Such populism identifies some of the primary correlates of Africa's difficulties but assiduously avoids relating these to structures of political economy; for instance, capitalism is never mentioned. While this publication transcends *dependencia*, it also refuses to recognize the class base of Africa's contradictions:

> . . . the threats to these fragile governments come not from the countryside but from the *Wabenzi* – the disaffected urban elites: the entrepreneurs, civil servants, police and armed forces . . . The policies of nearly all African governments favour the urban elite, by keeping food prices low . . . One of Africa's many vicious cycles is at work here: government policies degrade the rural resource base . . . (*Timberlake, 1985, 12*)

Similarly, although this publication makes many telling points, such as the distinction between causes and symptoms, it consistently avoids attributing causality and articulating theory. Rather, it merely asserts that:

> The book shows the relationships between Africa's environmental bankruptcy, its millions of environmental refugees, and the continent's political and social instability. Finally, it shows how development based on the under-used skills and ambitions of small farmers could help Africa to a more stable, famine-free future. (*Timberlake, 1985, 14*)

The analysis is not pushed beyond the political or the administrative factors to underlying structures. From a materialistic perspective, only 'intervening variables' are identified rather than basic contradictions:

> The African droughts and famines are not sudden natural disasters, nor are they simply caused by a lack of rainfall. They

are the end results of a long deterioration in the ability of Africans to feed themselves, a decline caused largely by mistakes and mismanagement – both inside and outside the continent. (*Timberlake, 1985, 7*)

By contrast, a more critical structural analysis would concentrate on the internal and external social relations of production which produce the features that erode famine resistance (Sandbrook, unpublished, and 1986). This is neither to overemphasize the colonial modes of production nor to blame 'imperialism' and 'dependence' alone. Rather, it is to recognize colonial inheritances and post-colonial constraints as well as to appreciate post-neocolonial transitions (i.e. from the early 1970s onwards); the vulnerability and marginality of the continent in a post-industrial, post-growth, post-Bretton Woods era.

The essential features of Africa's famine-proneness are (*a*) established urban- and export-oriented national economies (C. Young, unpublished) and (*b*) changes in the international economy in which colonial commodities have become less crucial, thereby making the neocolonial nexus vestigial (Shaw & Aluko, 1985).

In short, the assumptions on which African economies were established and flourished have been superseded; colonial-type economies, commodities, cities, and infrastructures have lost their *raisons d'être*. Thus, in a post-neocolonial period, self-reliance is an imperative. Only Africa can save itself. Because of its marginality, no one else is particularly interested in, let alone capable of, doing so.

Yet, self-reliance is only a prerequisite. Unless African political economies go beyond it to accumulate, reproduce and generate technology and transform exchange, they will at best merely survive. Survival alone is not, however, enough to rectify wrongs and to revive growth rates. Survival will always be problematic unless disaster-proneness is reduced and reversed. The basic deficiency of the *Lagos Plan* is that it was crafted by the same indigenous bourgeoisie that it needs to contain. Like Earthscan's *Africa in Crisis*, it eschews any radical analysis of the social relations of production, preferring instead to address the safer questions of trade and distribution. While the Economic Commission for Africa (ECA) has assumed the mantle of Economic Commission for Latin America as the most progressive of the UN regional organizations, it assiduously avoids offending member states, and particularly Africa's ruling class (Adedeji & Shaw, 1985; Shaw & Aluko, 1985). Thus, the social bases of self-reliance are not considered, nor are the possibilities of accumulation through 'primitive' production. Moreover, the state orientation of the OAU and ECA means that those organizations cannot incorporate the real character of self-reliance in contemporary Africa, i.e. the emergence of 'parallel' and 'illegal' economies (R. Lemarchand, unpublished).

The demise of the neocolonial nexus has meant that survival for displaced bureaucrats, traders, workers and farmers has involved a variety of 'informal' economic activities, from the 'smuggling' of commodity exports and consumer imports to the revival of traditional technologies and barter. Africa, like the rest of the world, has always had an active informal sector, which has never been completely incorporated into either the colonial or neocolonial nexus. With the passing of these economic relationships, the informal sector has undergone a revival. The only way that assorted African classes have survived during the last decade is by responding to 'market forces': smuggling of coffee from Uganda by entrepreneurs and cocoa from Ghana by peasants; importing consumer goods and spare parts whether by container or boxes into Nigerian ports or across the Sahara; the flow of 'illegal' laborers and currency notes across innumerable borders; and the whole 'black market' lifestyle of 'dash', *magendo* and *kalabule*.

Such unregulated activities are recognized and approved by neither Lagos Plan doctrines nor World Bank 'conditions', yet they constitute one of the few dynamic sectors keeping assorted African economies afloat. Although everyone knows they exist, no one incorporates them into official data and plans. Their presence is being tacitly recognized by the spread of state-sanctioned foreign exchange shops and by fluctuating exchange rates. But if the goal of the Lagos Plan and of the Berg Report is to release African dynamism, they should have recognized that it already exists in the distinctive form of self-reliance found in particular African political economies represented by *magendo* and *kalabule*. The expansion and recognition of such self-reliance is one scenario for the mid-term future with considerable potential for national redefinition and regional integration. Other futures, related to alternative plans and frameworks, are internal and external restructuring, regression and revolution (Shaw, 1983a; Adedeji & Shaw, 1985).

Alternative scenarios for Africa: reform, repression, regression and/or revolution

If unplanned self-reliance represents one possible future for the continent, the reform or restructuring proposed by the Lagos Plan and by the World Bank are others. Despite their dialectic qualities, the alternative OAU/ECA and World Bank formulations share at least a common concern with economic productivity and with basic needs (if not substructure); the social relations of production. At a similar level, ICIHI and Earthscan advocate populist, rural-based, appropriate strategies which transcend both the general crisis and particular famines. They call for longer-term Afrocentric responses:

> Better relief, however, only limits the consequences of Africa's vulnerability. The major thrust must be a longer-term re-establishing of a rural economy where the risk of famine is sharply reduced . . . Restoring self-reliance requires a restructuring of government policies and international aid so that they serve the interests of the rural majority in Africa more effectively. In particular, the rural poor, the current famine victims, must be economically enfranchised. (ICIHI, 1985, 132 and 137)

Despite the wisdom of its recommendations, ICIHI remains idealistic as well as populist. The only way in which more rural- and peasant-oriented policies will occur is through social pressure, not by 'economic enfranchisement'. This will involve going beyond all existing models of development toward environmentally sound policies that tend toward rural, peasant, basic needs and sustainable orientations.

Such orientation is, of course, already under way; how else would so many people in 'broken-back' states have survived over the last decade or so, in Chad, Ghana, Tanzania and Uganda (among others). The informal economy, however, will never replace the remaining formal sectors, unless it receives official recognition. For this to happen, peasants and other rural social groups have to organize (something they are not equipped to do) and to secure international support.

Likewise, the Lagos Plan and World Bank proposals lack popular support, their particular constituencies notwithstanding. The Lagos Plan's support derives from more progressive, state-oriented, technocratic elements within the bureaucratic bourgeoisie. Support for the World Bank's proposals comes from more conservative, economic, market-oriented members of the national bourgeoisie (Shaw, 1983a). It is precisely the continuing dominance of bureaucratic over national factions in the African bourgeoisie that favors the relative acceptability of the World Bank's proposals over those of the ECA. Just as such reforms cannot be taken out of historical or contextual situations, to what extent is the World Bank responsible for the very problems it is now trying to resolve? Thus, any effective 'conditions' have to be external as well as internal.

If Africa's crisis is in part a result of years of inadequate assistance combined with adverse global trends, then it can hardly help itself without external restructuring. Effective conditionality goes two ways, not one way (US Committee on African Development Strategies, 1985). To redevelop, Africa needs some kind of NIEO, rather than the emerging unequal division of labor (Ravenhill, 1986). As ICIHI (1985, 126) notes:

> . . . many of Africa's current problems originate in a depressed and inequitable international economy . . . A number of African governments are seeking to reform domestic policies. By contrast, no serious international economic reform is taking place.

Reforms within Africa may take place because of internal, continental, or global pressures. There is no such effective organization at the international level. Despite the pronouncements in *Cataclysm* (Clark, 1984), transnational terrorism has yet to force a restructuring of the global economy.

If reform of any kind is problematic, repression is all too likely, i.e. effective environmental or economic repression of new social forces through such practices as harrassment, arbitrary arrest, imprisonment, 'disappearance' and the like (Shaw, 1984). Given the revival of superpower competition and great power rivalries, such repression may be facilitated by external interests, particularly in 'sub-imperial' contexts, such as has happened, for example, in Angola, Egypt, Ivory Coast, Kenya, Nigeria, and South Africa (Shaw, 1983c).

If repression is most likely in such relatively privileged 'Third World' states, then in more peripheral 'Fourth World' situations, regression is even more likely. External and internal military–industrial complexes combined with manufacturing, distribution, tourist and communications sub-centers may come to revive a few African economies. By contrast, the burgeoning informal economies of the Fourth World will not be allowed to take over the shrinking formal sector, so further regression can be anticipated. *Triage* on a grand scale will occur, relieved only by real emergencies like 1984–85, to which external agencies respond.

Finally, a perspective not incorporated even in progressive texts on the African crisis, let alone in the reformist (populist or alarmist) literature, is that the African crisis may generate its own dialectic. It has already produced a range of short- and medium-term responses as well as alternative proposals and projections. Almost all of these, however, are at the level of distribution and consumption. All too few examine patterns of production, accumulation, and reproduction. In part, this is because of the paucity, formality and obscurity of much of the Marxist analysis of Africa which has yet to transcend its Eurocentric roots. On the other hand, a neomaterialist methodology, if not theory, may take one a good distance. Who suffers from famine? Who benefits from drought? Who decides alternative priorities to reduce drought-proneness? These are the bases for national and international reactions and coalitions.

In short, the African crisis is not universal. The more affluent, educated, urban, adult and male populations have suffered least, while the poorer, rural women, children and grandparents have suffered most. Likewise, the Third World has been less drought-prone than the Fourth World. Thus, many of these scenarios may coexist, but in different places and at different times.

African political economy: researches and relations for the next fifteen years

The current continental crisis has challenged prevailing assumptions of both the development and dependency schools of thought: neither capitalist 'miracles' (e.g. the Ivory Coast and Kenya) nor 'socialist' experiments (e.g. Ethiopia and Tanzania) have favorably withstood the decade of economic shocks. The current tendency toward market forces may yield its own set of contradictions (R. Sandbrook, unpublished). Meanwhile, established modes of analysis are also under attack (Shaw & Aluko, 1985).

The pervasive African political economy of export-oriented, urban-biased, import-dependent, bourgeois structure with inflated exchange rates is being dismantled rapidly under threat from external pressures (e.g. the IMF, the IBRD and bilateral agencies) as well as from internal imperatives (e.g. the demise of the neocolonial nexus, the parastatal system and the formal economy). In its place a new, less-structured system is evolving, one that incorporates foreign exchange shops, fluctuating exchange rates, parallel economies, black markets and intensified inequalities. The 'new' post-neocolonial Africa may display goods on the shelves but may be populated by people with insufficient income. Thus, the incentives for smuggling (informal 'transnational' exchange) increase.

The demise of the neocolonial nexus may seem to solve Africa's immediate difficulties, but whether it can lay the basis for sustained development is quite problematic. The privatization of parastatals and the liberalization of exchange may serve to revive Africa's development prospects. The costs, however, may be considerable. The question remains: Who will assist the mass of non-bourgeois people? The costs may also be cumulative. Will the indigenous bourgeoisie finally be allowed to accumulate and reproduce unhindered by leadership codes and progressive taxation? Meanwhile, external conditionality is a prerequisite for effective internal restructuring.

The often overlooked ingredient in the African crisis is the demise of the

North–South dialogue. During the last decade, Africa's development has been undermined by changes in the global economy. As Nyerere recently lamented, Africa has never had occasion to accumulate adequate capital. Thus, it has been most vulnerable to global shocks as well as to climatic anomalies:

> . . . I am tired of being told that Tanzania's present condition arises out of our own mistakes . . . that Africa's present condition is the result of African incompetence . . . that the solution to our problems is an agreement with the IMF . . . Every African country, and specifically Tanzania, has made mistakes of policy, and has areas of considerable inefficiency in the operation of good policies . . . Our mistakes have made an impossible situation worse; they do not account for the situation itself. (*Nyerere, 1985*b, *1–2*)

Nyerere documents his case by reference to declining terms of trade, increasing debt, reduced aid, and increased oil prices. Given the widespread character of the crisis, he suggests that its cause cannot be internal policies or politics alone. Rather, Africa has always been the most vulnerable continent:

> . . . the degree to which the malaise affects all African countries does suggest that internal policies are by no means the prime cause. Speaking from the standpoint of Tanzania, I suggest that Africa, being the poorest and least developed continent, has been the greatest victim of the malfunctioning world economic order. (*Nyerere, 1985*b, *2*)

Thus, as a prerequisite, Nyerere calls for new debates and decisions about a post-Bretton Woods order. As part of any new contract with Africa he reassures us that 'African states are themselves desperately concerned about their people's plight and doing what they can to deal with it' (Nyerere, 1985*a*, 2)

The new conceptualization of 'Africa First' is that the continent should take responsibility for any initial response, as recognized in the UN 'Declaration on the Critical Economic Situation in Africa': 'We are aware that African countries have the primary responsibility for their development and for addressing the present crisis. They have therefore undertaken and continue to undertake painful adjustment measures at very high social and political costs' (UN, 1984, 3).

The demise of the neocolonial nexus at the level of exchange, then, has its counterpoint at the level of rhetoric: Africa does not want charity, merely equality of opportunity. Its determination to reform itself is symbolized by the *Lagos Plan of Action* and the mood of self-reliance. To be sure, such a motif is something of an inevitability, given current straits. As Nyerere recognizes, self-reliance at all levels is imperative in response to global shocks: ' . . . we in Africa have no choice but to increase our own endeavors to be self-reliant, both on a regional and a national basis. Separately, we are all weak – all of us. Together, we would make Africa just a little less weak' (1985*a*, 3). But with the demise of neocolonial connections, the most prevalent type of self-reliance is that occurring outside the formal control of the state; that is, the ubiquitous 'illegal' and 'parallel' economies already mentioned.

Such informal sectors pose major problems for analysis as well as for policy; neither developmental, *dependencia*, nor materialist modes of analysis incorporate them readily. For the first, such sectors are counterproductive, as they detract from state-regulated industries and infrastructures. For the second, informal economies are essentially impossible as they exist outside center-periphery relations. For the third, they represent petty bourgeois aberrations that will be reincorporated into large-scale capitalist or socialist networks. Thus, a neomaterialist analysis is called for, one which recognizes the limits of the state sector for both production and accumulation and which incorporates treatment of social relations in both 'dual' economies. Likewise, any realistic policy framework must go beyond the formal sector and recognize that the informal sector is already privatized and 'de-parastatalized'. The real issue is whether such dynamism can be captured on a larger scale and whether the informal sector constitutes a basis for the accumulation of capital and technology. In short, until we are more confident of our mode of analysis, we should be ultracautious in advocating *praxis*. Any lasting solution to Africa's crisis involves going beyond ecological fallacies, orthodox regulations, and statecentric assumptions. The UN Declaration on the Critical Economic Situation in Africa really contains a fundamental challenge: 'We are convinced that unless urgent action is taken the rapidly deteriorating situation in Africa may well lead to disaster. We are therefore fully committed to supporting the efforts of African countries to meet the dual challenge of survival and development' (UN, 1984, 6).

References

ADB/ECA (1985). *Economic Report on Africa, 1985*. Abidjan & Addis Ababa: ECA.
Adedeji, A. (1985). *The African Development Problematique: Demography, Drought and Desertification, Dependency, Disequilibrium, Debt, and Destabilization*. Addis Ababa: ECA.

Adedeji, A. & Shaw, T. M. (1985). *Economic Crisis in Africa: African Perspectives on Development Problems and Potentials*. Boulder: Lynne Rienner.

Brandt Commission (1981). *North–South: A Programme for Survival*. London: Pan.

Brandt Commission (1983). *Common Crisis, North–South: Cooperation for World Recovery*. London: Pan.

Bush, R. (1985). Drought and famines. *Review of African Political Economy*, **33**, 59–63.

CIDA (Canadian International Development Agency) (1985). *The African Famine and Canada's Response: Recommendations*. A Report by the Honourable David MacDonald, Canadian Emergency Coordinator/African Famine, November 1984 to March 1985. Ottawa: CIDA.

Clark, W. (1984). *Cataclysm: The North–South Conflict of 1987*. London: Sidgwick & Jackson.

ECA (Economic Commission for Africa) (1983). *ECA and African Development, 1983–2008: Preliminary Prospective Study*. Addis Ababa: ECA.

Glantz, M. H. (ed.) (1976). *Politics of Natural Disaster: Case of the Sahel Drought*. New York: Praeger.

Hancock, G. (1985). *Ethiopia: The Challenge of Hunger*. London: Gollancz.

ICIHI (Independent Commission on International Humanitarian Issues) (1985). *Famine: A Man-Made Disaster?* London: Pan.

Jackson, H. F. (1985). The African crisis: Drought and debt. *Foreign Affairs*, **63**(5), 1081–94.

Lancaster, C. (1985). Update: Africa's food and development crisis. *ODC Policy Focus*, **8**, February, pp.1–7.

Luke, D. F. & Shaw, T. M. (eds) (1984). *Continental Crisis: The Lagos Plan of Action and Africa's Future*. Washington: University Press of America.

Mamdani, M. (1985). Disaster prevention: defining the problem. *Review of African Political Economy*, **33**, 92–6.

North–South Institute (1985). *After the Cameras Leave: The Long-Term Crisis in Africa*. A provisional conference report. Ottawa: North–South Institute.

Nyerere, J. K. (1985*a*). Speech at the UN Secretary-General's Emergency Conference on Africa, Geneva, March. Pamphlet published by the Government of Tanzania, 18p.

Nyerere, J. K. (1985*b*). Is Africa responsible? Address at the Institute of Social Studies, The Hague, March. Dar es Salaam. Pamphlet published by the Government of Tanzania, 13p.

OAU (Organization for African Unity) (1981). *Lagos Plan of Action for the Economic Development of Africa, 1980–2000*. Geneva: International Institute for Labor Studies.

ODI (Overseas Development Institute) (1984). Africa's food crisis. *Briefing Paper 1*, May.

Ravenhill, J. (ed.) (1986). *Africa in Economic Crisis*. London: Macmillan.

Review of African Political Economy (1985). Issue on war and famine in Africa. **33**, 1–115.

Sandbrook, R. (1986). *Africa in Crisis: State and Society Today*. Cambridge University Press.

Shaw, T. M. (ed.) (1983*a*). *Alternative Futures for Africa*. Boulder: Westview.

Shaw, T. M. (ed.) (1983*b*). Debates about Africa's future: The Brandt, World Bank and Lagos Plan blueprints. *Third World Quarterly*, **5**(2), 330–44.

Shaw, T. M. (ed.) (1983*c*). The future of the great powers in Africa: toward a political economy of intervention. *Journal of Modern African Studies*, **21**(4), 555–86.

Shaw, T. M. (1984). Unconventional conflicts in Africa: nuclear, class and guerilla struggles. *Jerusalem Journal of International Studies*, **7**(1-2), 63–78.

Shaw, T. M. (1985*a*). Africa prepares for economic summit. *West Africa*, **3533** (13 May), 948–9.

Shaw, T. M. (1985*b*). Neither tears, gifts enough. *Halifax Herald*, 28 May.

Shaw, T. M. (1985*c*). African solutions prepared in Addis. *Africa Now*, **50**, 16.

Shaw, T. M. (1985*d*). Not by relief alone. *Policy Options*, **6**(6), 33–6.

Shaw, T. M. (1985*e*). Africa: Beyond this famine. *International Perspectives* (July/August), 6–10.

Shaw, T. M. & Aluko, O. (eds) (1985). *Africa Projected: From Recession to Renaissance by the Year 2000?* London: Macmillan.

The People (1985). Mamdani is not a Uganda citizen, and Much ado about Mamdani. 5(17) (13 May), 1 and 10–11.

Timberlake, L. (1985). *Africa in Crisis: The Causes, the Cures of Environmental Bankruptcy.* London: Earthscan.

UN (1984). Declaration on the critical economic situation in Africa. New York: *UN Document A/RES/39/29.*

US Committee on African Development Strategies (1985). *Compact for African development.* Washington & New York: Overseas Development Council.

Wijkman, A. & Timberlake, L. (1984). *Natural Disasters: Acts of God or Acts of Man?* London: Earthscan.

World Bank (1981). *Accelerated Development in Sub-Saharan Africa: Agenda for Action.* Washington: World Bank.

World Bank (1983). *Sub-Saharan Africa: Progress Report on Development Prospects and Programme.* Washington: World Bank.

World Bank (1984). *Toward Sustained Development in Sub-Saharan Africa: A Joint Program of Action.* Washington: World Bank.

8

Linking and sinking: economic externalities and the persistence of destitution and famine in Africa

RANDALL BAKER

School of Public and Environmental Affairs, Indiana University

Introduction

> The best thing we could do for the debtors and for the underdeveloped nations generally, is to get clear in our own mind, what is required to make an economy succeed. (*Wall Street Journal, Editorial, 28 June 1985*)

The continent of Africa is trapped by its own history; a history largely shaped by external forces, some originating in the colonial period and some from the inertia which clothed the colonial model of relations in the guise of development once the colors on the flag had changed. That model has not served Africa well but most of Africa's external links were created and fashioned to serve *it*. This chapter examines the proposition that Africa may 'develop' itself into its own destruction if external sources of capital and its trading partners do not accept the need for fundamental, systemic change *now*.

It is, perhaps, an artificial exercise to isolate the external factors accentuating hunger, famine and destitution in Africa, since these operate through and with a complex of local policies, priorities and perceptions. To deal with the externalities in isolation is to deprive explanations of process and causation of most of their meaning. So, while concentrating on external factors in this chapter, I shall relate these closely to the more local circumstances throughout.

The second main area where a sense of balance is required is in consideration of the motives of the main participants in the development process. Much of the literature reflects a fairly radical polarization of views promoting ideas such as the perpetuation of dependency, the alliance of world classes, the Rousseauistic ideal, the virtues of modernization and so forth: something for everyone. Some of the literature has become polemical, verging in places on the use of conspiracy theory as an explanatory tool. I believe that there are valid elements of explanation in several of these hypotheses and have, consequently, cast the net fairly wide in a search for useful and practical realities. Ultimately, however, the touchstone is the subjectivity of my own experience. Increasing polarization does not help the immediate situation and makes dialogue between adherents to the various philosophies and the practitioners and policymakers in the field of development increasingly unlikely and less fruitful.

I work from the premise, albeit naive to many, that no one engaging in a serious search for a sustained improvement in the life of the world's poor is interested in a perpetuation or accentuation of policies and practices which deepen the crisis for those people. I have, therefore, tried to incorporate such realities as the fact that lenders also have to borrow and repay; that the majority of aid is bilateral and government to government and thus must work within a framework of local policies, power structures and sovereignty; and that many countries are caught in a serious trap of (*inter alia*) debt repayment, static or declining domestic revenues and a weakening of the revenue from the primary export-led development model. For those who subscribe to the conspiracy end of dependency theory this paper will have little to offer because to them the solution lies in a total restructuring of world power based on nothing less than revolution. This paper examines the historical roots of the present crisis in terms of a continuation of colonial policies – often largely through inertia – and how these relate to the prospects for the improvement of living standards for those on the margins of survival. Others may feel that a more overtly political stance is needed aiding only those countries with serious, poverty-oriented programs. Some element of this does appear in the final recommendations of the paper.

It is becoming less and less useful to talk about the Developing World as a concept. Africa seems to stand more clearly apart each year in the World Bank Atlas of indicators. It remains a continent in which per capita food supply continues to decline, [the average African in 1981 had 10% less food than he/she had in 1971 (Saouma, 1981)], in which dependence upon relief and food imports grows and becomes a regular and widespread feature of life, and where the future prospects for serious and sustainable growth along existing lines over the next decade look grim. Yet, in terms of natural resources

> . . . Africa has enough land for food self sufficiency. Even with the assumption of low levels of inputs, the combined potential productivity of all 51 countries could feed nearly three times the people in need . . . The specific results for the continent as a whole, estimate average potential population supporting capacities of 0.39 persons per hectare [pph] with low levels of

inputs, 1.51 pph with intermediate inputs and 4.46 pph with high levels of inputs. These potentials are respectively 2.7, 10.8 and 31.7 times higher than the average present population density of 0.14 pph. (Higgins *et al.*, 1981, 19)

In trying to reconcile the above quotation with the realities of famine it may be fair to say that the 'development' effort on that continent has not been successful insofar as so many have been left at the bottom of the heap and as the model, in general, seems to have run out of steam. Change is urgently needed. It is relatively easy to ascribe these problems on a temporary basis to dramatic oil price rises, declining commodity prices, adverse shifts in the terms of trade, growing debt, recession in the industrial countries and the absence of any 'green revolution' for the basic African staple crops. However, many of these factors – except the last – have been only marginally worse for Africa than for other parts of the world which have shown a better performance. Others point to the succession of natural calamities, principally the droughts, which struck Africa between 1968 and the present. There is little evidence, however, to suggest that the deepening crisis in Africa is a purely natural disaster, despite the unusual length of the drought in many places. The famine and the drought have served only to accentuate and dramatize a much broader and longer-term systemic malaise.

In this chapter, since the concern is with *famine* rather than *poverty per se*, we shall be concerned with the group at greatest risk and not just with those facing – albeit serious – deprivation. Lipton (1983) estimated the former group at around 15% of the total population and the latter at about 40%. Special problems arise when considering the role of external factors in relation to the 15% since they may well be, through the marginality of their land, their landlessness and/or their joblessness, outside some of the main avenues of development such as better international commodity prices. For these people there are basically only the following options:
— continued support from relief programs (institutionalized destitution);
— improving the traditional option (i.e. better strains of subsistence crops providing little 'return' on investment in the conventional financial or economic sense);
— greater incorporation of these people into the commodity sector (through access to land, labor-intensive agricultural programs, and better external prospects for crops);
— diversification: job creation favoring the poorest 15% in new sectors of the economy (incorporating the freeing of world trade and resisting trends towards protectionism).
It is clear that each of these paths involves a strong and determined local

policy backed by external understanding and support (e.g. research, more open trade environment, freeing up investment capital). In this effort the aid donors, the multilateral development banks, the International Monetary Fund and the UN agencies all have distinct but complementary roles to play.

To a great extent the problems in Africa derive from its basic models of African development which perpetuate policies from colonial times. This is not a polemic against colonialism, it is simply an acceptance of the realities of that period during which Africa was incorporated into a world system very rapidly, very thoroughly, very one-sidedly and very late. Furthermore, Africa was incorporated after an earlier period of slaving, which had all but beggared the pre-colonial African systems of production, society and exchange, and at a late stage in the development of the economies of those industrial nations which became the metropolitan powers (with the exception of the residual empires of Spain and Portugal). It is with this brief historical perspective that I would like to begin.

Africa and the World

Africa is locked into an economic system that obliges it to produce goods it does not consume and to consume goods it does not produce. (*Wall Street Journal, 28 April 1981*)

For most of the time that Africa has been drawn into the world economy the relationship has been a troubled one. It is to some extent the legacy of that period of integration which laid the foundation for many of the problems that presently afflict the poor.

From the time the Portuguese expanded beyond the conquest of Ceuta at the end of the fourteenth century, the African continent went into trauma. Earlier systems of culture, production and exchange were exposed to the ravages of centuries of slave trading by Europeans, Arabs and other Africans. A precondition for such a trade is the stimulation and perpetuation of anarchy, rivalry and internecine strife. Whatever may have been the course of Africa's evolution, this intrusion held it in check.

By the time of the 'Scramble' in the late nineteenth century, when the freebooters gave way to the administrators of the dependent territories, Africa already had a legacy of almost 250 years of total mayhem and haemorrhage. Those who came to Africa in the later phase may be characterized by certain qualities which have left their mark:

1) the missionaries who came to *replace* indigenous belief systems and cosmologies with the Christian ethic in one or other of its many guises;
2) the technologists and entrepreneurs who came to *transform* Africa into

a producer of tradable goods with an infrastructure to market and export these commodities;

3) the administrators who sought ways, through taxation and commoditization, to ensure that the dependent territories '*paid their way*';

4) the anthropologists who seemed to have, for much of their time in Africa, a preoccupation with the bizarre, the ritual, the deviant and form rather than function (Baker, 1984).

All these people have something in common: they were either there to effect change according to the principles and ideals they brought with them rather than to understand anything they found, or if they did try, they focussed in a descriptive fashion on the ways in which local people seemed to be outwardly different from 'us'. The approach was *unidirectional*, set upon change rather than on understanding.

To some extent this itself is understandable because the Africa they found had been ravaged and savaged and gave the outward appearance of unsophistication, chaos and darkness. The fact that this was not an endemic African trait but something wished upon much of the continent by an earlier wave of slavers seems not to have received the attention that it should have. Here again, let us not dwell on the barrenness of 'blame' for there is no value in that. Rather, let us try to understand the nature of the process of change and those (and their perceptions) who tried to effect the change.

Thus, there was little interest in ways the Africans had 'traditionally' managed to provide a living from the natural environment either by adapting to it or adapting it to them. From the earliest times of systematic European intervention, the ideal was to teach the 'dignity of labor' to a people who did not 'need to work' (Drummond, 1889); to give agriculture a commercial value and to incorporate Africa into the world system as it was then. Change, or 'development', became firmly established as something which came from outside: outside ideas, outside technology, outside capital and (often) outside management and control. This was the reinforcement of the unidirectionalism which has so long dominated the concept of development in Africa.

There were several consequences of this process shaping external links which are briefly outlined here.

1) Change was something overwhelmingly purchased or otherwise derived from abroad in the form of advice, technology, capital, etc., for which you needed the proceeds of foreign-exchange-earning exports.

2) Little attempt was made to research the indigenous subsistence crops or production systems, as the former were not marketed internationally

(particularly millet and sorghum) and the latter were considered to have no scientific basis.

3) The subsistence sector was largely left to its own devices and was actually encouraged to remain as it was because it subsidized the cost of labor to plantations, farms, mines, towns, etc. (Turok, 1979). It was, however, subject to pressures from the increased demand for land to grow cash crops, the confinement of population in the 'reserves' and the alienation of land for plantations.

4) A dual economy developed in which resources and attention were focussed on the 'modern' sector. This does not necessarily refer to plantations. Often in Africa the modern sector was simply the field of cotton or groundnuts that the farmer was required to graft onto the existing household garden. Thus, there was little substitution of cash crops for food crops, just an additional demand on the farmer to increase his or her labor and demand for land.

5) Increasingly, the export-crop earnings became essential nationally to support the often urban-biased development efforts. This was a process which became increasingly important after independence, as public sector expenditure rose and interventionism became widespread. The realities of urban power were such that not only was a tax extracted from the countryside to pay for the towns, but cheap food prices became widely established in order to subsidize the cost of urban labor and urban living. This sent a message of disincentive to the rural population and agriculture has remained the poor relation of development policy ever since. The stagnation of the rural areas then led to increasing urban migration, the increasing threat of political disruption among the growing disaffected urban poor, and the consequent necessity to retain and even accentuate the policy of cheap food. Thus, at home the towns predominated in policy, though with little productive return; the external economy was seen as the engine of development which, in turn, favored the towns (Lipton, 1977).

The neglect of the subsistence sector and the continued pressure to increase commercial production, accompanied by the growth in human population (among the highest in the developing world) have continued to subject the natural resource base to a process of accumulating pressure. It is doubtful whether the conventional wisdom regarding what has constituted 'development' in Africa is going to ameliorate or reverse this pressure. As a consequence, more of the same will result in the periods of stress, acting upon an ever less resilient and overtaxed environment and society, that is, more extreme events and collapse at the margins. What is needed is *a break with history and a re-evaluation of the development process*. As it now stands, an increasing proportion of the really poor population is not only excluded from the development process but also faces the prospect of a growing threat to its survival. Some elements of society, of course, do well from the established model so it may not yield without a struggle. Plans to revitalize the established primary export base – or to diversify it – are not going to reach the potential famine victims, unless a deliberate decision is taken, and practical means are devised, to reach them.

Recent external influences on Africa[1]

Despite the fact that Africa has been the subject of many economic shocks and reversals over the past decade or so, one of the principal problems derives not from oil or terms of trade (both of which are not much worse for Africa than for elsewhere in the developing world) but from the seemingly structural decline in the volume of traditional primary exports. This is particularly critical for Africa, because 32 major primary exports – with minimal value-added – account for about 70% of Africa's non-fuel exports to the industrial nations, compared to about 35% for other developing countries. Nineteen African countries showed negative rates of export growth during the 1970–80 decade and the continent's share of non-fuel trade from developing countries fell by more than half during that period. Growth in world trade in major African exports fell from 4.5% per annum in the 1960s to 1.5% in the 1970s, a trend which was then reinforced by the deepening recession in the developed world in the early 1980s. Minerals, on which some countries such as Zambia and Zaire place heavy reliance, declined by 7.1% per annum over the decade.

With the first oil price rises in 1973–74 there was a need in the short run to offset the extra foreign-exchange demands placed on African nations suddenly faced with current account deficits (excluding grants) which averaged 9% of gross domestic product between 1973 and 1982. This is approximately twice the level of other oil-importing developing countries. Reserves were depleted but the African countries were able to borrow from the international money market with relative ease, because commercial lenders were awash with money and they tended to see the situation immediately after 1974 as temporary, to be offset later by general recovery. However, this was not the case and the second oil shock (1978–79) and the subsequent deepening recession in the industrial nations produced the now-familiar scenario of heavy debt burdens, diminishing foreign exchange earnings to pay the debt interest, a rising dollar, revaluing the debt almost daily, and a fall-off in official development assistance (ODA) which plays so major a part in public-sector capital-investment programs in Africa. Thus, the existing foreign-exchange earnings had to be increasingly redirected toward the payment of debt or the countries had to (*a*) borrow more or (*b*) attempt debt rescheduling.

Although Africa is not normally singled out as one of the big debtors, this gives a false impression of the seriousness of the situation. By the late 1970s, and particularly the early 1980s, ODA failed to keep pace with the cumulative effects of the oil price rises, the accumulating debt burden, and the impact of the recession. The importance of this is seen in the fact that in eastern Africa ODA represented in 1980 78% of the regional capital inflow and 70% of the region's public capital expenditure. However, since 1980 there has been, at this most critical time, no increase in ODA in real terms. Indeed, between 1980 and 1982 ODA fell from $2.6 billion to $1.7 billion overall. Whereas, at the beginning of the 1970–80 decade ODA covered the current account deficits of most African countries, by the end of the decade it covered less than half. In West Africa's Sahel region about 85% of public financing comes from ODA. As I shall discuss later, this largely bilateral source is (at least in theory) of critical importance because much of it comes in the form of grants and should have a greater potential for use in such areas as improvement of subsistence conditions, sectoral restructuring, food-crop research, etc., where there is unlikely to be any immediate internal economic rate of return large enough to satisfy the commercial lenders and multilateral development banks.

Over the same period, the debt-service ratio (debts as a percentage of exports) of the African countries, which is one of their major links with the external world at the present time, doubled to 12.4% at the end of the 1970s and rose sharply again to 19.9% in 1983. Outstanding and disbursed debts for low-income Africa, taken as a percentage of the GNP (54.6%) are roughly double the percentage of the figure for all developing countries (and 7 times higher than the 8% figure for low-income Asia). As a percentage of exports, the value of debts for low-income Africa is twice that of the developing world and almost three times that of low-income Asia. Furthermore, the projections show the figure of debts outstanding as a percentage of export value rising from the present (1984) figure of 278% to 328% under the World Bank's 'Low' assumption (of global recovery) and only falling to 250% under the 'High' assumption. The declining economic performance, existing debt burden and the nervous reaction to debt have generally drastically reduced the prospect of further commercial lending. As more and more foreign exchange has to be directed toward servicing debts, less and less (especially with the decline in real ODA) is available for investment at home to ward off the perils of the future. Unfortunately, also, some of the earlier debt was incurred for consumption of large, non-productive public investments which generated little to no foreign exchange nor generated much in the way of an economic return. Compounding all this, of course, is the fact that the declining export situation produces a declining domestic savings situation, further reducing the capacity of the home government to dig itself out of its difficulties (domestic savings halved over the 1970–80 decade). This did little to help overcome the shocks that oil prices superimposed on declining export volumes, a 2.7% average annual decline in the purchasing power of exports and a rate of growth of GDP which fell steadily from 1973 through 1983.

In many ways Africa faces the same range and types of problems, externally speaking, as most developing countries, but with a much more serious impact (even if the sums borrowed are not as eye-catching as those in parts of Latin America). Briefly, the reasons are as follows.

— Africa has a narrower and much more traditional primary export base and is less able to spread its trade risks over a broader economic base than is the case in some Asian countries.

— The poorer end of the rural population has fewer opportunities to escape their poverty in any meaningful sense than do people in some other developing countries, because of the limited absorptive capacity of the secondary and tertiary sectors.

— The flow of ODA funds is particularly critical to Africa in light of its substantial contribution to public investment capital while private investment capital is very constrained by low income levels.

— There has been little to no research relevant to the improvement of the basic staple grains and other crops upon which Africa depends. The former colonial preoccupation with researching cash crops at commodity centers continued long after independence. Only around 1970 did any serious research begin on the largely dryland, rainfed enterprises (Norman, 1985).

— The continent continues to have one of the highest population growth rates in the developing world. Although interpretations vary, there are those who believe that people living so close to the margin of survival view numbers of children as one of the few forms of security in old age. There may well be, therefore, sound and rational reasons (for the individual) to increase the number of children. In a situation of increasing risk, the irony is that the natural inclination would be to increase the numbers even further. Individual rationality thus becomes collective suicide.

As outlined in the introduction, therefore, it is the interplay of external, local and historical factors which accounts for the acuteness of the African dilemma, rather than any one element in isolation. The rest of this chapter is concerned with looking at ways in which the external element may play a more meaningful role in the sustainable survival of Africa.

The future based on current trends

The facts of life at the present time are that Africa continues to show a growth rate of GDP below the population growth rate (3.0%) and the projections – even the optimistic ones – prepared by the World Bank do not present a rosy picture (Table 8.1). The high and low projections are based on assumptions regarding the prospects of (*a*) improved economic performance and the rate of inflation in the industrial world and (*b*) trends in protectionism in those countries. In fact (and ironically) Africa's vulnerability to protectionism is in many ways less serious than for some other, more diversified, developing economies, because the concern of most protectionists is with those manufacturing areas into which other primary producers have moved, such as shoes and textiles (so-called 'basic industries'). Even under the assumptions of the low protectionism, high recovery figure in Table 8.1 (the 'High' column) this would still result in a projected 0.1% annual per capita GDP decline over the period.

In its Annual Report for 1985, the World Bank (1985, 141) makes very plain the seriousness of the current situation:

> For many low income African countries the economic outlook is bleak. The Low simulation would mean another five year period of falling per capita incomes. Incipient economic reforms in many of these countries would surely fall victim to an international environment in which primary commodity prices would not

Table 8.1 *Average performance of developing countries, 1960–95*

GDP growth	1960–73	73–80	80–85	1985–1995	
				High	Low
Developing Economies	6.1%	5.5%	3.0%	5.5%	4.7%
Asia	6.0%	5.2%	6.4%	5.3%	4.6%
Africa	3.7%	2.7%	1.4%	3.2%	2.8%
(per capita)	(1.0)	(−1.0)	(−1.6)	(−0.1)	(−0.5)

World Bank (1985*b*, 35).

improve from the present very depressed levels, imports would need to be compressed even further, and additional aid flows would not be available. Unfortunately the High simulation holds out hope only *for the maintenance of average per capita incomes at the low levels to which they had declined in 1984 . . . Donors must . . .be willing to make adequate financial assistance, over and above that projected in the High simulation, available to support those low income African countries that are implementing substantial policy reforms* [emphasis added].

Table 8.2 shows that there has been some improvement in the terms of trade for Africa. Yet, even for some of the better-off parts, the general forecast regarding recovery within the conventional parameters are gloomy. In its annual report for 1984 the World Bank (1984, 83) states with regard to eastern Africa:

The adverse trade environment of the Eastern African countries is unlikely to be reversed radically in the 1980s . . . Nor are price prospects for Eastern Africa's export commodities bright. Even under optimistic assumptions about economic recovery in the industrial countries, prices of copper, coffee, cobalt, cotton, and sugar are not expected to increase in real terms by the late 1980s; in fact, they are likely to be lower than those prevailing in 1980 and about 15 percent to 20 percent lower than they were in the 1960s. As these commodities represent over two-thirds of the region's exports, there is *little prospect of any rapid expansion* . . . While efforts are and should be made to diversify exports, the areas into which diversification might take place are not easy to identify . . . [emphasis added].

In terms of improvements within the present framework of development, the prospects are bleak even for a return to the circumstances of the 1960s. Thus, we face the very real prospect of a continuing decline and increasing pressures on both the physical environment and the people at the lower end of the subsistence scale.

Table 8.2 *Changes in terms of trade: Africa, 1965–83*

1965–73	1973–80	1981	1982	1983[a]
−0.1%	−1.5%	−9.9%	−0.9%	4.6%

World Bank (1984).
[a] Estimate.

The emphasis still seems to be on fighting decline with the conventional armory of weapons: cash-crop diversification, sectoral diversification, marketing and price improvements, expansion of cash crops, etc., but now we have the additional problems of carrying a burden of debt repayments and a decline in the demand for Africa's basic exports which were not a major feature of earlier times. Like the White Queen in Lewis Carroll's *Alice in Wonderland*, Africa is faced with the prospect of having to run faster to stand still.

There is an additional problem, which is that many of the conventional development paths continue to neglect the food production sector (i.e. the subsistence sector). There are several reasons for this, deriving from external linkages and pressure:

— the need to realize more foreign exchange to meet debts and increased import burdens for consumption rather than investment;
— the inability of countries to raise foreign exchange from other sources, such as diminished ODA and commercial lending;
— the perceived need to maintain hard currency capital flow and credit in order to invest in the conventional arenas of development and growth for the future.

It must be reiterated that a growing proportion of Africa's population is falling outside the 'development' sector because of landlessness and joblessness. Thus, they lack the means to ensure their own subsistence either by growing food (except for illegal and ecologically devastating forays into forests and desert margins) or by purchasing it. Getting development to these people requires more than just rhetoric about 'aid to the poorest'. It requires programs specifically aimed at eliminating destitution, however economically 'sub-marginal' such programs may appear to be.

External factors and policy change

Even during some of the hardest periods of domestic food shortage, many African countries pursued policies of using agricultural land to earn foreign revenue at a time when they were on international food relief. For example, Mali's cotton production has increased 150% during the last 5 years and the production of groundnuts has risen 100% over the same period. Since these products compete for land with subsistence crops there is at least a prima-facie case for asking why this policy decision was taken, i.e. foreign exchange versus domestic food security: feeding the rich and not the poor.

It is too easy to blame the policymakers and their foreign advisers for this. These countries have little option but to raise the foreign exchange, as the impact of debt renunciation on the international monetary and banking system would be catastrophic once the tidegates opened. Furthermore, the response from the international community would very likely be recriminatory to the extent that there would be retaliation, leaving these countries in a situation of involuntary autarky. The political consequences of that for the African governments at home, especially from the more volatile and organized urban populations, would seal the fate of any government which attempted such an action. Yet, it is clear that the conventional models of 'development' are by-passing large sectors of the population (especially the growing landless population) which cannot afford to buy the additional food produced commercially. Modernization, as a way out, too often may mean a large-scale operation such as Zambia's state farm program with only limited capacity to absorb workers. Some countries, like Kenya, seem to be clearly divisible into two: those inside the development process and those outside it, scratching a living from an ever more run-down environment, resulting in the attempted coup of 1982; so, ironically, reproducing the civil disorders of the 1950s against the British over the same issue.

At the same time it is foolish to suggest that Africa should renounce *in toto* the conventional package, but a method has to be found for incorporating into 'development' the food production sector and food security for the poorest. This will not be easy, for the following reasons.

1. Governments are nervous about sinking large amounts of scarce investment capital into food-crop improvement which has no immediate, visible monetary return. In a wider sense the returns may be enormous but they tend not to figure high in conventional methods of cost/benefit analysis or on the revenue side of the treasury budget ledgers. Here it is worth noting that very often even the grant money given bilaterally and multilaterally is all too often appraised for 'economic' projects which favor those who already have a head-start.

2. Governments have very little appropriate research to apply in this field. Although work is starting now on better strains of rainfed sorghum, this is at a much earlier stage of development than other Green Revolution cereals. Furthermore, the Green-Revolution-type package does not have much meaning for those most exposed to hunger in Africa, because it involves costly inputs which would require the commoditization of the crops in order to pay for these inputs. In addition, hybrid seeds have to be multiplied and distributed, while the infrastructure in much of Africa for such distribution is notoriously bad. Furthermore, those without land cannot grow improved strains.

3. The multilateral development banks are in a difficult position. The most influential of these, the World Bank, has as one of its stated aims the alleviation of poverty. It has, however, found it easier to reach the poor than the extremely poor. Even though, since 1980, a requirement for poverty analysis has been incorporated into the initial appraisal for all agricultural projects, the World Bank has admitted that projects – even poverty-oriented ones – 'have provided few direct benefits for the landless, the tenants unable to offer collateral for loans, and for the near-landless farmer who finds it hard to borrow, acquire inputs and take risks' (1983, 13). It is easy to target the World Bank for criticism since it is rather more public about what it does than are many other organizations. Two things should, however, be kept in mind: the World Bank accounts for only 2% of foreign investment in developing countries and it is extremely difficult for it to secure the required economic rate of return from those with almost no resources at all. In poverty-oriented programs it has found that the economic performance has been at least as good as in the more conventional 'growth-dominated' investment packages; but mobilizing the potential famine victims, those with nothing, is extremely difficult. Although criticism of the World Bank still seems to be dominated by the gigantic project syndrome (*The Ecologist*, 1985), it is only fair to point out the changes in policy and orientation, the extreme problems of translating these into action and, bearing in mind that this refocusing came about only in 1980, the fact that it will take several years to see the impact of this in Bank project results and sectoral readjustment.

4. The IMF has built into its lending system pressures discouraging change within the African system. Basically the crisis atmosphere which predominates

> stress[es] the reduction of current account and public sector deficits at the cost of longer-term investments in, say, agriculture and education. In these circumstances, the real choice may be for short-term balance rather than for poverty alleviation *and* growth, although poor people in the poorest countries will suffer lasting damage. (*World Bank*, 1983, 32)

The IMF is one of the organizations which has a central role in permitting and shaping a restructuring rather than a recovery approach.

5. Money to help local governments to improve subsistence farming by strengthening research and extension, land conservation, etc., is more likely to be sought from grants than from bank borrowing. But, as we have seen, these grants and soft (mainly bilateral) loans have been

diminishing in real terms since about 1978. The main direction on the part of many governments, with the notable exception of Niger, has been to maintain business-as-usual on the commodity front and to rely on relief to deal with the crisis of extreme poverty. This, of course, will change nothing for the poor, except that they may live longer and extend their misery.

Prospects for change in the future

We have seen that there is something of a widening conflict between the demands of commercial export-oriented agriculture and the food security question, both of which compete for the same basic resource. The former does not easily answer the needs of the latter, as there is little effective linkage in many cases. While there does not seem to be much prospect for a country which turns its entire attention to ensuring subsistence, it is essential to move the question of national food security, especially for the very poorest, right up to the front. It is difficult for the countries concerned to get much in the way of constructive advice on this score from either the IMF or the World Bank. This is not a criticism of either of these organizations; one was created to deal with short-term balance of payments crises (not to grow more sorghum), while the other has an essential concern with securing an economic return from the funds it makes available for capital investment, including the International Development Association funds at concessional rates, and is, therefore, unlikely to adopt a mandate for eliminating absolute poverty and famine throughout Africa. Both institutions have a role since a 'balance of payments – debt crisis' perception reinforces the short-term view and the present economic system.

At the same time, the World Bank has an influence far beyond the size of its lending and is in a position to project a poverty focus into spending. But it will have to go beyond poverty. To the people we are talking about, poverty is a step *up* the ladder. Plus, of course, countries still need to aim for growth in concert with policy alleviation. The situation for the best-intentioned policymakers in Africa over the next few years is a very difficult one: to retain their links with the world monetary institutions means for the debtors that they dare not drop their primary concern with earning hard currency – despite roll-overs and rescheduling. In view of the dim prospects for 'recovery', however, in Africa's external economic relations and its vulnerability to a narrow range of primary exports, there has to be a breathing space to allow Africa to give priority to those on the verge of extinction and to secure the strategic base of survival. This policy may not immediately earn any dollars, except insofar as it eliminates imports of basic foodstuffs. It would seem perfectly reasonable to give top priority to countries with clear policies in this direction who will otherwise have a financial crisis. Even if some countries wish to do such things at the moment, they may be reluctant to advertise the fact because so few donors and lenders respond to the area of pure subsistence or of replacing export crops with food crops to some degree which is seen by them as going backwards. If investment in research and development in these areas is going to move money away from debt repayment, then so be it.

It is also essential, within *a clearly stated local food security policy framework*, to increase the amount of ODA. Now, more than ever, the continent is in need of such assistance if it is ever to free itself from the trap of the creaking, worn-out colonial economy and paying for the balance of payments crisis of the late 1970s and early 1980s. This is the time when ODA should be going up, not down. However, simple increases in ODA *per se* are not about to change the situation – they should only contribute to where there is a clearly defined path out of the present crisis. This may combine incorporation of more of the really poor into the cash-crop economy on the basis of fairer prices, land reform and less urban bias, the direction of attention to food as well as luxury export crops, and a much greater concern with agriculture in general which still remains, on the basis of budgetary allocations, the poor relation in many countries. Lastly, the above should be combined with as much diversification into manufacturing as is possible. Africa was a late starter in this respect. For this, the international climate should move away from the concept that industrial countries have a divine right to produce everything. As the economies of the industrial nations progress through innovation and technological change, they must be prepared to allow areas such as, for instance, textiles and footwear to be taken up by the developing world.

Thus, protectionism will ensure only that the Africans are held firmly on the land, where the crisis is located, able to sell only the same things as always into a diminishing and oversupplied market. It is all very well to accuse Third World countries of killing off basic industries in the developed world (e.g. shoes) by 'sweatshop' tactics, but this has to be seen in the context of the lack of any alternatives and the hopelessness of holding the continent of Africa in the colonial trap of exporting raw and unprocessed materials. They will not even be able to do this effectively while organizations such as the European Economic Community dump agricultural surpluses resulting from their own price anomalies into the world market. Even from the most practical and abstracted economic standpoint, an impoverished

Africa offers little as a potential market and affords the prospects for eternal political instability. But poverty is a blight and a scourge which must be eliminated and it is unlikely that a simple continuation of the past into the future is going to do much to reduce Africa's accumulating problem of hunger and destitution.

The lessons are as follows:

1. The continent of Africa has to be granted a period of recovery and rehabilitation. Simple rescheduling will not do, as the time horizon is too short and the prospects for conventional 'improvement' are too dim. Unless there is an understanding and expression of this emergency situation and there are programs to attack destitution and the man-made causes of famine, the roll-overs and rescheduling will simply delay the awful day and lead to a 'Ghanaianization' of the whole continent, i.e. the future consumed in paying off the mistakes of the past.

2. Aid donors and lenders will have to recognize that, under the straitened circumstances facing Africa over the recovery period, there is little to be achieved by holding rigidly to the division of aid/loans equal foreign exchange; local government equals recurrent costs. In some cases it has become evident that the countries do not have the capacity to absorb, or to make effective use of, some of the aid/loans being offered for capital projects. It is worth remembering that, as a proportion of GDP, savings in low-income Africa fell from 13.4% in 1970 to 5.9% in 1981.

3. There has to be a substantial increase in ODA on grant and concessional terms, insofar as that is directed toward clearly defined local policy to tackle destitution and the causes of famine. This consolidated policy direction is more important than the overwhelming preoccupation with *projects* which has dominated Africa and its donors/lenders over the last two decades. The International Fund for Agricultural Development has a key role to play because of its concern with the small, often conventionally 'uncreditworthy' farmer – yet this body is hard-pressed for funds. FAO has indicated that in real terms Africa will need a level of investment *two and a half times* that at present to achieve food self-sufficiency (Saouma, 1981, 23).

4. The lenders must balance the demands of their repayment schedules against real and well-substantiated claims to use interest payments for urgent food security investment and not for relief. Indeed, it may well be the only thing possible in view of the very poor predictions for Africa, *ceteris paribus*, to write off, rather than reschedule some of the

debt. This, of course, would only be done in the interest of agreed
and acceptable recovery programs.

5. The multilateral development banks (MDBs) should consider the wider
    incorporation of the poor into their rural projects. A lot of work is
    now being conducted into policy reform (e.g. the World Bank's
    Structural Adjustment Loans and Special Action Program). This target
    should be a major aim of structural adjustment lending of the World
    Bank and it may possibly require movement away from the
    conventional package of export crops. This might be seen to be in
    contradiction to the Compensatory Financing Fund of the IMF which
    tends only to fossilize the situation and incorporate the declining terms
    of trade more firmly into the system unless the compensatory lending
    is specifically designed to support systemic change (*Economist*, 15 June
    1985, 18).

6. The borrowers/recipients must demonstrate a serious policy shift
    towards agriculture and poverty elimination at the extreme bottom
    end. Declarations are easy to make (the UN Conference on
    Desertification showed that), and so too are plans of action. Serious
    programs with budget lines are something else altogether. At the time
    of writing the OAU summit was taking place in Addis Ababa (18–20
    July, 1985). Once again, we have mention of a plan: a 5-year plan for
    'greater economic cooperation, agricultural reform and food self-
    sufficiency', much like the earlier Lagos accord. Western diplomats
    at the meeting frequently alluded to the sense of déjà-vu, pointing out
    not only earlier experiences but the fact that the declaration (the 'Addis
    Declaration') would not be binding on member states. The summit
    also called for an international conference to discuss Africa's debt
    problem, but this is unlikely to be responded to positively by the
    creditors (who generally do not like 'clubs'). There is no universal way
    to achieve the aims of these various declarations. Most likely, a
    combination of factors is needed: the creation of more paid
    employment; the development and distribution of better strains; a
    strong rural bias and access for all to the means of eventually ensuring
    their own survival through their own efforts. Without demonstrable
    evidence of will and a practical local policy line, donors and lenders
    are under no obligation to help, since such help would only serve to
    compound the problem.

7. At the beginning of this chapter I quoted a study by Higgins *et al.*
    (1981) on the potential productivity of the African continent. The
    general potential, which is optimistic, is countered by the division of

the continent into nations, 11 of which, the report states (mostly in cool highland Africa and the Sahel) 'have resources insufficient to provide for the year 2000 population, even if the intermediate levels of input were to be applied by that date' (p. 21). This is a suggestion for a stronger intra-African policy to realize that potential for all who live in the continent.

So, to survive, the people of Africa – or rather the really poor people of Africa – rely on us to open our markets to them and use our technological know-how to keep our own diversification moving while allowing them also to diversify, and to give assistance only where there is some serious prospect of giving the recipients (not the country or the government) some real prospect of acquiring the means of secure subsistence directly or indirectly (i.e. food or wages). It is time for Africa to break free from the legacy of its recent history.

Notes

1 All figures cited in this section, unless otherwise noted, are from the World Bank. The following brief notes enlarge on some of the institutional references made in the text.

Structural Adjustment Loans (World Bank) are intended to help countries with deep-rooted balance of payments difficulties to reform their policies. They provide foreign exchange to enable the countries to make the transition (i.e. shifting out of parastatals).

Special Action Program (World Bank) financial measures and policy advice help countries implement adjustment measures needed to restore 'growth and creditworthiness'. The thinking on restructuring is much influenced in the African context by a belief that state intervention in many areas of production and marketing of primary products has been a negative force and would be better replaced by a more market-oriented economy. To a large extent this is what is meant by readjustment. A good example is provided by the writings of Balassa: ' . . . successful adjustment policies through export promotion and import substitution permitted market-oriented countries to accelerate their economic growth and limit their reliance on foreign capital . . . By contrast, the interventionist states lost market shares and were not able to offset the loss through import substitution . . . [p]rivate market-oriented countries experienced an improvement, and socialist-oriented countries a considerable deterioration in their economic performance . . . The above conclusions indicate the importance of policy choices in sub-Saharan Africa and the *need for policy reform towards greater market orientation*' (Balassa, 1984, 337). This, of course, is not the same as alleviating poverty or ensuring food security, but it is a means of increasing the government's wherewithal to act. At the same time the World Bank has pointed out that 'Poor people can benefit from the freeing of markets and similar actions only if concomitant structural changes provide the access, assets or information the poor need to compete in the marketplace' (World Bank, 1983, 29).

The International Fund for Agricultural Development (IFAD) Since its inception in the 1970s, IFAD has been remarkably successful in its goal of reaching the smaller, generally 'uncreditworthy', farmer with small loans. Though this seems to be the one organization that usually opposing parties agree is doing a good job, its future is threatened at the time of this writing. Originally, IFAD's funding was to have been split 50/50 between the Developed World and the OPEC countries, but these latter have requested that their share be reduced to about 40%. The United States has vigorously opposed this (since OPEC has done little to earn the love of the USA over the last decade), and there is strong reason to believe that there are many senior people in government who would accept the demise of IFAD with equanimity. This is partly because they are often uncomfortable with multilateral lending where it is difficult to tie the flow of funds to US national policy interests.

Acknowledgments

I would like to thank Professor William Siffin and Catherine M. Sokil of Indiana University for their assistance in the preparation of this chapter.

References

Baker, P. R. (1984). Protecting the environment against the poor: The historical roots of the soil erosion orthodoxy. *The Ecologist*, 14(2), 53–60.
Balassa, B. (1984). Adjustment policies and development strategies in sub-Saharan Africa. In *Economic Structure and Performance*, ed. M. Syrquin, L. Taylor & L. E. Westphal, pp. 317–40. Orlando: Academic Press.
Drummond, H. (1889). *Tropical Africa*. London: Hodder & Stoughton.
Higgins, G. M., Kassam, A. H., Naiken, L. & Shah, M. M. (1981). Africa's agricultural potential. *Ceres* (Sept.–Oct.), 13–21.
Lipton, M. (1977). *Why Poor People Stay Poor: Urban Bias in World Development*. London: Temple Smith.
Lipton, M. (1983). African agricultural development: The EEC's new role. *Development Policy Review*, 1(1), 1–21.
Norman, C. (1985). The technological challenge in Africa. *Science*, 227(4687), 616–17 (quoting Edmund Hartmans, Director of the Nigeria-based International Institute for Tropical Agriculture).
Saouma, E. (1981). A new food order for Africa. *Ceres* (Sept.–Oct.), 22–6.
The Ecologist (1985). Special issue on the World Bank: global financing of impoverishment and famine. 15, 1–2.
Turok, B. (ed.) (1979). *Development in Zambia: A Reader*. London: Zed Books.
World Bank (1983). *Focus on Poverty*. Washington, DC: World Bank.
World Bank (1984). *Annual Report*. Washington, DC: World Bank.
World Bank (1985a). *Annual Report*. Washington, DC: World Bank.
World Bank (1985b). *World Development*. Washington, DC: World Bank.

PART III

Case studies

9

Drought, environment and food security: some reflections on peasants, pastoralists and commoditization in dryland West Africa

MICHAEL WATTS

Department of Geography, University of California, Berkeley,

Social relations are given to a historical subject . . . as realms of possibilities, as structures of choice. Adam Przeworski (1982, 311)

The space of death is crucial to the creation of meaning and consciousness . . . We may think of [it] as a threshold yet it is a wide space whose breadth offers positions of advance as well as those of extinction. Michael Taussig (1984, 497)

Famine has been represented to us in the United States almost daily over the last year. One such representation appeared in the *New York Times Magazine* (17 November 1985): a photograph of one particular moment of agony and human suffering in an Ethiopian relief camp. Like many such moments the arresting quality of the image is dispersed by a sort of personal inadequacy. As John Berger puts it, the moment 'becomes evidence of the general condition . . . [it] accuses nobody and everybody' (1980, 40). In some respects this inadequacy coupled with a type of diffuse causality, precisely captures the current crisis in Africa; agrarian stagnation and, in some cases, collapse on an unthinkable scale, famines of almost biblical proportions, and the recurrence of drought in the context of fragile ecology, fuelwood shortages and soil erosion. For much of semiarid West Africa and the Sahel, 1984 was 'an apocalyptic year of famine' (Derrick, 1984, 281) emerging from the ashes of 10 years of rehabilitation and rural development – Kissinger's plea to 'roll back the desert' comes to mind – led by a phalanx of donor agencies and development organizations. When one situates agriculture on the larger canvas of debt repayments, unfavorable world markets for many primary commodities, and fiscal austerity associated with the hegemony of monetarist policies at the core, the development scenarios are bleak and the economic prospects grim if not altogether hopeless.

I wish to take a somewhat different tack, however, starting from Berry's prescient comments on the dangers of generalizing about agrarian collapse on the basis of quite problematic statistics (1984, 59). Rather than to deny the actuality of accumulation and social change in the rural sector – as is sometimes implied by reference to stagnation, collapse, withdrawal and so on – I wish to emphasize precisely the dynamic, changing character of pastoral and peasant production systems in semiarid West Africa and the new and unanticipated forms of differentiation that are a central aspect of the agro-pastoral landscape. In focusing on change, and on the social relations of agriculture and pastoralism, I wish to examine drought, environmental deterioration and famine in a specific, albeit partial way. First, I will look at land-use management and various responses to environmental perturbations, such as drought, against the backdrop of a large body of work which suggests that pre-capitalist communities are fashioned, broadly speaking, around the dual foci of risk and subsistence security (see Scott, 1976; Watts, 1983). Second, I will examine the ways in which these social and technical relations, constituting what Edward Thompson (1978) calls a 'social field of force', have been transformed and are changing with the growth of commodity production and the deepening of the market.

Following a brief overview of the ecological context of Sudano-Sahelian West Africa, I discuss what Paul Richards (1985) refers to as an 'indigenous agricultural revolution'; namely the human ecology of resource use and the adaptive flexibility of agriculture in reaction to environmental uncertainty – in short a sort of 'peasant science' – in the semiarid savannas of West Africa. Then, I try to show how environmental problems in the Sudano-Sahelian regions can be understood in relation to the way in which political–economic change has fundamentally shaped, limited and ruptured certain patterns of land use; to argue; in other words, that peasant knowledge and practice is overridden by forces that compel peasants and pastoralists to 'destroy their own environment in attempts to delay their own destruction' (Blaikie, 1984, 29). In the third part, using evidence from Nigerien and Nigerian Hausaland, I show how these same forces can also provide insights into the nature of food insecurity in peasant communities in dryland areas; here I specifically target the process of commoditization and the social context of the development of markets among differentiated rural producers. Finally, I speculate briefly on what the implications of this local level work might be for the analysis of famines.

Geographical and ecological context

The geographic context for this chapter is an area I have somewhat idiosyncratically referred to as Sudano-Sahelian West Africa, a zone centered on latitude 15° N and embracing both the southern fringe of the Sahara (the Sahel proper) and the northern reaches of the Sudanic savanna woodlands. There are no universally accepted standards for delimiting climatic or vegetational zones in West Africa, but following the recent report by the National Research Council (NRC, 1983), the Sudano-Sahelian region as denoted here would include an area of thorn scrub steppe dominated by annual grasses, between 200 and 600 km wide, bracketed by the 100 mm and 600 mm isohyets. Covering some 3 million square kilometers across nine states, this zone subsumes three phytogeographic communities: a grass steppe (Sahelo-Saharan) between the 100 mm and 200 mm isohyets in which the tree component is slight and annual grasses such as *Panicum turgidum* predominate: a tree steppe (the Sahel) situated between the 200 mm and 400 mm isohyets which is more heavily vegetated including such characteristic species as *Acacia laeta, A. nilotica, A. tortilis, A. ehrenbergiana* and annuals such as *Aristida funiculata, A. adscensionis*: and the Sudano-Sahelian borderlands (shrub savanna) in which the vegetative cover may reach 10–12% on the sandy loams and over 60% on the silty soils. The grassy component is usually identified with *Andropogon gayanus* and *Zornia glochidiata*, while tree species such as *Acacia albida, A. seyal* and *Combretum glutinosum* predominate.

The predominant climatic aspect of this biome is a pronounced, short wet-season of 2–4 months, the remainder of the year being completely dry. Total precipitation and the probability of 'abnormal' dry years diminishes as one moves from the Saharan borderlands toward the savanna grasslands. As a general rule, however, climate is distinguished by extraordinary variability; the onset, duration, termination and spatio-temporal distribution of rainfall varies quite dramatically from one year to the next. As Nicolson (1981, 1983) notes, climatic fluctuations are abrupt and extreme; rainfall in one decade may persist on time scales nearly double that of the following, while the alternation of dry and wet episodes can create a false sense of 'normal' conditions. In this regard the drought years of the late 1960s and 1970s were not unique; they were part of a recurrent pattern and as such, in palaeoclimatic terms, fell within the expected range of variability both for the past several centuries and, in all likelihood, for the last 12 millenia. Recent research on climatic perturbations during the late Quaternary reveals both periods of desert incursion into savanna lands (for example the arid episode between 20 000 and 12 000 BP) and five millenia of much more humid conditions (9000–5000 BP) associated with greater riverine discharge and larger lake volumes. For the past two and a half millenia, however, the climate of the Sahel has been relatively stable, subject to periodic short- and medium-term oscillations toward drier or more moist conditions.

As in other semiarid regions, for example Rajastan in India (see Mann, 1982), prevailing systems of land use in the Sudano-Sahel include a variety of forms of animal husbandry integrally linked to, or combined with, sedentary rainfed agriculture, pump irrigation, and water-management systems such as flood retreat cultivation. It is clear that the rigid evolutionary and typological distinctions often made between Sahelian pastoralists and farmers are erroneous not simply because the two economies are fundamentally linked – few pastoral systems, for example, can sustain themselves without the acquisition of cereals from farming communities – but also because the system boundaries are fluid and permeable.

There are three major systems of animal husbandry: first, nomadism proper in which herds and households remain in the northern Sahel throughout the year following a peripatetic existence in search of water and pasture; second, transhumance in which many pastoral groups – for instance the Fulbe described by Sowers (1983) in southeastern Upper Volta (now Burkina Faso) – increasingly adopt rainfed agriculture of millets and sorghums and simultaneously maintain herds of cattle, sheep and goats which are taken to northern pastures during the rains and return to permanent villages and water sources in the savannas immediately after harvest, in order that animals can browse on crop residues; and third, sedentary peasant farming communities for whom small ruminants and cattle are important sources of manure, draft power and accumulation. These sedentary, and often ancient, farming communities are part of a rainfed cereal complex – millets, sorghums and pulses in particular – characteristic of much of the Sudano-Sahel and sustained by a bewildering variety of complex shifting, short-fallow and permanent cultivation systems.

In some areas – for example among the Hausa in northern Nigeria and in the Mossi heartland in Burkina Faso – intensive cultivation of cereals based on high inputs of organic fertilizer supports population densities in excess of 150 per km^2 and, in the case of the Hausa, have supported such densities for several centuries. In the closely settled zones, land scarcity and relatively high population growth have combined to create a shortage of pasture, the possibility of conflicts between herders and peasants associated with access to grazing 'commons', and the incursions of cattle onto standing crops, as well as new forms of interrelationships between sedentary cattle owners – for whom cattle have become in recent years a source of considerable wealth and speculative profits – and nomadic shepherds to whom animals are in effect subcontracted (see Delgado, 1978; Bassett, 1984).

Although the question of environmental change in semiarid West Africa was at the center of Sahelian debates throughout the 1970s, there was an identical and equally volatile debate conducted half a century earlier on the question of Saharan expansion. Stebbing (1935) and Bovill (1920) both maintained, in quite dramatic terms, that desertification and soil erosion threatened the productivity of the northern savannas, as a result of population pressures and inept agrarian practices. Stebbings' work in particular was granted a good deal of legitimacy in the 1980s and appears with striking regularity as evidence of the recursive quality of environmental degradation and secular climatic change in the Sahel. Yet Renner (1926), Brynmor Jones (1938) and Stamp (1940) were all deeply skeptical of Stebbings' claims and, not least, his supposition that a 'creeping Sahara' was already 3 miles north of Maradi by 1934. Stebbing, a forester with little tropical experience, visited Niger and northern Nigeria during the dry season and his report catalogued extensive dune systems, falling water tables and a marked deterioration in ground cover. Yet his report was roundly demolished by the experienced District Officers in northern Kano and Katsina Provinces who argued that Stebbing was overwhelmed by the aridity and apparent impoverishment of dry season ecology (cf. Chambers, 1983).

In actual fact, the water table was rising in northern Katsina, the dune systems had been stable for several thousand years and the close settled systems in the region had supported extremely high population densities for several centuries (see National Archives in Kaduna, Nigeria, ref. NAK Sokprof 2/1 #4094). This local verdict was subsequently upheld by the Anglo-French Forestry Commission in 1936–37 and a report on sylvan and land-use conditions in Katsina, Daura and Kano (Fairburn, 1937). In both cases any mechanical advance of the desert was fully discredited, although attention was drawn to the 'potential danger' of unregulated shifting cultivation and deforestation for fuelwood. Indeed, Fairburn was above all impressed by the relative *absence* of moisture deficits and the apparent ecological resiliency of savanna vegetation (1937, 44).

I raise these somewhat prosaic issues for the following reasons. First, environmental problems can be used for political ends and, hence, their extent and severity accordingly magnified, as for example also happened in the 1920s and 1930s in Central and Southern Africa (William Beinhart, personal communication). Second, a good deal of the discussion of Sahelian degradation remains hopelessly entangled in unproven generalities and dubious, though often unstated, assumptions concerning the ecology of grasslands, patterns of overgrazing, and agronomic practice. To say, as Picardi (1974, 55) does, that half a century of work by explorers and range ecologists and extensive satellite coverage during the 1970s, all show that 'desertification (has) existed for a long time', is to say very little. As Spooner (1982, 22) points out, a measure of desertification, conventionally defined, is an inevitable consequence of human appropriation and transformation of nature in order to survive. But there are important co-evolutionary processes of mutual adaptation in all such systems so that environmental change – for example, the obvious expansion of the cultivated area during the period prior to the 1970s famine in the Sahel – cannot be taken as synonymous with *degradation*. To cite several instructive cases: Bernus & Savonnet (1973, 117) using a careful study by Toupet in Mauritania point out that the period 1945–65 was clearly a period of Saharan *recession*, and that deserts can and do retreat, often spectacularly. Detailed local studies by Valenza (1979) in the Senegalese Ferlo establish rangelands in good condition in spite of heavy use, in which modifications in plant cover are minor but vary continuously in relation to the enormous (and recursive) variability in precipitation. Finally, Warren & Maizels (1977, 1), using Hollings' (1973) notions of resiliency and stability in ecological systems, point out that until much more basic research has been undertaken on Sudano-Sahelian grasslands, we might assume that sandy rangelands can recolonize and recover very quickly.

The fundamental dilemma in any debate on the nature of environmental degradation in the Sahel is the paucity of local data through time in an area in which synchronic studies can be quite spurious. In an important article, for example, Hare (1977, 340) believes that in northern Nigeria, where population densities exceed 150 per km^2, there is definite evidence of desertification, though living standards are not deteriorating and a disaster is not imminent. Yet, what are we to make of this? To take the example of the densely settled area around Kano, where even short-term fallowing has long disappeared, demographic growth has certainly resulted in a decline in the average size of family holdings. Between 1932 and 1964,

> the number of separately occupied plots on the 448 acres which were surveyed increased by 42% to 185 . . . During the same period the cultivated area increased by 26 acres [of] mostly marginal land . . . Of all plots registered in 1932, 41% had been subdivided by 1964 while only 16% had been consolidated . . . Fragmentation is also increasing. The average plot decreased in size by 22% between 1932 and 1964, [and] the average holding by 11%. (*Mortimore 1970, 385*)

Yet Mortimore observes that the permanence and productivity of agriculture in Kano has not been achieved through ecological deterioration. A colonial assessment report in a peri-urban district (Dawaki ta Kudu) reveals that densities in excess of 150 per km^2 were sustained by impressive grain yields from heavily manured holdings (Table 9.1). The sandy loams received at least five tons of animal droppings per hectare provided by the 40 000 animals in the district, almost 12 per household. More recent work by Schultz (1978) near Zaria substantiates the significance of organic manures for agrarian and ecological stability in northern Nigeria. I would argue that the evidence for desertification in Nigeria is far from clear. Yet, changes within the rural economy associated with the development and impact of commodity production and state intervention have not necessarily reduced the risks of a disaster (see Watts, 1983).

This is not to insert a completely Panglossian view of ecological conditions in the Sudano-Sahel. Careful empirical work along the lines of that undertaken by Talbot & Williams (1979) in Niger on cycles of erosion and deposition, or by Hurault (1975) on overgrazing in north-central Cameroon, show clearly that meticulous local studies can identify concrete processes of degradation. Rather, I wish to emphasize the following: first, that 10 years after the drought-famine we are still in a state of relative ignorance – it is only recently, for instance, that the critical significance of nitrogen and phosphorus rather than rainfall has been established for rangeland productivity in the southern Sahel (de Vries & Djiteye 1982; Bremen & de Wit 1983) – and this should warn against global generalizations on ecological degradation. Second, that many analyses of ecological change rest on problematic assumptions – and this is all that they are – concerning the purported rationale for, and impact of, large herds or poor farming practice on Sahelian range or arable lands, and such analyses do not make for penetrating social analysis on dryland ecology, of the sort demanded by

Table 9.1 *Millet yield, Dawaki ta Kudu district, Kano Province 1937*

Soil Type	Millet yield (kg per ha)	Manure (loads per ha)[a]
Jangargari or Jigawa	1647	138
Dabaro or Tsakuwa	1145	74
Shabuwa	926	67
Fako	564	25

NAK SNP 17 30361/1938.

[a]One load is roughly 54 kg.

Spooner & Mann (1982). Third, that desertification and environmental degradation are ultimately *local, place-specific processes* and must be understood as such, which requires a careful analysis of the interdigitation of determinate ecological and political-economic forces (Blaikie, 1984).

The human ecology of agricultural and pastoral systems: a paradox?

> Rural people's knowledge is often superior to that of outsiders. Examples can be found in mixed cropping, knowledge of the environment . . . and results of rural people's experiments . . . .
> (*Robert Chambers, 1983, 75*)

If there has been one recurrent intellectual trend in the study of African agro-pastoral systems over the last 15 years, it is the belated recognition that farmers and herders are expert practitioners of their respective modes of livelihood and are particularly sensitive to reproduction of the ecological systems of which they are part (see Richards, 1983). The question of the ecological rationality of stock-rearing economies, for example, has been the subject of careful local studies (see Dahl & Hjort, 1976; Dyson-Hudson, 1980; Raikes, 1981); all of these works point to a plethora of adaptive strategies – for instance herd mobility, species diversity, a sophisticated ethnoscientific understanding of local ranges – which reveal how herders cope with the vicissitudes of semiarid ecosystems and adapt to changing environmental and economic pressures. The recent spate of detailed ethnographic work also emphasizes the intense concern of all herders with the possibility of overgrazing and range deterioration. Dupire (1962, 69), for example, refers to the desire among Wo'daa'be Fulani to regulate tightly the length of dry season interlineage festivities 'for the stockmen fear the destruction of pastures from too great a concentration of animals'. Likewise, Bernus (1974) refers to a case in which Illabakan Tuareg in Niger actually turned off a mechanical pump because the new source of water severely exacerbated interethnic relations, the regulation of which was a key element in the control and maintenance of pastures. Such arguments need not imply a timelessness to pastoral adaptation but rather, in contrast to the view that pastoral systems inevitably lead to overgrazing, stress the longevity of such systems of livelihood – quite literally for millenia – in savanna biomes.

A similar argument can be made for agricultural resource use in the semiarid African tropics where questions of soil moisture conservation and erosion control are fundamental land-management issues. To simplify vastly a considerable body of human ecological research on dryland agriculture in West Africa, one can identify four broad areas or principles of indigenous resource use (see also Richards, 1983, 24–9). First, a variety of techniques frequently practiced by peasant farmers to conserve or improve the *physical properties of the soil* by manuring, mulching, ridging, minimum tillage, rotation and terracing. Second, the *complex land-use combinations* by which farmers exploit a variety of microenvironments, combining upland and lowland ecotones. In northern Nigeria, for example, Hausa farmers utilize local variations in savanna ecology and soil properties to cultivate expanded combinations of a variety of long- and short-maturing cereals and pulses (Watts, 1982). Third, *intercropping* in which multiple species or many varieties of the same species are interplanted in one field in the course of which the erosive potential of tropical rains is minimized, disease infestation reduced, nutrient use and the utilization of sunlight maximized, and weed infestation contained (see Igbozurike, 1977; Norman, 1977). Fourth, *adaptive capability*, the capacity of farmers to respond to perturbations in the physical environment – for example drought or soil deterioration – through complex decision-making sequences based on what one might call 'peasant science', that is to say, a tradition of agronomic knowledge and experimentation (see Chambers, 1983). Once again, then, to invoke the 'backwardness' of African shifting cultivation or the lack of any concern for ecological management among African farmers, is quite specious. Rather, this body of revisionist work speaks directly to continued inventiveness and adaptability among arid-land farmers.

The salience of this work can be seen in my own study of drought and peasant agro-ecology in northern Nigeria, an area of intensive, rainfed cereal cultivation (Fig. 9.1). There are three obvious attributes of semiarid

Fig. 9.1. Study area: northern Katsina, Kaduna State, Nigeria.

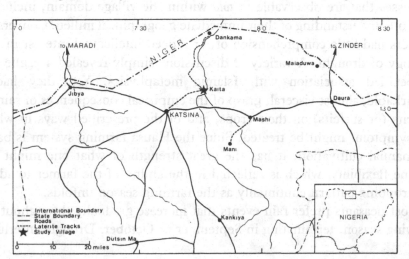

ecosystems such as the northern Nigerian grasslands; first, precipitation is so low that water is the dominant controlling factor for biological processes; second, precipitation is highly variable through the year and occurs in discrete, discontinuous packages or 'pulses'; and third, spatio-temporal variations in precipitation has a large random (unpredictable) component. In such areas, climate is bimodal; precipitation is almost never 'normal' over the short- or long-term, it is almost always considerably above or below any mean statistic. Superimposed upon the seasonal cycle of wet and dry are epicycles of enormous variability; periodically the onset of the monsoon is delayed, the totals may be poor, monthly distribution uneven, or the termination of the rains unusually early. In view of its latitude (North 13° 05′), the onset of the farming season in the village in which I worked north of Katsina is quite late, usually commencing sometime in late May with a pronounced monomodal distribution, peaking in July and August. While movement accounts for this distribution, the Intertropical Convergence Zone (ITCZ) does not act like a temperature front, for there is no direct correspondence between its surface and rainfall. Hence, the advance of the ITCZ is irregular, proceeding in a series of surges, halts and retreats. For the Hausa farmer, then, the regular round of the seasons is constituted at the local level as irregularity, by discontinuity, false starts and extreme patchiness. To talk of averages or mean statistics is, in other words, somewhat obfuscatory.

To the extent that drought is recurrent throughout the Sudanic savannas and that rainfall variability is part of the climatic order of events, it is to be expected that those who depend directly on the land for their livelihood demonstrate a sound knowledge and judgment of climatic variability and environmental risk. Hausa farmers appear to have a firm grasp of local processes that are observable *in toto* within the village domain, including an acute understanding of their immediate geographical milieu. Conversely, farmers had little comprehension of, or indeed intellectual interest in, the etiology of drought. A variety of discussions simply revealed a vague and ill-specified association with Islamic metaphysics. Yet, they had a remarkable, almost visceral, grasp of the empirical consequences of rainfall deficits (or surfeits) on their crops, and of the prescribed ways in which the symptoms might be treated. Since the Hausa farming system is based on manual cultivation, it has the great strength of what one might call on-line flexibility, which is reflected in the ability of the farmer to adjust his cropping pattern continually as the farming season unfolds.

Most peasants prefer rain events that increase gradually throughout the growing season, terminating in September or October. Drought interludes may have a serious impact, especially if they occur in early June (germination of the cereals), late July (heading of millets), or September (heading of sorghum). Drought spells at other times may not be problematic and indeed some farmers cherish a dry spell in early September which can facilitate the millet harvest. Particularly intense rain, conversely, can result in waterlogging and arduous farmwork. Of course, at the farm level, as all farmers know, there is no model wet season but only empirical constellations of rainfall generating specific crop conditions. Peasants have their own cognitive assessments of the likelihood of more or less wet or dry years, and it can be assessed in relation to four typical 'states of nature'. In Kaita village, a sample of 20 farmers revealed that the frequency of 'very bad' years (like 1972–73) was estimated at 8.3% and that of 'very good' years (such as 1978) at 12%. 'Good' and 'bad' years were assigned probabilities of 44.3 and 35.4% respectively (see Watts, 1983).

Rainfall, then, is highly variable, and Kaita peasants are indeed preoccupied with the nature of this variation. In Hausaland, I believe that the prevailing agricultural system can be viewed in relation to the limiting factors of low and uncertain precipitation, high rates of evaporation and low biological productivity. Specifically, farming behavior is best seen as *sequential adjustments* (or response strategies) to time-honored adaptations. The uncertainty of each year determines that Hausa farmers make daily decisions as the farming season develops. This close supervision and management revolves around two critical moments: the start of the rains and the first weeding. The process of adaptive management can be grasped by dividing each growing season into five somewhat artificial stages (see Fig. 9.2), which roughly correspond to discrete decision-making segments (World Bank, 1981). At stage 1, the farmer has little idea of what state of nature will prevail, except for subjective estimates based on past experience. Some farmers may undertake the risky venture of dry planting (*bizne*) during this first stage in the hope of germinating large acreages of millet early on and, hence, bringing forward the harvest date. At stage 2 (planting of millets), the farmer still cannot predict rainfall and generally plants millet followed later by sorghum, possibly altering spacing and sown area in response to the lateness of the planting rains to compensate for the possibility of reduced yields. On this information farmers can begin to form definite expectations of their cropping patterns. Poor germination demands replanting or resupplying and perhaps crop substitution; this may occur into stages 4 and 5 for later-maturing crops such as sorghum and cowpeas. Conversely, if germination is good, the farmer can be more confident that some or all subsistence requirements will be met since the crop mix will probably be able to withstand a mid-season drought. This stage model is an obvious simplification of an almost infinitely complex decision-making process. Figure 9.2 gives a vivid illustration of this sequential complexity, though it refers only to the operations on an individual field in Kaita village.

The second dimension of Hausa dryland farming is the prevalence of *polycultures or intercropping*. The research undertaken at Ahmadu Bello University has documented the close correspondence between the incidence of crop mixtures and the length of growing seasons (and, hence, the variability of rainfall). The proportion of cultivated area in double and triple crop mixes clearly increases as the probability of drought is amplified. Crop mixtures provide a secure and dependable return in the face of a capricious climate, though Norman and his colleagues have demonstrated that risk reduction may also be compatible with profit maximization and high returns per man hour (Norman, Pryor & Gibbs, 1979, 62–3). However, the practice of intercropping is inseparable from the sequential patterns of farm management since plot crop mixes emerge lexically as the season unfolds.

Fig. 9.2. Planting strategies by stages within the farming season: Kaita Village, 1978.

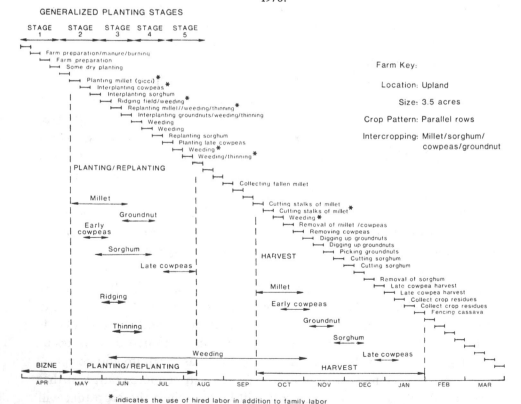

* indicates the use of hired labor in addition to family labor

In Kaita, the arrangement of crops in mixtures varies considerably between farmers. Indeed, a World Bank survey in one part of Kaduna State in 1979 identified at least 300 different crop combinations during one season librarian. Within this diversity, however, it is possible to identify three broad patterns (World Bank, 1981, 66–7): first, interplanting of the same rows which includes the placement of later crops between stands of earlier seedings but on the same ridges; second, inserting at regular intervals a row of one crop between two crops of another (for instance groundnuts between millet); and third, mixtures in which the second or third crops are planted between the ridges of the earlier crop, either in the furrow or on the side of the ridge. These latter mixtures, especially in the more humid areas, can become extraordinarily complex if six or seven different cultivars are involved.

In all upland systems in Kaita the *gicci* system prevails, which consists of sowing wide-spaced early millets in the furrows of the previous year's ridges immediately after the first rains, subsequently splitting the ridges and sowing other crops between the millets, usually sorghum and cowpeas when the rains are established. The rationale of *gicci* is to produce an early millet crop, as soon as possible after the long dry season, to defray or offset expenditures on highly inflated grains. Early sowing would not be possible, unless plant populations were sparse enough to minimize competition for the scant soil moisture during the early wet season. The wide spacing thus permits subsequent interplanting of less drought-tolerant cultivars, when the probability of 'stop-start' rains is less likely.

The simple intercropping model developed by Norman and his colleagues implies an immutable 'standard farm plan' that is apparently instituted *grosso modo* every farming year. Crop mixtures appear as static, ready-calibrated agricultural programs that are run automatically year after year. Yet, I have argued that what characterizes Kaita agriculture above all else is its flexibility, and a crucial property of most Hausa farming systems is their ability to be continuously adjusted and fine-tuned. Kaita farmers attempt wherever possible to preserve this flexibility and do not adopt a rigid mini-max strategy. They adjust agricultural practice in accordance with the objective local level (farm) occurrence of rainfall.

To return to the stages of Fig. 9.2, we can infer specifically that through adaptive flexibility, farmers gradually switch from subsistence concerns to income and cash demands. By stage 3, farmers (*a*) are able to devise concrete expectations based on the prevailing state of nature, and (*b*) estimate with some confidence the food production outlook. Based on these calculations, farmers will also decide whether to sell domestic grains from the previous years (at relatively favorable seasonal prices), assuming of course that such surfeits exist. What concerns me here, however, is the sort of flexibility that farmers possess in this sequential decision process in relation to drought.

Responses to rainfall variability generally consist of two broad strategies, (*a*) the control of microclimate and (*b*) the sequential use of crop varieties (polyvarietal strategies) and management of lowland microenvironments.

Control of microclimate and physical soil structure

As one would anticipate in an area where rainfall is erratic and evaporation rates high, the majority of drought responses refer to the control of microclimate and the preservation of local soil moisture. Of particular significance is the increase in ridging (*huda*), especially cross-ridging (*kadada*), to conserve surface moisture, and an intensification of weeding to reduce weed–crop competition. Respondents[1] reported that ridging (68%) and increased weeding (84%) were preferred strategies during a period of erratic rainfall. As a general rule, plant spacing tends to be much more compact on the heavily manured holdings near settlements, but three-quarters of the farmers interviewed believed that stands should be thinned and spacing widened when rains appear patchy or badly distributed.

The other category of adjustments attempts to reduce moisture requirements, principally through the replacement of long-maturing cereals by drought-tolerant varieties. Sixteen percent of the sample took the precaution of planting cassava (*rogo*) during July, as a 'hunger breaker', if the upland crop prospects looked bleak by the mid-rains. The limited number who resorted to cassava is largely a reflection of land limitations, inability to procure cassava cuttings, and a general demand for early cereals to defray the cost of grain purchase. On the other hand, the polyvarietal crop substitution of drought-resistant millets and sorghums was much more widespread, being preferred in slightly over 70% of all cases. While almost no research has been undertaken on the varietal makeup of indigenous cereals and pulses in Hausaland, it is evident from the ethnoscientific terminology alone that farmers have a long-standing record of genetic experimentation, and a subtle knowledge of crop types (particularly millets) in relation to moisture tolerances, pedological requirements and maturation rates. Many of the northern Katsina varieties pre-date the European presence, while others are complex hybrids of local and introduced high-yield grains. The most commonly planted cereal varieties have drought tolerances that are carefully determined; millets are generally able to withstand a 20-day drought when about 1 month mature (15–23 cm high) in contrast to 15 days for sorghum and 19 days for beans. By mid-season when the plants are 90 cm or more high, drought tolerances among the cultigens are much more equitable, usually between 25 and 30 days.

Polyvarietal crop strategies and management of lowland microenvironments

The inventory of drought adjustments constitutes only one dimension of the human ecology of drought. In this respect, the environmental surface on which Kaita farmers conduct their agronomic affairs is not homogeneous but consists of three cognitively discrete microenvironments; a broad terrace adjacent to the river is devoted to rice and sorghum, the riverine basins of clay-loam (*laka*) soil to wet-season sorghum or dry-season irrigation, and the levees to tobacco or market gardening (Fig. 9.3). The precise manner in which these niches are exploited and articulated each year depends in some measure on the pattern of local precipitation. As a general rule, cultivation of the *fadama* (low wetlands) expands during drought years to compensate for the decline in food availability (see Watts, 1983).

All of this work leads to a paradox, however: namely, that in view of this new corpus of work the purported existence of widespread environmental degradation along the desert edge becomes highly problematic. In a sense, degradation should be posed as an *exceptional condition* and what has to be explained, therefore, are the conjunctural forces which disequilibriate a resilient system in large measure calibrated to the specific ecologic constraints of semiarid lands. I do not want to imply that all climatic explanations should be jettisoned, since it is clear that since the mid-1960s rainfall has been relatively low – but still within statistical expectation – for the entire Sudano-Sahelian zone. Similarly, there is no need to idealize peasant agronomy or herder-management tactics. But one must nonetheless still explain the peculiar environmental severity of the recent conflation of

Fig. 9.3. Farming microenvironments, Kaita Village, Nigeria.

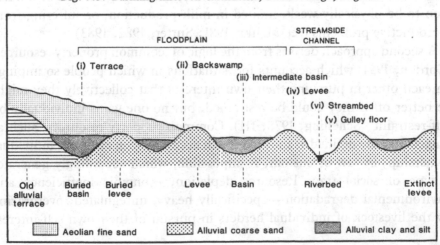

poor rainfall, ecology and land use in light of what we know of indigenous systems of resource manipulation.

How, then, is this paradox to be explained? I would like to discuss three explanations frequently espoused in the literature, though they clearly do not exhaust all of the theoretical elaborations currently in vogue. The first is broadly Malthusian in the sense that the rapid growth of animal and human populations is seen to have exceeded the carrying capacity of extant resource systems. This argument is of somewhat limited utility for human populations, insofar as population densities for most of the Sudano-Sahel remain quite low while the rural-urban exodus which has created some of the highest urbanization rates in the world has contained – with the exception of a few closely settled zones where densities exceed 150 per km^2 – rural demographic pressures.[2] Much more pervasive is the view that the number of animals greatly exceeds the rangeland's carrying capacity; the increase in livestock density is attributed in part to the unintended consequences of post-World-War-II inoculation programs and well development but also to an intrinsic propensity among herders to accrue animals on the grounds of status or as a means of savings (i.e. mobile capital). This is, of course, a sort of essentialist argument common to Malthusian forms of analysis. Yet, the imputation of herd maximization and low culling rates – in particular a belief that animals are kept until they are old irrespective of their productivity or worth – is seriously undermined by pastoralist research over the last decade (see PPS, 1979). On the one hand, there are labor constraints on herd size (reflected in increased disease, reduced watering and care, greater predation beyond a critical size threshold) and on the other hand, a good deal of evidence for higher off-take rates associated with the necesary and invariant pastoralist demand for cereals (assuming millet or sorghum is not cultivated by the 'herders' themselves), and the regular sale of animals seen to be physically weak, limited in milk production, or suffering from reproductive problems (see Maliki, 1981; Sutter, 1982, 1983).

A second approach derives from the logic of 'common property resources' (Gordon, 1954) which accounts for 'situations in which people so impinge on each other in pursuing their own interests that collectively they might be better off if they could be restrained, but no one gains individually by self-restraint' (Schelling, 1978, 111). Common property resources, such as rangelands, are defined as those to which access is open and free; individual ownership, by law or custom, is impossible, which accordingly generates a type of social trap. Resource depletion, economic inefficiency and environmental degradation – specifically heavy, unregulated, overgrazing by the livestock of individual herders in pursuit of their own self-interest

– inevitably ensue, what Hardin (1968) calls 'the tragedy of the commons'. Despite the intuitive appeal of a commons approach, it is far from clear whether the preconditions of self-interest through open access are actually met in pastoral systems. First, ethnographic research documents the numerous local, ethnic, and jural claims which herders may exercise with respect to a rangeland. Such claims often bring herders and farmers into direct conflict in the southern Sahel where sedentary cultivators move into and cultivate lands 'traditionally' considered by clans, lineages or households as their legitimate use-values. The most compelling case of such regulations is the *dina*, a codified system dating back to the early nineteenth century, governing access and use of pastures in the interior delta of the River Niger (Gallais, 1972) among perhaps almost 200 000 Peul nomads and 1.5 million cattle (excluding sedentary populations and their livestock). Furthermore, pasture use is functionally determined by access to water, specifically wells, which is invariably an individually or group-owned resource (see Horowitz, 1979; Sutter, 1982; P. Reisman, unpublished). In sum, common property resources are Janus-faced, embracing a form of *laissez-faire* utility with locally constructed forms of exclusive use. Parenthetically, a second problem with the commons view is that it has a built-in logic of accumulation which in the herder case assumes the form of a compelling motivation for herd growth, a condition which is – as I have already suggested – highly problematic.

Conflict is certainly endemic in the Sudano-Sahel – over water, pasture, land rights, cattle tracks and so on – but it would be an act of faith to see such tensions and their ecological consequences as the mechanical result of a Hardin-like structural logic of open access and individual use. Rather, one might reflect on the origins of common law and sentiment in western Europe which was, of course, rooted in the class struggles between plebs and patricians over access to wild game (see Thompson, 1975). Such dissent was naturally part of a mores upheaval in social and political relations associated with the development and deepening of capitalism and this might, therefore, point toward the significance of new global and market forces among Sahelian agro-pastoral communities for an understanding not only of jural conflicts but of ecological processes as well.

A final perspective is *social*, an embracing but indeterminate referent, a characteristic which speaks directly to both its strength and weakness. An exemplary case of a social approach is the book edited by Spooner & Mann (1982, 1) which investigates the 'social dimension of desertification . . . [to] deepen our understanding of it . . . [and to] enlarge the number of factors that we consider in relation to it'. Expanding the intellectual horizons of deterioration in drylands is most admirable, because local ecological processes are, in complex ways, embedded in webs of social, political and economic relations which frequently extend beyond the geographic confines of the degraded regions themselves. Yet, it is ironic that nowhere in that volume is the question of what actually constitutes 'social' adequately addressed. As a consequence, social perspectives on dryland ecology degenerate into a pluralist, grab-bag of ideas embracing everything from land tenure to international political organizations.

The National Research Council (NRC, 1983, 24) report is illustrative of this theoretical untidiness, when it simply inventories nine anthropogenic forces central to the modification of Sahelian ecosystems: bush fires, trans-Saharan trade, settlement sites, gum arabic trade, agricultural expansion, proliferation of cattle, firearms, transportation networks and urbanization. It is not that such agents must be incorporated into a grand, holistic 'social theory' but rather any serious discussion of social relations must begin with a notion of structure, that is to say, social practices that are reproduced through time (Giddens, 1981). In many discussions of the social context of desertification it is unclear what the structure of society is and how one might go about examining it.

In my view, which I take to be social, I shall begin with the work of Wright (1978, 114) who distinguishes between two broad categories of persons in social systems – the direct producers and non-producers – which correspond to two categories of labor:

> *Necessary labor* constitutes the expenditure of human activity for production of the means of livelihood of the direct producers.
> *Surplus labor* represents the human activity which produces a surplus beyond the requirements of simply reproducing the direct producers themselves, a surplus . . . appropriated by the non-producing classes.

What distinguishes social structures, then, is (*a*) the manner in which social relations conjoin varieties of producer and non-producer and (*b*) the social mechanisms for surplus extraction. Such a view is of some assistance in the case of Sahelian West Africa, because it behooves one to begin with notions of the various forms of household production, differentiation between these entities, and the manner in which surpluses are extracted from the basic units of production. This is akin to Blaikie's (1984) analysis of soil erosion and land use where he begins with fundamental differences between decision-making units – in his case peasant households in Nepal – most particularly inequalities in asset-holding and income-earning opportunities. Whether one is studying herders, peasants, or some syncretism of the two,

I would like to argue that the *social relations of production and exchange* are central to understanding not only the complexities of land-use decisions but also in broaching the paradox of why – and for whom – the problem of environmental change arises at all. There is, then, on the other hand, the sphere of production (at the level of the household) and exchange addressing such issues as the management and organization of resources and domestic labor in farm or herder production systems, the prices of inputs and outputs, the extent of commodity production, and the intersection of the market with use-value production; and on the other hand, the social relations in each sphere, the manner in which households of different economic and social status interact in the course of social reproduction, how labor is mobilized and how surpluses are extracted in the context of intra-household and extra-household relations.

From the local Sahelian vantage point, then, pastoral or farming households reproduce themselves through time in quite differentiated ways. Productive units vary markedly in terms of their size, off-farm income-earning capability, landholding, herd size, and in their capacity to accumulate, save and invest. Households are in no sense homogeneously situated in relation to the production process and the circuits of exchange: in other words, their reproductive strategies differ considerably. This is not simply a recognition of an on-going process of household growth, fission or fusion associated with the demographic cycle, but rather an identification of complex patterns of accumulation, impoverishment and differentiation through time. Implicit in such a statement is the realization that commodity production – and by extension the market and the state – has become, and has been for some time, an integral part of the cycle of household reproduction among farmers and herders alike throughout the Sudano-Sahel.

In two brilliant papers, Bernstein (1978, 1979) has opened up one avenue along which one can readily move from questions of social relations or production and exchange among peasants – and also herders, as I will subsequently seek to demonstrate – to concerns of environmental degradation. Bernstein identifies what he calls 'a simple reproduction squeeze' among peasant households for whom the production of commodities for exchange has become, in varying degrees, an integral part of their survival. Specifically, a deterioration in the terms of trade between commodities produced for the market and items of necessary domestic consumption acquired through the marketplace is transmitted to the household economy in terms of reduction in consumption, an intensification of commodity production or both. Hence, falling export prices (for example, for cotton) may result in reduced domestic consumption (there is less money available for food purchase, for example) or a deepening of commodity production (through an intensification of labor time and a further 'mining' of small landholdings). Such perturbations in the price mechanisms can result, then, in a sort of superexploitation of land and labor; working harder, consuming less and squeezing more from the land in order to remain at the same level (i.e. barely subsistent). Central to Bernstein's exegesis is both *differentiation* and *social relations*: those who suffer from fluctuations in the terms of trade are poor peasants strapped by limited access, assets and limited income-earning possibilities. Likewise, the necessity for commodity production to acquire money capital – and more generally, the limited opportunities for accumulation among this class of producer – must be situated in the context of poor peasants' indebtedness to wealthier farmer-traders, and their sale of wage labor (to cover seasonal food expenditures, for example) to other households; in other words, the social relations of surplus extraction (through rent, usury or the wage relation).

By concentrating on social relations, one can begin to understand the connections between material circumstances and ecological conditions by starting, as it were, from the bottom up: from the perspective of the domestic herder or peasant-household economy and the nexus of relations in which it is enmeshed. More concretely, one can derive five conclusions which really are little more than starting points for any social theory of dryland ecology. First, the social relations of household inequality allows one to move, via the terms of trade argument, to land-labor intensification and the risks of ecological deterioration among poor households. Second, differentiation also points to the *location* of ecological problems: location in a class sense – poorer households seem especially vulnerable – yet also geographically, since marginalized peasants often find themselves marginalized in space, working peripheral (i.e. distant) or low fertility holdings (Blaikie, 1984). Third, it appears that surplus extraction – whether through usury, rents, or unequal exchange – is concretely transmitted in some social settings to the physical environment since farmers/herders accordingly must extract a surplus in ecological terms. This vulnerability which compels intensification in use on family holdings may also be extended to village commons, on which the rural poor depend heavily, since they may also be 'mined' as a response to the limited access to privately owned means of production. Fourth, an understanding of the material conditions leading to a simple reproduction squeeze can perhaps account for the increasing demise or irrelevance of adaptive strategies; namely, immediate gratification (i.e. survival) overrides conservation. As Blaikie (1984) notes, hard-pressed cultivators (or herders) may not be able to afford the luxury of conservation or ecological stability. Fifth, as is now well known, it is precisely the existence of such high-risk economic conditions among poor sections of rural communities which accounts for the persistence of high fertility among farming households (Mamdani, 1972; Cain, 1981). Yet the utility of children for immediate household survival only deepens and exacerbates problems of carrying capacity and pressures on often fragile ecosystems.

Social relations and ecological deterioriation in the Sahel

In this section it is not possible to develop many of the linkages referred to in the previous discussion. Rather, I simply seek to initiate a dialogue which may go some way toward elucidating the recondite connections between differentiation and the political economy of peasant or herder economies, and what was referred to earlier as the paradox of degradation. My examples are drawn from research among Hausa peasants in northern Nigeria and southern Niger, and pastoral communities in several parts of the Sudano-Sahel, most especially Niger and Senegal. It needs to be reiterated, however, that the entire region is anything but stable and unchanging at the level of agro-pastoral economy. Recent work continually emphasizes the increased involvement by herders in farming with the result that orthodox polar types – sedentary–non-sedentary, or farmers–herders – are of limited utility. Work by Toulmin (1983) north of Segou (Mali) in a Fulani-Bambara region documents growing competition and new forms of interaction between two ethnic groups in the context of a perceived deterioration in rainfall and an increased involvement by Fulani 'herders' in rainfed millet cultivation. In her words:

> The conflict that occurs between, in particular, the Bambara and Fulani communities in the region studied should not be seen primarily in terms of farmer versus herdsman, since this obscures the real point at issue, that is, who should have rights to settle, to cultivate land and to dig wells within what is traditionally conceived of as the Bambara village's territory. The question of rights to dig a well are of special importance since a livestock owner without a water supply of his own must cede control over the manure produced by his animals to the owner of a water source during the dry season. The question of who can get access to a supply of manure and on what terms is of growing importance within a context of (a) a deterioration in rainfall conditions in comparison with the pre-drought period, and (b) an increased

> involvement by pastoral groups in cultivating sufficient grain for
> their own food needs. This has meant that access to particular
> types of land – sandy soils are currently much in demand – and
> ownership of a well capable of watering livestock have become
> assets of great importance to both the Fulani and Bambara. This
> is because of the relatively high yields of short-cycle millet that
> can be harvested from manured, sandy soils in comparison with
> yields from the same variety of millet grown on clayey . . . soils.
> (*Toulmin, 1983, 20*)

In citing this work – and related research by Sowers (1983) and Delgado
(1978) – I seek to emphasize the continual fluidity and changeability along
the desert edge, as farmers and pastoralists alike respond to complex changes
and pressures to which they are subject. Further, it is entirely unclear what
such ongoing adaptations imply for the question of the physical
environment. The cultivation of grains by herders is an important vehicle
by which they can circumvent the risks of a volatile local grain market
which, as I show below, can be a major inducement to inflate herd size.
Equally, in Toulmin's work, the presence of Fulani pastoralists and the
expansion of well-digging provided the means by which Bambara farmers
could expand and intensify their production of millet under conditions of
limited rainfall and yet not exert ecological strains on a fragile upland
ecology through an unregulated shortening of fallow. While animal numbers
are increasing, in part because cattle are seen as important sources of local
accumulation for farmers, for the area studied by Toulmin in Mali the
availability of northern pastures seems to have minimized the threat of
wholesale environmental deterioration both on the rangelands and the arable
terroir. Similarly, it is often noted that the northward movement of largely
Hausa farmers in Niger Republic in search of new farmland has contributed
significantly to the marked degradation of borderland ecological systems
that are, even in good times, of dubious value for sedentary agriculture.
Yet, it appears that the poor harvest of 1983 throughout the Sudano-Sahel
has been instrumental in the recent dissolution of such immigrant
communities, at least in Niger, which obviously is of some consequence
for the recovery of these purportedly threatened ecosystems.

Ecology, herders and terms of trade

In coming to terms with the question of herd size and animal populations
in the Sahel, a focal point of research has been the area of herd reconstitution
in the face of environmental vulnerabilities (Dahl & Hjort, 1976). The
impact of severe drought, for example, on herd dynamics and restitution
can be dramatic – loss rates of 80–90% are not uncommon – and the capacity

for the herder household to re-enter the pastoral economy in the aftermath of such a perturbation is of great significance for herd size. A minimal formulation, then, must be a herd capable of withstanding 'normal' shocks while maintaining recuperative capability. Incursions on herd size, however, originate not only from drought or disease but also from the marketplace. According to Dahl & Hjort (1976), during the dry season with poor pasture and low lactation rates, a herd of 593 animals would be required to sustain an average herding household (5–6 persons)! As a consequence, herders are necessarily projected into the grain market, and the sale of animals and off-take rates more generally become centrally related to food needs. The cattle-millet terms of trade, in other words, is of great consequence for the reproduction of the pastoral economy (see Baier, 1974; Haaland, 1977; Watts, 1983).

Sutter (1982) has shown how the millet-cattle terms of trade fluctuate wildly for Wo'daa'be herders in Niger; the millet equivalent received for an export bull varied from 360 kg in 1949 to 1938 kg in 1977. The terms of trade argument is of specific interest during crisis periods such as drought, when herders flood the market with weak and dying animals while millet prices are simultaneously escalating; in the 1972–73 drought-famine, for instance, cattle prices often fell in excess of 60%, while grain prices doubled (Watts 1983). Conversely, in the post-drought period, the terms of trade move favorably for the herders since cattle prices rise quite dramatically in relation to millet. Sutter's work shows, however, that since the mid-1970s millet prices have been consistently high in Niger and the grain market extremely volatile. Furthermore, not only grains but other cash demands – for clothes, tea, sugar and so on – have become socially necessary for herder households, thereby necessitating animal sales and deepening commercialization. Yet, in 1978, millet prices doubled with the result that nomads were compelled to sell more animals to cover cereal costs and in some cases to sell *productive animal capital* such as cows and heifers. Put differently, unfavorable terms of trade transmitted through the grain market can act to decapitalize the herder economy. Incorporation into regional or national markets, then, is itself a source of extraordinary risk and this provides the basis for a new dynamic for increasing herd size to buffer the possibilities of price perturbations.

To grasp the relations between herd size, risk, terms of trade and the implications for local ecology is, however, to realize that one must understand social and economic phenomena that transcend the parochialism of the local herder household; namely, the organization of the grains trade, the role of the market, the impact of livestock traders and cattle-fattening programs, and the utility of cattle for speculative profits not least among wealthy farmer-traders. I cannot discuss these issues here (for a preliminary discussion see Raynault, 1980; Clough, 1981; Watts, 1983); rather, I seek to establish a thread of causality between the potential for ecological deterioration through herd size, to the question of household or herder risk, transmitted through commodity markets (in this case, grain) and the plethora of external factors (the state, sources of accumulation, traders) that enters in their functioning.

The possibility of herd decapitalization has been a major incentive for herders to take up cultivation of grains to circumvent the market altogether, while other households are quite literally forced out of the pastoral economy by unequal exchange, having to liquidate all of their animal assets. The cultural premium placed on the pastoral *genre de vie* has naturally been a major incentive for herders to retain their classically cattle-based mode of operation. Accordingly, large herd size is a rational adaptive response to the vicissitudes of the market and the climate. The increase in the number of wells, which has reduced the labor constraints on herd maintenance and regulation, has clearly provided the basis for such a growth of herds, as has the increase in commercial demand for meat, and the fact that there are few other sources of productive investment for herders outside of cattle.

However, this explanation remains only partial because the historic changes confronted by herders in the course of their incorporation into ever-larger circuits of exchange have dramatic implications for the *internal structure* of the pastoral economy itself (Samatar, 1985). Sutter (1983), working among Fulani in the Ferlo of Senegal, has documented the marked stratification between households on the basis of herd size (see Table 9.2); the structure of differentiation, moreover, corresponds to differing off-take rates, sources of non-pastoral income, and prices received for animals. Sutter suggests that there are new patterns of cumulative privilege and disinvestment which have critical implications for the ability of certain segments of pastoral society to survive. To talk, then, of increase in herd size – with all the implications for range deterioration and trampling in the vicinity of wells – is spurious. Sutter's exciting work suggests that current ecological concerns are inseparable from the current realities of the restructuring of the herding economy and the emergence of new forms of accumulation and impoverishment.

Peasant differentiation and ecology in Hausaland

Like peasant villages everywhere, farming communities in Hausaland (embracing southern Niger and northern Nigeria) are marked by pronounced economic inequality. Though commonly measured in landholding, such stratification based on acreage also corresponds to distinctive positions in relation to the circuits of production and exchange. Poor households, holding less than 2 hectares and barely self-sufficient in cereals (millet, sorghum and beans), are also those most likely to engage in the sale of wage labor, have limited and unremunerative sources of non-farm income (collecting firewood, petty trade), few cattle and small ruminants (and hence less manure), almost no modern agricultural inputs and are frequently heavily indebted. A wealthier class of farmers holding in excess of 15 hectares are generally part of larger extended households, employ wage labor and ox plows, invest in cattle and are not infrequently large-scale rural traders. They are, needless-to-say, self-sufficient in staple foodstuffs from their own domestic production.

Rural population densities are quite high in Hausaland (in excess of 150 per km^2 in some of the close-settled zones) and long-term fallowing systems have disappeared almost everywhere. Rainfed agriculture is in many instances permanent, through the intensive application of manure and intercropping strategies, both of which act to maintain the physical structure of the soil and to reduce erosion. Uncultivated bushland is extraordinarily scarce. Traditional agricultural production was based on two types of land-use: extensive and intensive. The village land-use system contained

Table 9.2 *Differentiation and off-take among Fulani pastoralists in northern Senegal*

| Variables | Herd Size (by number of cattle) | | | | |
	I (0–9)	II (10–24)	III (25–49)	IV (50–100)	V (100+)
Number of households	34	44	36	27	22
Average number of people/ household	7.6	7.9	8.3	9.2	9.4
Average number of cattle/ household	3	15	33	71	146
Average number of goats & sheep/household	37	32	40	47	104
Cattle off-take rate	26%	18%	15%	12%	7%
Sheep/goat off-take rate	32%	32%	34%	29%	28%

Sutter (1983, p. 8, p. 27)

gradations of farming intensity that corresponded to a von Thunen-like pattern of concentric rings around the village. The closest area (*karakara*) was permanently cultivated through manure application; a second ring (*maiso*) was cultivated extensively through long- or short-term fallow; and a peripheral zone was bush or commons devoted in part to livestock. Population growth has, of course, transformed this pattern at least in the sense that bush (*daji*) has disappeared and fallows have been reduced through intensification.

What, then, does all this imply for ecology? In spite of the intensive manuring systems, in some cases soil quality has declined and yields seem to appear especially low (Watts, 1983). Recent work by Raynault (1980) in Maradi shows, however, that the discussion of ecological stress must be seen through the optic of inequality. Raynault demonstrates that households vary in terms of landholding and animals per capita and also in terms of yield. In addition he has determined a spatial aspect, since the poorer production units are precisely those with peripheral holdings (see Table 9.3). Strapped by limited labor power, limited manure, and unrenumerative off-farm income, these families have to travel further to their fields which, though permanently used, are not heavily manured.

Social relations, household differentiation and food security: case studies from Hausaland

Peasant households vary markedly in terms of life chances of their members, which in part reflects the quantity of, and access to, resources each may command in his/her lifetime. Landholdings and income tend to vary with the demographic structure of households; the large farm units till more land and earn higher incomes than the smaller households. Income and farm statistics on a per capita basis, however, tend to reduce the equalities that emerge from more aggregate figures based on the household. Most village research in Hausaland (in both Niger and Nigeria) has not failed to notice the pronounced differences in economic and material welfare between households. Simmons' (1976) work in Dan Mahawayi (Zaria), for example, estimated that 6 households spent between N1000[3] and N4800 per annum, while the remaining 34 in her sample spent between N80 and N680. Matlon (1977) reported the annual net income of six households in southern Kano averaged N2715 or eight times the mean income per household and three times the average per resident. The degree of interhousehold landholding equality is markedly reduced when a per consumer equivalent statistic is computed, but it remains evident nonetheless that there are quite high levels of landholding concentration, particularly in the densely settled zones of Kano, Sokoto, and Katsina.

In her seminal work, Polly Hill (1972, 1977) distinguishes four distinct economic strata, which she defined by 'their ability to withstand the shock of a very poor harvest'. The rich are able to make loans, distribute gifts and perhaps even prosper through trade and off-farm income. The poor, conversely, borrow, incur debts, migrate, decapitalize and perhaps sell farms. Hill shows that wealthy farmers are more likely to be older men with large households and sons in extended patrilineal domestic units

Table 9.3 *Space, location and differentiation: Gurjae Village, Niger* [a]

	B		C
	1=14.8 ha		1=16 ha
	2= 3.9 ha		2= 4.3 ha
	3=18%		3=17%
	4=615 kg		4=440 kg
	5=210 kg/ha		5=135 kg/ha
	6=0		6= 0.7
	7=16		7= 12
3 ha	A		D
	1=10.2 ha		1= 5.9 ha
	2= 2 ha		2= 2 ha
	3=10%		3=11%
	4=460 kg		4=270 kg
	5=260 kg/ha		5=140 kg/ha
	6= 2.5		6= 0.1
	7=21.5		7= 6.5
0		1200 m	

Land holding per adult (vertical axis)

Distance from village

Raynault (1980, 44).

[a] Key: 1: cultivated area per household; 2: cultivated area per adult; 3: percentage in fallow; 4: cereal production per adult; 5: yields per ha; 6: number of cattle per household; 7: small ruminants per household.

(*gandu*); they own and farm more land per household, per working male and per resident. They manure their land more intensively and are more likely to buy land and hire labor than smaller farmers. The wealthy produce more grain and groundnuts and are more likely to cultivate high-value crops such as tobacco, to own plows and groundnut decorticators and to keep cattle. They generally engage in intervillage trade in grain and other commodities, store grain for resale and lend money. Each of these resources and activities provides the means to accumulate other resources and to permit the wealthy to reproduce and consolidate their advantaged position.

The poor, conversely, are most likely to be found among the old and in households headed by young men on small, lightly manured holdings. The poor have smaller households and their married sons are less likely to remain in *gandu*. Heads of poor families usually engage in menial, often transient, occupations that yield low returns. They engage in wage labor and borrow heavily to buy grain in the rainy season. The worst off are those with no creditworthiness; at best, they may borrow a little land to farm, or use cornstalks to make beds for sale. It is common for poor men to sell both their farm manure and their farms. The poor are not only short of land but also lack the means (cash, manure and family labor) with which to farm effectively. They dispose of their few resources to meet their reproductive needs, trapped in a cycle of impoverishment.

In simple *quantitative* terms, these dimensions of socioeconomic differentiation are dramatically indicated in Table 9.4, which compares patterns of household inequality in three Kano villages in Nigeria (Matlon, 1977) with a community in Nigerien Hausaland close to the Nigerian border (Sutter, 1981). While there is a relative equivalence between strata, as regards landholding acreage per resident or per consumer equivalent, there are important differences between classes in terms of farm location (the poor households invariably holding peripheral farms 2–4 km from the village) and manure application/field quality (the rich conversely having well-manured, central holdings). This is given expression in the huge yield and productivity differences between Classes I (wealthy) and III/IV (poor). The stratification of Hausa communities translates directly into equally significant typological distinctions for food availability and security (Table 9.5). The data from Kano and Yelwa highlights this even more starkly; on average, the poorest 20–5% of farming households were deficient subsistence grain producers, often purchasing nearly 100% more grain than they sold. Lower-income households tended to fall well below estimated minimal caloric requirements; deficits of 20% were computed for the poorest quintile in the Kano villages, even after food purchases were considered (Matlon,

Table 9.4 *Dimensions of household inequality: Yelwa and Kano*

	Class:	Yelwa (Niger)				Kano (Nigeria)		
		I (Rich)	II	III	IV (Poor)	I (Rich)	II	III (Poor)
Sample size (*n*)		9	11	12	10	10	20	10
Household (HH) size		11.2	6.9	6.9	3.9	9.3	5.8	3.7
% of HHs in *iyali* (nuclear families)		44	82	67	80	50	55	50
% of HHs in *gandu* (extended families)		56	18	33	20	50	45	50
Total cultivated area per HH (ha)		13.1	9.6	6.3	4.5	3.2	2.4	2.5
Cultivated area per resident (ha)		1.17	1.39	0.91	1.15	0.37	0.41	0.37
Cultivated area per consumer equivalent (ha)		1.82	2.07	1.38	1.58	0.80	0.59	0.54
Cattle per HH		3.4	2.0	0.83	0.3	0.36	—	—
Sheep per HH		10.6	4.5	4.3	2.5	32.0	15.6	11.4
Gross harvest value	CFA[a]	21 100	20 400	17 200	14 200	₦603.3[b]	451.6	496.4
Millet produced per HH (kg)		1 405	683	731	285	332.6	212.8	177.4
Sorghum produced per HH (kg)		583	397	350	243	1337.6	870.1	391.8
Food purchases/consumer equivalent (value)	CFA	8 800	10 500	12 000	14 600	N27.9	19.0	22.4
Food sale/consumer equivalent (value)	CFA	600	900	800	900	N6.7	4.2	1.84
% off-farm income		n.d	n.d	n.d	n.d	36.8	14.3	31.9
Off-farm income per HH	CFA	50 400	22 100	14 300	16 700	n.d	n.d	n.d
Hired labor as % of all agricultural tasks		n.d	n.d	n.d	n.d	51.0	38.0	30.8
Total number of days of hired labor sold per HH		0.5	3.7	6.4	7.7	n.d	n.d	n.d

Matlon (1977) and Sutter (1981).
[a] CFA 100=US$ 0.36.
[b] ₦=US$ 0.99

Table 9.5 *Patterns of grain sale, purchase and production: Yelwa and Kano[a]*

	Yelwa				Kano[a]		
	I (Rich)	II	III	IV (Poor)	I (Rich)	II	III (Poor)
(1) Millet and sorghum production per consumer equivalent (kg)	277	252	238	182	585	357	142
(2) Minus 10% seeds and loss[b]	249	227	214	164	527	321	127
(3) Minus 24% bran[c]	189	173	163	125	400	244	97
(4) Grain flour available per consumer equivalent per day (kg) (#3 ÷ 365)	0.52	0.47	0.45	0.34	1.09	0.66	0.26
(5) Calories of grain flour available per day (#4 × 3450)[d]	1794	1622	1553	1173	3760	2277	897
(6) Estimated daily calorie needs for consumer equivalent supplied by on farm production (%) (#5 ÷ 2575)[e]	70	63	60	46	146	88	34
(7) Millet and sorghum bought per consumer equivalent (kg)	42.9	56.8	83.8	103.5	26.1	20.9	30.7
(8) Millet and sorghum sold per consumer equivalent (kg)	10.52	15.78	21.62	24.32	49.9	48.5	16.1
(9) Total grain availability/ capita/day (calories)[f]	2106	2005	2128	1891	3557	2031	1035
(10) Millet and sorghum sales as % of total production (#1 ÷ #9)	3.75	6.26	9.1	13.3	8.5	13.5	11.26

Adapted from Sutter (1981) and Matlon (1977).

[a] Classes I, II and III refer to decile 10, quintile 3 and decile 1 respectively in Matlon's (1977) stratification.
[b] Estimated from quantitative and qualitative interviews by Sutter (1981).
[c] Based on eleven observations of the weight of grain as it comes in from the stalk and the weight of the bran resulting from processing by Sutter (1981).
[d] Calories per kilogram of flour from FAO, *Provisional Food Balance Sheets* (Rome, 1977).
[e] Daily calorie needs of a male 16–60 years old (which closely reflects our consumer equivalent measure) are based on FAO calorie requirements adjusted by temperature and demographic characteristics to Zaria, Nigeria (Simmons, 1976, 72).
[f] Computed from #5 − (#7 − #8) and does not include gifts and loans in grain.

1977, 286). In those low-income classes, the gross value of farm production could not cover the purchasing power required to obtain the food necessary for minimum caloric intake. Yet, to assess these sheer quantitative distinctions adequately, as significant as they may be, one must examine the social relations between households and, more generally, what I refer to as the intersection of commodity markets. Stated differently, the stratification debate in relation to food security must be seen structurally to illuminate how contrasting cycles of household reproduction (between strata) actually intersect. While I cannot do this adequately in this chapter, I may at least give a strong sense of food security in relation to household inequality by examining, first, the cycle of reproduction and, then, what one might call the social relations of trade, particularly seasonal sequences of grain sale and purchase, and the critical role of debt.

In a nutshell, by examining quantitative differences *relationally* one can begin to appreciate how food supply for certain rural strata is uncertain and risky every year, particularly during the critical months before harvest, and how patterns of debt, grain sales and wage laboring can simultaneously (*a*) devalue their household production, (*b*) limit the time devoted to domestic production with deleterious consequences for farm productivity and (*c*) create seasonal liquidity and cash-flow crises that compound their indebtedness and render them vulnerable to the vicissitudes of a volatile grain market.

In Hausaland the seasonal cycle of household reproduction has its own unique timbre (see Chambers, Longhurst & Pacey, 1981). Toward the end of the dry season, the labor and energy requirements for fetching water, gathering food and off-farm activities to earn cash to purchase food tend to increase. Food becomes scarcer and less varied as domestic granaries run down and millet prices climb steadily in the marketplace. Poorer families, strapped by their small holdings and limited familial labor, begin to suffer earlier than other households. They have less food in view of the lower productivity of their farms and their labor, a shortage of capital and miserly non-farm occupations. Some migrate seasonally, others undertake casual and poorly paid wage labor. The rains denote a crisis in the reproductive cycle. Heavy and urgent agricultural demands have to be met promptly and the timing of labor mobilization is of the essence. Intense physical labor – for land preparation, planting, and weeding – comes at the time when food is most scarce. Many of the lower income groups are in a negative energy balance, losing weight as their work output exceeds caloric intake; the quality of nutrient intake declines. The wet season is also the least healthy period of the year with maximal exposure to infections. Future food supplies and the securing of subsistence for a new cycle of reproduction depend upon the farmer's ability to work or hire in wage labor during the crisis period. Afflicted by sickness, birth and pregnancies, food shortage, poor diet and high grain prices, the 6 or 8 weeks of the 'soudure' prior to harvest represent quasi-crisis conditions. The pressures to borrow, mortgage crops or sell labor are necessarily acute and the seasonal energy crisis shifts the terms of trade against labor. When the harvest comes, debts have to be repaid along with tax. Food is abundant, prices are low and debts are repaid in a buyer's market. As the dry season progresses, conditions improve, risk of infection declines, caloric intake rises and diets improve.

In short, the period prior to the first millet harvest constitutes, even under normal conditions, a major configuration in the reproduction cycle. A poor harvest and the onset of a food crisis simply expands and intensifies what is, in any case, a period of some anxiety and stress. This pre-harvest crisis is readily appreciated through an examination of labor mobilization. According to Norman (1967), an average of 241 man hours are spent on the family upland farm during the peak wet-season month, more than 80% above the mean monthly input. The four busiest months from June to September account for 53% of the total annual labor.

Subsistence production and grain availability should be situated within these cycles of labor mobilization and seasonality. A considerable body of work shows the reduction of food availability during the farming season in contrast to the immediate post-harvest period (Schofield, 1979). Protein and caloric malnutrition among children rises significantly and adult body weight declines prior to the harvest. In spite of limited nutritional data, however, work on seasonal under-nutrition in Nigerian Hausaland is hardly equivocal. Simmons (1976), working in villages near Zaria, computes an average daily per capita intake for the whole year of 2264 calories, varying from a low of 1949 calories in December and January, after the main upland harvest, to a high of 2458 calories in April and May, the first 2 months of the planting season. Protein intake was similarly distributed, peaking in August and September with a trough in December and January. Average protein intake generally met recommended requirements.

A rather different image is brought into focus, however, when consumption and sale data are disaggregated. Simmons made a crude economic distinction between rich and poor, predicated on cattle ownership. On the basis of this dichotomy, it is clear that during the critical pre-harvest months, grain consumption per capita per day was almost 30% higher among cattle owners than among those without cattle. Further, the average grain purchase (in kilograms per month) between April and September, when prices were 90% higher than the harvest level, was at least one-third greater among poor farming households. By the January following the 1970 harvest, poor households had already disposed of 100% of their total annual grain sales at roughly 5 kobo per kilogram. The rich, conversely, sold 49% of their sorghum between April and August 1971, during which period it fetched, on average, 6.9 kobo per kilogram.

These data suggest that the social relations of trade specifically generate conditions in which considerable grain 'surpluses' circulate in the food economy, fueled by a complicated interdigitation of debt, cash hunger and the seasonal timing of cereal sale and purchase (Raynault, 1977*a*, *b*). Of course, food prices are unstable in Hausa markets; in Hays' (1975) study of 17 urban markets, wet-season peak prices were on average 30–32% higher than immediate post-harvest levels. In Kaita village the 1977–78 season was unusually freakish in this regard; following a mediocre harvest, the price of a *tiya* (bowl) of millet leapt from 60 kobo in September 1977 to a record price of ₦1.20 in July 1978 (Watts, 1983). By May 1978, almost 80% of Group II and III households reported empty granaries. I have suggested that a critical element in the local food economy is the timing of crop purchase and disposal in relation to class position. There is, in fact, a direct inverse relationship between cash outlay on millet and economic situation; Group III households spend at least three times as much per consumer as their Group I counterparts. Much of this buying activity in 1977–78 did not occur in the marketplace at all but represented private grain transactions between peasant and village-based food merchants, usually during June and July. Households projected onto the grain market procured millet in a variety of ways, but short-term cereal loans (*falle*) repaid in kind at high interest and the sale of small livestock were of signal importance. These same households had paradoxically sold some of their millet and cowpeas after the harvest of 1977, principally to cover tax, for debt repayment and necessary social expenditures, most critically house-repair and consumer items for ceremonial gift exchange. The complementarity of post-harvest distress sales (at low prices) and wet season purchase (at inflated prices) defines the structural parameters of a cycle of impoverishment in which poor households must operate (Fig. 9.4). Many of the better-off Group II households also contributed to the post-harvest millet glut, because of heavy social consumption associated with ceremonial (especially marriage) costs. However, they sold proportionately less and were generally able to borrow more and to finance the hiring of wage labor in the rainy season, principally from income from their irrigated market garden agriculture.

The majority of the wealthy farmers consciously withheld grain in 1977–78 for the wet-season price rise; those with surpluses sold millet in June, July, and August in order to buy wage labor and invest in cattle, which (since the early 1970s) have become sources of speculative profits. Several in Group I are grain dealers who are active in wet-season lending operations. During 1978, a bundle of millet borrowed in May was repaid in kind at harvest time with a very minimum of 50% interest. Most *falle* loans characteristically demand two bundles for every one borrowed (see Clough, 1981). In September and October, then, debt repayments in kind provide a means by which rurally based wholesalers can channel large stocks of undervalued millet, which are promptly evacuated to urban consumers in Katsina, Kano and Daura. In Kaita, six large grain traders were agents for urban-based merchant capital; the urban merchants were huge food wholesalers who fronted capital to their rural clients to facilitate post-harvest buying operations.

Many poor households find themselves deeply imbricated in market relations and commodity production yet are incapable of fulfilling their own food needs. Such households are often compelled to sell grain at harvest,

Fig. 9.4 Simplified model of household differentiation, Kaita Village, Nigeria.

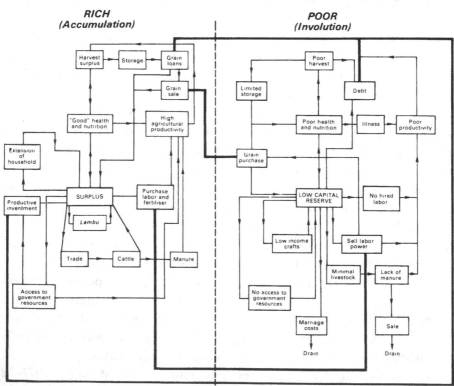

when prices are lowest, in order to cover necessary cash outlays (for tax, to repay debts) yet are projected back into the marketplace 6 months later when their domestic granaries are empty. Prices during the following wet season, prior to the new harvest, are at least 50% higher than during the period of sale. Middle or wealthy peasants conversely are not subject to the same pattern of bimodal distress sales. These same land-poor households also engage in wage labor during the farming season, diverting labor from their own farms at critical periods (seeding, weeding). In short, they are caught in a cycle of poverty that originates in the specific social relations of production and exchange in which they are enmeshed (Watts, 1983). An understanding of these historical and political economic specificities is also instrumental, I would suggest, for the analysis of famines.

Understanding famines in social terms

[It] is by no means clear that there has ever occurred a famine
in which all groups in a country have suffered from starvation,
since different groups typically do have very different
commanding powers over food . . . (*Sen, 1981, 43*)

Famines are enormously complex social and biological phenomena and it is quite hopeless to expect a generic theory of their origins or consequences, particularly since good empirical case studies tend to be quite rare (Torry, 1984). It has become commonplace to recognize, nevertheless, that the causal powers joining drought, or other natural perturbations, and famine are particularly attenuated even under conditions in which productivity and farm technology is low or limited. However, to say that famines are 'acts of man not of nature' (Timberlake, 1984; Owen & ICIHI, 1985) hardly goes far enough, since such a proposition demands quite careful analyses of prevailing social conditions, of their possibilities and their constraints. In this regard Sen's (1980) work has almost certainly done more than any other publication in theorizing these social relations in relation to immediate or short-term causes and consequences of famine. Sen demonstrates of course that food crises can occur without a decline in absolute food availability and that demand structure or changes in patterns of entitlement can be constitutive elements of mass starvation. Yet even his work has its lacunae. He tends to focus on immediate or proximate causes of famine, neglecting long-term structural patterns in agricultural development – in other words, larger crises of social reproduction – and he tends to gloss over intrahousehold entitlements as opposed to intracommunity entitlements. Hence, it is clear that during famines, children and females are especially vulnerable (Vaughn, 1984; Whitehead, 1984) and that units such as the 'household' must be disaggregated and handled with extreme care as units of analysis.

The general point, nevertheless, is that entitlement patterns are highly variegated with the inevitable consequence that famines can, and often do, contain intricate causal relationships and generate complex effects. Some people profit from famines; some people die; other people end up in refugee camps. Hence it is the aim of research and scholarship to deconstruct notions of a generalized food crisis and to establish its specific character, extent and effects for particular classes, strata and gender groups in particular places, regions and localities (H. Bernstein, unpublished). I believe that a useful starting point for such analyses would involve, following D. F. Bryceson (unpublished), a need to distinguish between malnutrition, seasonal hunger and famine. Malnutrition is a problematic notion as debates between nutritionists over genetic as opposed to adaptive or homeostatic models of nutrition testify (Payne & Cutler, 1984), but one can distinguish between chronic states of under-nutrition as aspects of the human energy costs of production, and seasonal or cyclical reductions in dietary intake causing or exacerbating states of malnutrition. Both of these conditions are clearly distinct from famine as large-scale caloric deprivation associated with the terminal points of a famine cycle (Currey, 1984).

In my discussion of drought and environment, I have tried to emphasize three sorts of concerns that bear upon this simple taxonomy. First, that malnutrition and seasonal hunger are parts of 'normal' cycles of reproduction in peasant or pastoral communities, though the precise form and extent of each is an empirical question. It follows that a theorization of the mechanisms through which such cycles are produced and reproduced in differentiated communities can be helpful in understanding (*a*) how malnutrition/seasonal hunger can, if triggered by a drought-induced harvest shortfall, become a full-fledged famine, and (*b*) what the internal dynamics of such a famine might be. Second, the work in Hausaland shows clearly that hunger and famine must be situated on the larger canvas of household differentiation and specifically how commoditization can produce particular patterns of vulnerability. I emphasized how the intersection of quite undeveloped commodity markets – for instance of grain and debt – can have deleterious consequences for some households. This is not to suggest that commoditization or the market always produces famines or vulnerable households, wage laborers, pastoralists and so on; indeed Cowen (1983) has shown how the reverse was the case in Kenya. I wanted to emphasize how it is the social context of partial or incomplete market development (commoditization) that generates certain weaknesses in the rural food system. In this regard, I share Spitz's (1983) view that drought, for example, cannot be conceived in isolation from the quite specific character and state of food systems as constituted in specific social systems. In this sense it is nonsense to charge, as Torry (1984) does, that such analyses are simply 'dependency theory' when my own work – and Spitz's on India – offers on the one hand a specific critique of dependency theory, and on the other purports to show the unexpected and uneven forms and consequences of differentiation in peasant societies.

Finally, even a penetrating analysis such as Sen's must be placed historically not only in terms of immediate conjunctures (Bush, 1985). In other words, short-term changes in entitlements occur in the context of larger transformations in agriculture, in particular sorts of states, in terms of specific patterns of peasant differentiation and market development, and as part of wider processes of ecological change and deterioration. These developments are, moreover, not merely additions to the famine equation but are fundamentally constitutive of a larger *crisis of social reproduction*. It is simply not enough to say, as does, for example, Torry (1984) that 'soft' states or corruption are relevant to famines, because it is precisely these characteristics that have to be theorized and accounted for in the 'new theoretical insights' that he calls for insistently, but does not himself provide. I believe that current debates on state–peasant relations (the macro level) and attempts to understand the almost infinitely complex patterns of decomposition and recomposition of peasantries (the local level) in Africa (see Berry, 1984; Watts & Bassett, 1985; H. Bernstein, unpublished paper delivered to the Institute of Commonwealth Studies, London) are in this respect critical for famine studies. Yet it is so often the case – for instance in Torry's purportedly synthetic work on famines – that such grand historical political economy of the so-called agrarian question in Africa is conspicuously missing. It is exactly these silences and absences which render many studies of famine so incapable of assessing the likelihood of adopting long-term solutions to food crises rather than short-term palliatives.

I began this chapter with a brief note on the way in which images of starvation can accuse everyone and no-one, which is a treacherous quality, because famines are so deeply imbricated in the constellations of social and class forces of African states. Indeed, it is salutary to recall in this volume on famine and drought in Africa that at the very time of its preparation Mamood Mamdani has been stripped of his Ugandan citizenship because he dared to suggest that famine is fundamentally a social and class phenomenon and that solutions must 'revive the creativity and initiative of the people' (Mamdani, 1985, 96).

Notes

1. This research was based on a sample of more than 30 farmers. See Watts, 1983, for further information.
2. Some scholars have argued that pastoral populations are indeed growing quite rapidly – in excess of 3.0% per annum – and this indirectly contributes to the growth of animal populations along the desert edge which is, according to this view, the direct cause of forage shortage and range degradation (see Bremen & de Wit, 1983).
3. One Nigerian naira (₦) consisting of 100 Kobo is currently equivalent to US$1.00.

References

Baier, S. (1974). African Merchants in the colonial period: a history of commerce in Damagaram 1880–1960. PhD thesis, University of Wisconsin.
Bassett, T. (1984). Food, peasantry and state in the northern Ivory Coast. PhD thesis, University of California, Berkeley.
Berger, J. (1980). *About Looking*. New York: Pantheon.
Bernstein, H. (1978). Notes on capital and peasantry. *Review of African Political Economy*, **10**, 60-73.
Bernstein, H. (1979). African peasantries: a theoretical framework. *Journal of Peasant Studies*, **6**, 420–43.
Bernus, E. (1974). Possibilités et limites de la politique d'hydraulique pastorale dans le Sahel nigérien. *Cahiers d'ORSTOM*, ser., Sci. Hum., **11**, 119–26.
Bernus, E. & Savonnet, G. (1973). Les problèmes de la sécheresse dans l'Afrique de l'ouest. *Presence Africaine*, **88**, 113–138.
Berry, S. (1984). The food crisis and agrarian change in Africa. *African Studies Review*, **27** (2), 59–112.
Blaikie, P.(1984). *The Political Economy of Soil Erosion*. London: Methuen.
Bovill, E. (1920). The encroachment of the Sahara on the Sudan. *Journal of the African Society*, **20**, 259–69.
Bremen, H. & de Wit, C. (1983). Rangeland productivity and exploitation in the Sahel. *Science*, **221**, 1341–7.
Bush, R. (1985). Drought and famines. *Review of African Political Economy*, **33**, 59–64.
Cain, M. (1981). Risk and insurance: perspectives on fertility and agrarian change in India and Bangladesh. *Population and Development Review*, **7**, 435–74.
Chambers, R. (1983). *Rural Development: Putting the Last First*. London: Longman.
Chambers, R., Longhurst, R. & Pacey, A. (eds) (1981). *Seasonal Dimensions to Rural Poverty*. London: Frances Pinter.
Clough, P. (1981). Farmers and traders in Hausaland. *Development and Change*, **12**, 273–92.
Cowen, M. (1983). The commercialisation of food production in Kenya. In *Imperialism, Colonialism and Hunger in East Africa*, ed. R. Rotberg, pp. 199–224. Lexington, Mass: Heath.
Currey, B. (ed.) (1984). *Famine as a Geographical Phenomenon*. Boston: Reidel.
Dahl, G. & Hjort, A. (1976). *Having Herds: Pastoral Growth and Household Economy*. Stockholm: Studies in Social Anthropology, University of Stockholm.

Delgado, C. (1978). *Village Livestock Intensification in Southeastern Upper Volta*. Ann Arbor: University of Michigan, Entente Studies.

Derrick, J. (1984). West Africa's worst year of famine. *African Affairs*, 83 (332), 281–300.

de Vries, P. F. & Djiteye, M. (eds) (1982). *La Productivité des Pâturages Saheliens*. Wageningen: Pudoc.

Dupire, M. (1962). *Peules Nomades: Etudes descriptive de WoDaaBe du Sahel Nigerien*, Paris: Institut d'Ethnologie, 1962.

Dyson-Hudson, N. (1980). Strategies of resource exploitation among East African savanna pastoralists. In *Human Ecology in Savanna Environments*, ed. D. Harris, pp. 236–54. London: Academic.

Fairburn, W. (1937). *Report on Sylvan Conditions and Land Utilization in Northern Nigeria*. Kaduna: Ministry of Agriculture.

Gallais, J. (1972). Projet de développement d'élevage dans la région de Mopti, Annex A. Paris: SEDES.

Giddens, A. (1981). *Central Problems in Social Theory*. Berkeley: University of California Press.

Gordon, H. S. (1954). The economic theory of a common property resource. *Journal of Political Economy*, 62, 124–42.

Haaland, G. (1977). Pastoral systems of production. In *Land Use and Development*, eds P. O'Keefe & B. Wisner, Report #5, 179–193. London: International African Institute.

Hardin, G. (1968). The tragedy of the commons. *Science*, 163, 1243–8.

Hare, K. (1977). The making of deserts: climate, ecology and society. *Economic Geography*, 53, 332–45.

Hays, H. (1975). The marketing and storage of food grains in northern Nigeria, *Samaru Miscellaneous Paper*, No. 50. Zaria: Ahmadu Bello University.

Hill, P. (1972). *Rural Hausa: a Village and a Setting*. Cambridge University Press.

Hill, P. (1977). *Population, Prosperity and Poverty: Rural Kano 1900 and 1970*. Cambridge University Press.

Hollings, C. (1973). Resilience and stability in ecological systems. *Annual Review of Ecology and Systematics*, 4, 1–23.

Horowitz, M. (1979). The Sociology of Pastoralism on African Livestock Projects, *A.I.D. Discussion Paper*, #6, Washington, DC: USAID.

Hurault, J. (1975). Surpâturage et transformation du milieu physique. *Etudes de Photo-Interpretation de l'Institut Geographique Nationale*, No. 7, Paris.

Igbozurike, U. (1977). *Agriculture at the Crossroads*. Ile-Ife: University of Ile-Ife Press.

Jones, B. (1938). Desiccation in the West African colonies. *Geographical Review*, 91, 401–23.

Maliki, A. (1981). Ngaynaaka: l'élevage selon les Wo'daa'be, Niger-Range and Livestock Project. Washington: USAID.

Mamdani, M.(1972). *Myth of Population Control*. New York: Monthly Review.

Mamdani, M. (1985). Disaster prevention: defining the problem. *Review of African Political Economy*, 33, 92–6.

Mann, H. (1982). The Central Arid Zone Research Institute. In *Desertification and Development: Dryland Ecology in Social Perspective*, eds B. Spooner & H. Mann, 295–303. London: Academic.

Matlon, P. (1977). The size, distribution, structure and determinants of personal income among farmers in the north of Nigeria. PhD thesis, Cornell University.

Mortimore, M. (1970). Population densities and rural economies in the Kano Close Settled Zone, Nigeria. In *Geography and a Crowding World*, ed. W. Zelinsky; L. A Kosinski; R. M. Prothero, 380–8. London: Oxford University Press.

Nicholson, S. (1981). The historical climatology of Africa. In *Climate and History*, ed. T. Wigley, M. Ingram & G. Farmer, pp. 249–70. Cambridge University Press.

Nicholson, S. (1983). The climatology of sub-Saharan Africa. In *Environmental Change in the West African Sahel*, pp. 71–92. Washington, DC: National Research Council.

Norman, D. (1967). An economic study of three villages in Zaria Province: Part 1, Land and labour relationships. *Samaru Miscellaneous Papers*, No. 19. Zaria: Ahmadu Bello University.

Norman, D. (1977). The rationalization of intercropping. *African Environment*, 2/3, 3–21.

Norman, D., Pryor, D. H. & Gibbs, C. J. (1979). Technical change and the small farmer in Hausaland, Northern Nigeria. *African Rural Economy Papers*, No. 21. East Lansing: Michigan State University.

NRC (1983). *Environmental Change in the West African Sahel*. Washington, DC: National Research Council.

Owen, D. & ICIHI (1985). *Famine: A Man Made Disaster* (the Independent Commission on International Humanitarian Issues). London: Pan.

Payne, P. & Cutler, P. (1984). Measuring malnutrition. *Economic and Political Weekly*, (25 August), 1485–91.

Picardi, A. (1974). *A Systems Analysis of Pastoralism in the West African Sahel*, Annex 5. Cambridge, MA: MIT Centre for Policy Alternatives.

PPS (1979). *Pastoral Production and Society*. Cambridge University Press.

Przeworski, A. (1982). The ethical materialism of John Roemer. *Politics and Society*, 11(3), 289–313.

Raikes, P. (1981). *Livestock and Policy in East Africa*. Uppsala: Scandinavian Institute for African Studies.

Raynault, C. (1977a). Circulation monétaire et évolution des structures socio-économiques chez les haoussa du Niger. *Africa*, 47, 160–71.

Raynault, C. (1977b). Aspects socio-économiques de la circulation de la nourriture dans un village hausa (Niger). *Cahiers d'Etudes Africaines*, 17, 569–97.

Raynault, C. (1980). *Recherches Multidisciplinaires sur la Région de Maradi, Rapport de Synthèse*. Bordeaux: DGRST, Universite de Bordeaux II.

Renner, G. (1926). A famine zone in Africa: the Sudan. *Geographical Review*, 16, 583–96.

Richards, P. (1983). Ecological change and the politics of African land use. *African Studies Review*, 26, 1–72.

Richards, P. (1985). *Indigenous Agricultural Revolution*. London: Hutchinson.

Samatar, A. (1985). The state, peasants and pastoralists: agrarian change and rural development in northern Somalia 1884–1984. PhD thesis, University of California, Berkeley.

Schelling, T. (1978). *Micromotives and Macrobehavior*. New York: W. W. Norton.

Schofield, S. (1979). *Development and the Problems of Village Nutrition*. Montclair, N.J.: Allanheld & Osman.

Schultz, J. (1978). Population and agricultural change in Nigerian Hausaland. PhD thesis, Columbia University.

Scott, J. (1976). *The Moral Economy of the Peasant*. New Haven: Yale University Press.

Sen, A. (1981). *Poverty and Famine*. London: Oxford University Press.

Simmons, E. (1976). Calorie and protein intake in three villages of Zaria Province, *Samaru Miscellaneous Paper* No. 55. Zaria: Ahmadu Bello University.

Sowers, F. (1983). Fulani Resettlement and Livestock Production in the Southern Sudan Zone of Upper Volta, Ouagadougou. Washington, DC: USAID.

Spitz, P. (1983). *Food Systems and Society in India*, 2 volumes. Geneva: UNRISD.

Spooner, B. & Mann, H. (1982). *Desertification and Development: Dryland Ecology in Social Perspective*. London: Academic.

Stamp, L. D. (1940). The southern margin of the Sahara. *Geographical Review*, 30, 297–300.

Stebbing, E. (1935). The encroaching Sahara: the threat to the West African colonies. *Geographical Journal*, **85**, 506–24.

Sutter, J. (1981). Economic integration and peasant economy: a case study of two Hausa villages in Niger Republic. PhD thesis, Cornell University.

Sutter, J. (1982). Commercial strategies, drought and monetary pressure: Wo'daa'be nomads of Tanout Arrondissement, Niger. *Nomadic Peoples*, **11**, 26–61.

Sutter, J. (1983). *Cattle and Inequality: A Study of Herd Size Differences Among Fulani Pastoralists in Northern Senegal*. Berkeley: Institute of International Studies.

Talbot, M. & Williams, M. (1979). Cyclic alluvial fan sedimentation on the flanks of fixed dunes, Janjari, Central Niger. *Catena*, **6**, 43–62.

Taussig, M. (1984). Culture of terror – space of death. *Comparative Studies in Society and History*, **3**, 467–97.

Thompson, E. P. (1975). *Whigs and Hunters*. London: Allen Lane.

Thompson, E. P. (1978). Eighteenth century English society: class struggle without class? *Social History*, **3**(2), 133–65.

Timberlake, L. (1984). *Natural Disasters: Acts of God or Acts of Man*. London: Earthscan.

Torry, W. (1984). Social science research on famine. *Human Ecology*, **12** (3) 227–52.

Toulmin, C. (1983). Herders and Farmers or Farmer-Herders and Herder-Farmers. *Pastoral Network Paper*, No. 15. London: Overseas Development Institute.

Valenza, J. (1979). Les pâturages naturels de la zone sylvo-pastorale du Sahel Senegalais vingt ans après leur mis en valeur. Bamako: ICLA. Cited in Horowitz, M. (1979). The sociology of pastoralism on African livestock projects, *AID Discussion Paper*, No. 6, Washington, DC: USAID.

Vaughn, M. (1984). Poverty and famine: 1949 in Nyasaland. Mimeo. Cambridge University.

Warren, A. & Maizels, J. (1977). Ecological change and desertification. London: *University College, Paper No. A/CONF. 74/7*, UN Conference on Desertification.

Watts, M. (1982). On the poverty of theory: Natural hazards research in context. In *Interpreting Calamities*, ed. K. Hewitt, pp. 231–62. London: Allen & Unwin.

Watts, M. (1983). *Silent Violence: Food, Famine and Peasantry in Northern Nigeria*. Berkeley: University of California Press.

Watts, M. & Bassett, T. (1985). Politics, the state and agrarian development: a comparison of the Ivory Coast and Nigeria. *Political Geography Quarterly*, **5** (2), 101–25.

Whitehead, A. (1984). Gender and famine in West Africa. Mimeo. Sussex: University of Sussex.

World Bank (1981). *Agricultural Technology Adoption in Northern Nigeria*, Vol. 1. Washington, DC: World Bank.

Wright, E. (1978). *Class, Crisis and the State*. London: Verso.

10

Drought, food and the social organization of small farmers in Zimbabwe

MICHAEL BRATTON

Department of Political Science, Michigan State University

Introduction

Africa is a continent of smallholders. The key actors in the Africa food crisis are the farming and herding families who seek a living from the land. At present, there is unprecedented interest in the African countryside by state and market institutions, both national and international. The solution will not, however, come from outside. Governments and donors now agree that the future welfare of the continent rests in the hands of the small family farmer.

The basic assumption of this chapter is that autonomous action is required by smallholders to organize themselves for increased agricultural output. I therefore use a broader-than-usual interpretation of the notion of 'social impact'. I do not see social organization simply as a passive structure upon which drought and other exogenous factors have a formative, usually disruptive, effect. I view social organization also as a vital independent force – every bit as important as technological innovation and economic incentive – that can actively shape and stimulate production. In short, people can *have* an impact, as well as being impacted *upon*. By considering the impact of farmer organization on food production and consumption, I hope that this chapter can highlight a practical method of 'denying famine a future' which the readers of this volume might otherwise overlook.

The chapter is divided into four sections. The first briefly describes the main forms of social organization in the African countryside and indicates where farmer organizations fit in. The second presents background information on Zimbabwe: its climate, agricultural potential and peasant farming systems. The third part analyzes how, in Zimbabwe, farmer organizations have helped smallholders to emerge as major food producers. Finally, the paper examines the effects of the three-year drought of the early 1980s on collective action among farmers and the sources of household food supply. On the basis of the Zimbabwe case, the conclusion is drawn that farmers have the capacity to act autonomously even under conditions of drought, but especially in less stressful times.

Social organizations in rural Africa

The literature on peasant agriculture in Africa almost always identifies the *household* as the key unit of production and consumption. Even where social obligations are observed by a widely extended family, a core group of relatives can be distinguished who live together, work the same fields, and eat from the same pot. The household, however, has severe limitations as a development institution. It is often critically short of key production resources. Labor, for example, may be lacking where the household is embedded in a larger labor migration economy; draft power may be a constraint where the household is too poor to purchase cattle or equipment. The scale of production in the household is also an inhibiting factor, usually being too small to make economical use of new technologies or commercial services.

These limitations have led practitioners of agricultural development in Africa to look beyond the household for institutions of greater scope and capacity. In the past, official thinking put the onus on the *state* as the responsible authority. Yet, it is clear that the span of control of even the most unitary government rarely extends in any complete way into the countryside (Russell & Nicholson, 1981, 4). The state has often failed to penetrate the periphery, let alone transform the structures of production it encounters there. A fundamental organizational problem is to reach a large number of small producers in scattered and remote locations and to deliver services in a timely manner at precise intervals across an agricultural season.

Now, the current orthodoxy in development policy is that the *market* 'has been neglected or suppressed despite evident capacity to do many of the tasks [of] government' (Berg, 1984, 45). It is not immediately apparent, however, that private sector solutions are available for smallholder production problems. In most African countries, the market has even flimsier institutional foundations than the state. The purchasing power of a peasantry is extremely limited and entrepreneurs are not naturally attracted into rural areas by market opportunity. Moreover, because smallholdings are usually the least accessible of all production units, peasant farmers must always pay a premium for the delivery of any good or service priced by the market.

As a result, all types of central institutions enjoy limited competence in Africa, more limited, perhaps, than on any other continent. Attention must, therefore, turn to what small farmers do for themselves. Since the public and private sectors fall short, and since the household is small and poor, the alternative is to search within the rural social fabric for institutions that have development potential.

I contend that in some African countries there are local organizations that can mediate transactions between the household on the one hand and the state and market on the other. I am referring to *voluntary associations of farmers* that coalesce of their own accord on the basis of shared economic interests and cultural values. Organized collective action among smallholders has been an element in the agricultural development of most advanced economies (Esman & Uphoff, 1984, 31–5) and currently plays an important role in the development of irrigated agriculture in Asia (Korten, 1982; Uphoff, 1985). The time to examine its potential role among dryland farmers in Africa is overdue.

The distinguishing feature of collective action among small farmers is that it is '*self-managed*'. Local organizations of farmers, unlike a household or a clan, are operated by a management committee – chairman, secretary, treasurer – of elected officers. And, unlike institutions of the state and market, they do not draw leaders from a cadre of professionals but from among their own ranks. This sets farmer organizations apart, for example, from cooperative societies that are regulated by a government department or which operate as a firm in the market. The principles of collective action are voluntary membership, government by agreement, and social control by peer pressure. When this kind of organization arises among farmers, it may take many forms: mutual aid parties, special interest clubs, primary cooperative societies. Whatever the form, however, all lie within the 'informal sector' of popular local organization, a field of enterprise which development studies have yet to address comprehensively.

Background on Zimbabwe

Peasant agricultural systems

Until the end of the nineteenth century, peasant farmers in southern Africa relied for food on shifting cultivation. The staple crop was millet grown on mounds in ash-enriched plots, and the basic agricultural tools were the axe and the hoe. Major changes occurred in the twentieth century: the introduction of the ox-drawn steel plow and the ascendance of white maize as the dominant food and cash crop. These innovations have penetrated all but the most remote reaches of what is now Zimbabwe. The Tonga-speaking people of the northwest are among the few remaining groups to cultivate by hand and to rely for food principally on small grains. The majority of farmers now plant single stands of maize in rows, on the flat, according to extension service recommendations. Intercropping, usually maize with curcubits (pumpkin family), can be found in homesite gardens, alongside small plots of groundnuts, bambara nuts, sweet potatoes, and vegetables. By 1980, maize accounted for 80% of national grain production by volume, with millets and sorghum lagging behind at 11% and 3%, respectively.

Despite the wide range of natural conditions in the peasant farming areas – known in Zimbabwe as the 'communal lands' – it is possible to make some simple generalizations about the farming systems that prevail. The objectives of small farmers are broadly similar wherever they live. A first priority is to produce or acquire sufficient food to meet household subsistence throughout the cycle of seasons. Once household food security is assured, a secondary (but increasingly important) objective is to earn a cash income by producing agricultural surpluses and moving them to market. Peasant farmers in Zimbabwe are more thoroughly integrated into a cash economy than their counterparts in most of the rest of Africa. The use of purchased inputs in crop production, notably improved seed, but increasingly fertilizer and insecticide, is widespread. Where natural conditions permit, new cash crops like tobacco, sunflower, and particularly cotton, are being rapidly adopted.

All farmers in Zimbabwe face the same set of farm management decisions, namely, how to allocate the limited resources at their disposal among crop production, livestock production, and off-farm employment. The proportional mix of enterprises is likely to vary according to local conditions, with livestock being relatively more important in dry regions and off-farm employment ascendant in rural areas close to towns. The point is, however, that the elements in the production system – crops, livestock and wage labor – are identical in all cases.

These elements are intimately interconnected and form an integral whole. Crop production, for example, cannot be understood in isolation from livestock husbandry and wage remissions. A common view among agricultural economists in Zimbabwe, for example, is that the output of crops is a function of the size of a household's holding in cattle (Collinson, 1982; Shumba, 1984). Cattle are central to the life of small farmers in Zimbabwe, not only in their own right as a source of food, cash and prestige, but also for the draft power and manure they provide to arable agriculture. Another common view is that food crop output is a function of the availability to the household of wage remissions from a family member in town.

Households which enjoy such a source of ready cash are more likely to afford the purchase of the modern inputs on which dramatic gains in food crop productivity depend (Bonnevie, 1983; Callear, 1984).

Maize would appear to be a poor choice of staple crop for Zimbabwe, because a reliable annual yield can be achieved only in high rainfall areas. Yet, small farmers attempt to grow maize even in the driest regions where surpluses are attainable in only 1 year out of 4. This seemingly irrational behavior illustrates how household food production takes place within a broader political-economic context. The consumer preference for maize as a staple food reflects the transfer of urban tastes and standards to the countryside. The production of maize also requires less labor than small grains, particularly for crop protection and grinding, an advantage much appreciated in small or female-headed households who have members absent in town for all or part of the year. Finally, peasant farmers in labor migration economies may be more entrepreneurial and less averse to risk than their counterparts in countries where the family plot constitutes the sole source of subsistence. Producers in Zimbabwe are able to take a chance on maize, because they know that they have alternative sources of food or cash (e.g. urban relatives) in the event of crop failure.

Whatever the crop, however, 'overall productivity is determined to a great extent by how effectively farmers make use of limited amounts of water' (Norman, Baker & Siebert, 1984, 211). Timeliness of operations is a key management factor in plant germination and growth. Farmers who are able to plow and plant with the first rains, or to weed for the first time before maize has reached knee-height, are able to take full advantage of available moisture. The prospects of a peasant household in agriculture, therefore, depends partly upon the ecological factor of rainfall and partly on the organizational factors of access to, and management of, production resources. My hypothesis, tested below, is that farmers who are organized into groups are best able to mitigate the effects of low rainfall and drought.

Rainfall and drought

As in other subtropical areas, the passage of the seasons in Zimbabwe is marked less by changes in temperature than by variations in rainfall. The annual pattern is a short, wet 'summer' from late November to late March, followed by a long, dry 'winter' from April to October. General rains occur as a result of the southward incursion of the intertropical convergence zone (ITCZ) over the territory of Zimbabwe. Local showers and thunderstorms are common late in the day as a result of convection.

Zimbabwe has a dry climate with rainfall inadequate for agricultural production over much of the country. In terms of national distribution, there is a close association between rainfall and altitude. The heaviest rainfall occurs in the mountains of the eastern border (above 1000 mm per year) and the lowest levels are recorded in the valley basins of the Zambezi and the Limpopo Rivers and their tributaries (below 600 mm per year). Between these areas lies the central 'highveld' plateau which runs from west to east at an altitude of 1400–1700 m and which enjoys moderate rainfall (600–1000 mm per year).

Water is the major limiting factor in agricultural production and the greatest source of uncertainty for the farmer. In general, rainfall tends to decrease from north to south and from east to west. As well as being limited, rainfall is also unreliable with variations from year to year. Figures for mean annual rainfall do not necessarily constitute a sound prediction of what may

Fig. 10.1.

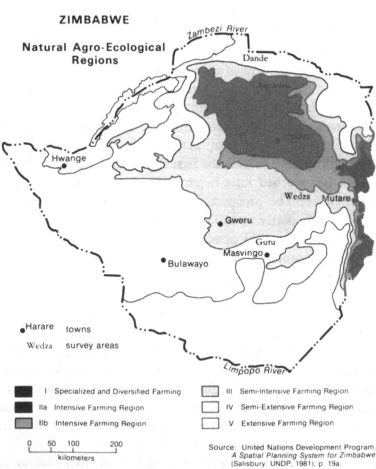

ZIMBABWE

Natural Agro-Ecological Regions

Zambezi River

Dande

Hwange

Wedza Mutare

Gweru

Gutu
Masvingo

Bulawayo

Harare   towns

Wedza   survey areas

Limpopo River

■	I Specialized and Diversified Farming	▨	III Semi-Intensive Farming Region
■	IIa Intensive Farming Region	□	IV Semi-Extensive Farming Region
▨	IIb Intensive Farming Region	□	V Extensive Farming Region

0   50   100   200
      kilometers

Source: United Nations Development Program.
A Spatial Planning System for Zimbabwe
(Salisbury: UNDP, 1981), p. 19a.

be available in any given rainy season (Peake, 1952, 1). The areas of low rainfall are also those with the least reliable distribution, ranging from 20% variability in the north of the country to 45% variability in the south. There is also a regular occurrence of mid-season dry spells in January and February precisely at the time that grain crops, especially maize, have the greatest need for moisture (Lineham, 1972, 2).

In the first 5 years of independence in Zimbabwe, 1980–85, the distribution of rainfall was variable in the extreme. The seasons unfolded as follows:

- 1980–81 saw wet conditions reminiscent of the mid-1970s. Rainfall totals were above average over the country as a whole and some areas on the plateau received 150% of normal.
- 1981–82 was the beginning of the drought. While all areas were short of rain, there were marked regional differences. The plateau generally received 80% of normal, whereas parts of the lowlands, especially Matabeleland, received 40% or less.
- In 1982–83 severe drought was experienced throughout the country. In terms of total rainfall it was the fifth worst season this century. The major problem was the rainfall distribution with most of the useful heavy rainfalls occurring early in the season, followed by prolonged mid- and late-season dry spells. The Bikita-Buhera area was worst hit with deficits of over 400 mm by the end of the season.
- A third year of drought occurred in 1983–84. The pattern of partial impact reappeared with the lowlands in heavy deficit and the parts of the uplands nearer normal. Here, despite mid-season dry spells, there were useful late rains in February and March.
- The drought broke in 1984–85. Mean annual rainfall totals were exceeded at most meteorological stations, with the highest surpluses in the south. Although more rain fell in total in 1980–81, the overall distribution in 1984–85 was better for crop production, without excessive rainfall in February.

Together, the three dry years comprise the longest sustained period of drought on record in Zimbabwe. For purposes of analysis, in this paper I will refer to 1982–83 as the year of 'severe drought' and 1981–82 and 1983–84 as years of 'partial drought'.

Agricultural potential and performance

At the national level, Zimbabwe's record in agricultural production is outstanding. The country is self-sufficient in most foodstuffs and in favorable years is able to export maize to neighboring territories. In order

Table 10.1. *Zimbabwe: classification of natural agro-ecological regions*

Natural region	Mean annual rainfall (mm)	Annual rainfall reliability	Farming system	Agricultural potential
I	above 1170	Rains every month	Specialized, diversified	Tea, coffee, fruit, forestry, etc.
II	710–1170	Infrequent mid-season dry spells	Intensive, mixed	Food and fiber crops, livestock
III	570–710	Fairly severe mid-season dry spells	Semi-intensive	Livestock and crops
IV	450–610	Severe dry spells and periodic drought	Semi-extensive	Livestock and drought-resistant crops
V	below 510	Frequent seasonal drought	Extensive	Livestock only

Vincent & Thomas (1961).

to meet national needs, farmers must produce the 900 000 tons of maize per year in excess of home consumption, a volume attained in 8 out of the last 10 years. Zimbabwe has a larger national cattle herd than all other SADCC (Southern African Development Coordinating Conference) countries combined and is also a net exporter of beef. Agricultural exports account for 33% of the country's foreign exchange earnings. As the drought ended in 1985, record crops were anticipated of tobacco, cotton, wheat, soya and millet. Even during the drought, in 1982 and 1984, farmers delivered more maize than expected to the Grain Marketing Board, in part because forecasters were unable to predict rising levels of output from peasant farmers.

The potential for agricultural production varies markedly from one part of the country to another. Vincent & Thomas (1961) have classified Zimbabwe into five natural agro-ecological regions, principally on the basis of 'the adequacy and efficiency of the rainfall' (1961, 3). Each natural region has a distinct agricultural potential in terms of the intensity of land use and the types of commodities that can be economically produced (see Table 10.1). In the eastern highlands (region I), it is possible to grow a diverse range of specialized crops like tea, coffee, fruit and timber. The plateau (regions II and III) is best suited to a mixed crop and livestock system of an intensive or semi-intensive nature. Almost all of Zimbabwe's three principal crops – maize, cotton and tobacco – are produced here, as well as much of its beef and dairy products. In the drier parts of the country (regions IV and V), crop production – even for the drought-resistant millets and sorghums – is a risky proposition. Intensive farming is impossible except under irrigation; the recommended farming system is ranching or herding, usually of cattle, on relatively extensive holdings.

The agricultural economy is dualistic, with a large-scale commercial sector standing alongside small-scale family farms in the peasant sector. This structure reflects the colonial history of Zimbabwe in which European settlers appropriated large tracts of land for agricultural purposes. As of 1985, the commercial lands constituted 33% of the total land area with an average holding size of over 3000 ha. The 'communal lands' of the peasant farmers constituted 42% of the total land area with an average holding of 3 ha of arable land, plus grazing. The critical point is that the agricultural potential of the land held by peasant farmers is extremely low, because of limited rainfall and sandy soils. Whereas 30% of the commercial areas lie on prime land (regions I and II), only 9% of communal areas do so (see Table 10.2). Indeed, a full 75% of the peasant farming areas falls in the regions of lowest agricultural potential (regions IV and V). By virtue of geographical location, therefore, the smallholder is more vulnerable than any other type of farmer to variations in rainfall and the incidence of drought.

Under these circumstances, it is remarkable that peasant farmers were able to make significant gains in agricultural output in the period 1980–85. At independence the communal areas contributed only 8% to the total marketed output of the national food staple, maize; by 1985, the share had risen more than five-fold, to 45% (see Table 10.3). A similar meteoric rise occurred for cotton. In just 5 years, and against a background of drought, small farmers were able to effect a shift in the structure of agricultural production in Zimbabwe and to begin to challenge the dominant position previously occupied by the commercial farmers.

This does not mean, however, that all peasant farming areas have grown and benefited equally. In 1983–84 more than 80% of peasant maize sales originated from the 26% of the communal land that lies in natural regions I–III. Even during the good rains of 1980–81, only 57% of peasant farmers on average nationwide sold maize. The remaining 43% were producing at subsistence levels or below. The figure for non-surplus households was highest in areas of low potential and in years of low rainfall.

Because the capacity to produce food is not available in equal measure to all households, it is not surprising that undernutrition is a public health problem in rural Zimbabwe. While the country can normally produce ample food to meet the nutritional needs of the population, food is not always available in the right place, at the right time, and at an affordable price. One estimate is that, at any moment, 20–30% of young children in the rural areas are undernourished (Republic of Zimbabwe/FAO, 1983, i–ii). The majority of the population subsists on two daily meals of maize porridge

Table 10.2 *Natural region by land tenure*

Natural region	Zimbabwe	Commercial land[a]		Communal land	
I	1.8%	2.7%	29.8%	0.8%	8.5%
II	15.0%	27.1%		7.7%	
III	18.7%	22.4%		17.2%	
IV	37.8%	25.5%	47.8%	44.9%	74.2%
V	26.7%	22.3%		29.3%	

Chavunduka (1982).
[a]Includes both large- and small-scale commercial areas, plus resettlement areas.

(*sadza*) with vegetable relish. Young children are fed a diluted mixture of the porridge, often without added protein or fat. In normal times, only a small proportion, 1–2%, becomes severely malnourished, but upwards of 10% may be at risk in times of emergency such as drought.

The impact of drought, 1982–84

The drought of the early 1980s had an immediate and severe impact on aggregate agricultural production. From a bumper crop of over two million tons of marketed maize, following the good rains of 1980–81, national output dropped to 640 000 tons in 1982–83. As shown in Table 10.3, this was the only year in which the secular rise in the peasant sector's share of marketed maize was interrupted. Since the national economy is based on agriculture, the shrinkage of agricultural output had a ripple effect elsewhere, notably in the agro-industrial and transport sectors. Gross domestic product declined by 3% in 1984 (Republic of Zimbabwe, 1984). For the first time ever, the supply of maize and wheat from the Grain Marketing Board to the milling companies was rationed in October 1983. Sorghum and wheat were imported in that year, and by 1984 stocks had dwindled to the point where maize imports became necessary.

Table 10.3. *Zimbabwe: official maize sales, 1979–85*

Season	Total Sales (tons)	Percentage sales by type of farmer[a]	
		Large-scale commercial	Communal lands
1979–80	825 563	87.9%	8.1%
1980–81	2 013 759	82.0%	14.4%
1981–82	1 451 827	70.4%	21.8%
1982–83	639 747	72.6%	21.4%
1983–84	941 591	58.6%	35.5%
1984–85[b]	2 000 000	50.0%	45.0%

Calculated from producer delivery records, Producer's Registry, Grain Marketing Board, Harare, Zimbabwe.
[a] Totals do not add up to 100% because data for resettlement and small-scale commercial farmers are not shown.
[b] Estimate, Minister of Agriculture, *The Herald*, Harare, Zimbabwe, 20 March 1985.

As might be expected, the effects of drought were not consistent at the local level; in the driest areas (natural regions IV and V), the maize crop was a total failure throughout the drought years. Maize plants were stunted and withered at waist height and no farmer could report a surplus for sale. In some parts of Chibi, for example, farmers did not even get a crop in the ground for three consecutive years because of the failure of heavy planting rains to arrive.

In the intermediate zone (natural region III), the maize crop was entirely written off only in the year of extreme drought; in the years of partial drought, the crop was patchy, ranging from 'severely affected' to 'above average', depending on location. In Wedza communal land, for example, rainfall was so localized in 1983-84 that some farmers attained maize yields of over 5 tons per hectare within a 15 km distance of others who yielded a bare 300 kg (Bratton & Truscott, 1985, 2).

In the most-favored parts of the country (natural region II), there was scant evidence of drought at all. The proportion of peasant farmers marketing maize in Chipuriro communal land actually rose slightly during the drought (from 67% to 78%) and the tonnage delivered remained consistently high. Even in the disastrous 1982–83 season, Chipuriro farmers were able to deliver an impressive total of 17 500 tons of maize, an average of 2.35 tons per household to the Grain Marketing Board.

The human toll of drought was heavy and was felt earliest in the chronic food deficit areas of the lowlands and among the most vulnerable segments of the population. As early as April 1982, the Ministry of Health reported that up to 70% of children were undernourished in parts of Matabeleland and Masvingo provinces. Older residents recalled that the drought of the early 1980s was worse than the drought of the late 1940s when 'at least children had a bit of green maize and cattle had grass on which to feed' (*The Herald*, 2 February 1983). Teachers reported that pupils were arriving at rural schools without having eaten and were fainting in the classroom.

In the absence of a supply of grain, some villagers resorted to traditional or unconventional foods. In Mberengwa, wild fruits commonly used for stockfeed were dried and ground into a powder in order to make porridge. In Binga, Batonka women gathered grass seed which, once pounded and cooked, served the same purpose. In Tsholotsho, desperate individuals gathered maize waste around the community grinding mill. In lower Rushinga, where pressure on the natural environment was intensified by the presence of refugees from Mozambique, people had no alternative but to subsist on the roots, fruit, and leaves of the baobab.

The most immediate countrywide effect of drought was the desiccation of domestic water supplies. It was not uncommon for village women to walk up to 15 km each way – 20 km in Filabusi – to seek an alternative source. In Sabi, women and livestock had to stand all day waiting for overburdened wells to recharge with water. There were reports from Zabagwati of 'women exchanging blows and scratches . . . [after accusing one another] of filling too many containers' (*The Herald*, 28 April, 1982). Even in the wetter areas, domestic water supplies ran dangerously low in places of population concentration. Water rationing was introduced in urban centers and rural schools were forced to shut down during the dry season.

Perhaps the most serious long term effect of drought on the economy of the communal lands was the decimation of livestock herds. While it is safe to say that no person in Zimbabwe died as a direct result of starvation, cattle deaths between 1978 and 1983 are estimated at 36% of the communal land herd (Republic of Zimbabwe, Department of Veterinary Services, 1971–1984). In Tsholotsho, farmers said, 'we have lost count [of cattle deaths] . . . and we have stopped skinning the dead' (*Herald*, 13 September 1982). Cattle died at dried-up dams and waterholes when they got stuck in the mud and were too weak to struggle out. In many areas farmers refused to dip their cattle for fear that, in their weakened condtions, they might drown in the dip.

The social organization of production

Farmer organizations in Zimbabwe

How, in the face of drought, can we account for the fact that peasant farmers grew and sold more maize in the first 5 years of independence than during the previous years? What factors were at work to counteract the failure of the rains? We recognize that surplus production emanated principally from areas of highest potential where the drought was least severe. Nonetheless, expansion of food crop output is most unusual under drought conditions and begs an explanation.

To an important extent, changes in national agricultural policy helped to release a smallholder supply capacity that had been pent up by colonial economic restrictions. Producer prices for maize were raised by 124% between 1980 and 1985, an increase sufficiently large to outstrip rapid inflation in the cost of fertilizer and other inputs. Crop packs of seed and fertilizer were distributed as a measure to re-establish farmers on the land in the aftermath of war and drought. Efforts were begun to reorient Zimbabwe's marketing system – a mixed system of private input supply and public control of crop sales – towards the smallholder.

But the contribution of social organization should not be forgotten. In

1980 people were keen for peace to return to the countryside. There were fresh political ideas abroad about community cooperation, and opportunities were available, often for the first time, to obtain agricultural services. Farmers responded with a grassroots impulse to create new collective organizations and revive old ones. By 1983, 44% of farmers said that they belonged to some form of voluntary agricultural association.[1] Some of these organizations were products of independence, but most had roots going back to the early 1970s and before.

One of the features of farmer organizations in Zimbabwe is their diversity. In Zimbabwe, small farmers have had considerable contact with agencies from the public, private, and nongovernmental sectors. For example, the national agricultural extension service has promoted 'Master Farmer Clubs' and 'group development areas'; fertilizer companies have encouraged 'cash groups' and 'savings clubs'; and the rural missions, notably of the Catholic Church, have given support to 'mushandira pamwe' (working together) groups. Farmers groups are most common in the parts of the country where services from the various economic sectors are readily available. The Zimbabwe case suggests that it is not geographical isolation or neglect by policymakers that prompts small farmers to organize for self-help. Rather, autonomous farmer organizations are most likely to arise where effective state and market institutions provide stimulation and support. It is also vital that the policy regime allows sufficient 'space' for unaffiliated and autonomous organizations to exist.

I asked farmers, 'What activities does your farmer group engage in?' The most common activity (60% of group households) is the *attainment of knowledge*, with farmers assembling to listen to an extension worker or simply sharing ideas during the course of daily production. The second most common activity is *pooling of labor* (54%), with farmers getting together on one another's fields for collective work parties (*zvikwata*). Draft animals and farming equipment are often shared at the same time. Thirdly, farmers make *bulk orders of farm inputs* (47%), with the objective of achieving substantial savings in the cost of these expensive items. Finally, farmers organize to *sell in bulk* (36%), usually by hiring a truck to transport their commodities to market.

Farmer organizations in Zimbabwe tend to be composed of 'middle peasant' households. Group members have holdings in land, cattle and labor that are slightly above average, but have not developed to the full their potential for productivity and income. For a variety of reasons, the very poorest households generally do not join: they have no time or resources to contribute, they lack confidence in themselves, or they are not interested in improved farming. The richest individual farmers also stand back, either because their farm enterprises are sufficiently large to be economic, or because they are unwilling to share household assets with their less fortunate neighbors. About two-thirds of the members of farmer organizations are women. Women tend to be most numerous in groups that engage directly in production, for example, through the pooling of labor; men dominate the groups that engage in money transactions in central markets.

Collective action and food production

In order to assess the effect of social organization of food production, I recorded maize output figures for individual and group farmers in the 1981–82 season. This was a drought year in the south, but merely a drier-than-average year in the north. I found that farmers in groups *consistently outproduce* farmers working alone. For example, in Chipuriro households that belonged to farmer organizations, 'group households' produced an average of 4.25 tons of maize per hectare in 1981–82, compared with 2.59 tons for individuals, as shown in table 10.4. This group advantage holds true under every set of rainfall conditions, whether in well-watered Chipuriro (region II) or in drought-prone Gutu (region IV). Indeed, the positive effect of group organization seems to increase as natural conditions worsen: the productivity of group farmers is 137% higher than individuals in Gutu, compared with 64% in Chipuriro. The group advantage also holds true regardless of the measure of maize output used – whether productivity of land, total household production, or total household sales. The most pronounced advantage, however, lies in the amount of maize sold, with high sales providing a direct boost to the household incomes of group farmers.

What is the explanation for the superior production record of farmers who work in groups? High output is not achieved solely by virtue of economic status, that is, because group members are 'middle peasants' who are relatively 'well off' to begin with. This can be demonstrated by introducing a statistical control for the wealth of farmers, measured in this case by household holdings in cattle. While the production advantage of group farmers is slightly reduced by this procedure, the significant difference between individual and group farmers remains. Indeed, we discover that group farmers consistently outproduce even the stratum of 'rich' individuals who stand aloof from groups. The explanation of production, therefore, seems to rest, not so much with the resource base of the household, as with collective social action as undertaken within farmer organizations.

Table 10.4 *Zimbabwe: maize output by small farmers, 1981–82 season*[a]

	Natural region	Individual farmers	Group farmers	Ratio of group to individual
Productivity (Tons per hectare)				
Chipuriro	II	2.59	4.25	1.64:1
Wedza	III	0.95	1.75	1.85:1
Gutu	IV	0.54	1.28	2.37:1
Production (Tons per household)				
Chipuriro	II	3.65	6.91	1.89:1
Wedza	III	0.97	2.15	2.22:1
Gutu	IV	0.54	1.47	2.72:1
Sales (Tons sold)				
Chipuriro	II	2.58	5.86	2.27:1
Wedza	III	0.36	1.21	3.36:1
Gutu	IV	0.09	0.66	7.33:1

[a]From all survey areas except Dande which is excluded because the conditions there are not well suited to maize production. Farmers prefer to grow cotton as a cash crop and purchase their food requirements in maize.

The objective of collective action is to mobilize resources for agriculture. Peasant farmers are poor and household output is limited by resource constraints. We would, therefore, expect self-help organizations to center on attempts to guarantee to member households a reliable supply of production resources. Collective action can be conceptualized as occurring at two levels. The first is at the *level of production*, that is, where group households turn to one another for resources, for example, by pooling of labor or draft power. The second is at the *level of exchange*, where group households organize to improve access to services from outside agencies of state and market. An example here is bulk buying or selling of agricultural commodities.

I have found that, in Zimbabwe, farmer organization at the level of exchange has a greater impact on agricultural performance than at the level of production. In other words, farmers achieve larger gains in output by organizing to enter the central cash economy, than by organizing to mobilize resources at village level. This argument can be supported as outlined below.

Collective action at the *level of production* has an uneven and limited effect on the access of group households to agricultural resources. For example, group membership does not improve the ability of the household to get access to the basic resource of *land*. Not only do very few farmers lend and borrow land (12%), or establish collective plots, but there is no difference whatever in the behavior of individual and group farmers with regard to the sharing of land (see Table 10.5). Similarly, with *draft power*, while a large proportion of farmers do borrow oxen to plow (44%), this form of collective action is as much, if not more of, a traditional practice among family members as it is an organized function of farmer groups. The only exception is with labor groups organized for collective work (82%) in which draft pooling is a basic function. Group organization at the level of production is felt most strongly in terms of access to *labor*. Labor pooling, too, has traditional roots in *nhimbe* (beer party), but now has been institutionalized as *majangano* (reciprocal labor exchange). A written roster of work is drawn up in labor groups to rotate around the fields of members in a planned series of improved practices. Households in groups (46%) are significantly more likely than individuals (21%) to get extra labor in the form of a reciprocal work party. Group members receive four times as much labor from this source as do individuals, an average 28 versus 7 extra person-days per year. In sum, it should be noted that, at the level of production, group organization is most effective in relation to labor, less so in relation to draft power, and not effective at all in relation to land.

A quite different and more consistent pattern of group organization prevails at the *level of exchange*. Across the board, group farmers enjoy higher levels of access than individuals to agricultural services from central agencies. They are significantly more likely to be recipients of extension advice, seasonal credit and inputs like fertilizer, and to be clients of official crop marketing boards. This relationship holds consistently true, whether the service is relatively abundant, like *extension* (55% of all farmers), or extremely scarce, like *credit* (18% of all farmers). For services for which there is a user charge, like *fertilizer* or *transport*, group farmers regularly pay lower prices due to the economies of scale associated with bulk supply. For example, in 1984 in Mashonaland, group farmers paid between 13% and 33% less for a bag of compound maize fertilizer than did their individual counterparts. They are no more likely than individuals, however, to be able to guarantee timeliness in service delivery. When services are considered in

Table 10.5 Zimbabwe: access to agricultural resources by small farmers

	All farmers (%)	Individuals (%)	Group members (%)
Level of production			
Borrow extra land from another household	12	11	13
Get extra labor from a communal work party	38	21	46%***[a]
Borrow draft oxen and implements	44	48	40%[b]
Level of exchange			
Get advice from agricultural extension worker	55	31	86%***
Have a loan from a credit scheme	18	7	32%***
Buy fertilizer	61	48	77%***
Sell crop(s) through an official market	39	25	57%***

*** p=<0.001.
[a] 57% *** in labor groups.
[b] 82% *** in labor groups.

'packages' – a good combination is advice, fertilizer and market – the advantages of group organization are revealed most clearly. Group farmers are four times as likely (57%) as individuals (14%) to get a full, complementary package.

Effects of drought on collective action

A central question of interest in this chapter is whether prolonged drought has an impact on the propensity of farmers to organize themselves for production. Do conditions of extreme environmental stress lead to the intensification of social collaboration or a retreat into individualism? Commentators on this subject have offered opposite interpretations. In an article on social insurance in 'primitive' societies, Richard Posner argues that 'greater uncertainty in food supply increases the demand for the principle of reciprocal exchange' (1980, 18). The argument here is that social obligations of mutual assistance exist to address, and are triggered by, situations of hardship. On the other hand, correspondent Henry Kamm reported from Chad that the long drought of the 1980s had led to a 'weakening of the social structure' and that 'mutual help . . . traditions . . . [are] showing signs of strain' (*The New York Times*, 4 February 1985; see also Watts, 1983, 53). The implication is that, when hardship is generalized, then no individual household is in a position to help any other. Obligations may be recognized under moderate stress but cannot be activated when everyone is desperately short of resources.

In order to judge the validity of these contradictory propositions, we can begin with evidence on farmer attitudes in Zimbabwe. A subsample of farmers in Wedza was re-interviewed after an interval of 2 years, during which the drought peaked. Whereas in 1982 (after 1 year of drought) 93% of farmers reported that they planned to plant more crops in the following season, in 1984 (after 3 years of drought) only 76% had the same positive attitude. The decline in optimism in agricultural prospects occurred equally among group and individual farmers. After 1 year of drought, 82% of group members felt that their farmer group was getting 'stronger', and after 3 years of drought only 63% felt this way. Almost all the increase in negative responses came from the dry areas where maize and groundnut crops were completely wiped out. The reasons given by those who reported that their groups were 'weaker' are instructive. The most common response was that the group was new and, because of the drought, had never been able to demonstrate the concrete benefits of cooperation to its members. Even members of well-established groups reported a decline in group cohesion due to the fact that 'we no longer work together in the fields because of the lack of rainfall'.

Beyond farmer attitudes, we can measure the propensity of collective action actually to occur, with and without drought. Two short case studies will be taken from farmer group activity, one each from the level of production and the level of exchange.

At the level of production, the drought had a clear negative impact on the ability of farmer groups to pool *draft power*. Apart from the shortage of rainfall and along with the shortage of good-quality land, small farmers in Zimbabwe cite the scarcity of draft oxen as the main resource constraint inhibiting the production of food crops. A full 58% of households are either short of draft power or are effectively draftless. In order to plow, cultivate and to transport agricultural materials to and from the fields, they must borrow oxen and equipment from other households. Draft pooling, if organized within labor groups, permits even draft-deficient households to plow the land in winter and to plant summer crops early at the time of the first rains (Bratton, 1984). Following the good season of 1980–81, more than half of all households (53%) were able to borrow draft oxen; by 1983–84, however, following the severe drought, there was a significant decrease (to only 35%) (see Table 10.6).

The main reason for the disruption of collective action was that drought intensified the shortage of draft animals. Many oxen died or had to be sold during the drought, and those that remained were weakened for want of water and forage. Even young, trained animals were unable to pull a plow, and those under the yoke were sometimes known to expire from the effort. Under these circumstances, owners became increasingly reluctant to lend out their oxen, even to close relatives. There was no evidence of a retreat from organized groups into the family, because the frequency of lending to 'relatives' and 'strangers' remained the same before and after the drought. Interestingly, however, the means of payment did change: cash transactions for draft services declined significantly and more transactions were made free of charge (60%) or in return for labor (25%). So, while the drought tended to impair the frequency of draft pooling, it also brought out the more altruistic and less commercial aspects of farmer organization.

Let us shift now to the level of exchange and inquire into the effects of drought on the collective action of farmers. In this case the example will be credit services. We already know that in Zimbabwe group farmers (32%) are more likely to have a loan for seasonal inputs than are individual farmers (7%) and that all loans go to farmers in high potential agro-ecological zones (mainly region II).

I found that farmers in groups are usually more likely to repay their loans (Bratton, in press). To understand this important facet of collective action, and the effects of drought upon it, it is necessary first to appreciate the various methods of loan liability that apply in the peasant sector in Zimbabwe. First, there is *individual liability*, as illustrated by the terms of loans issued under the Small Farm Credit Scheme of the Agricultural Finance Company (AFC) in Zimbabwe. Loans are made to and recovered from individuals, and liability rests with the one farmer whose name appears on the AFC loan register. Second, *voluntary joint liability* is practiced by non-government organizations, for example, in the Revolving Loan Scheme of Silveira House (a rural development training institute operated by the Catholic Church). Loans are given only to active members of proven groups and, if any member fails to repay, the group as a whole loses eligibility for credit in the following season. Finally, terms of *mandatory group liability* are applied by the Group Lending Scheme of the Agricultural Finance Corporation. Under this arrangement, responsibility for repayment rests with the group as a whole, and the debts of recalcitrant members are automatically deducted from the accounts of those who pay.

Overall repayment rates are rather low, as can be seen in Table 10.7. Nonetheless, before drought struck in the north, farmers who accepted group liability were significantly more likely (between 72% and 92%) than individuals (53%) to pay off their loan obligations. Mutual social pressure within the group seems to have been the mechanism lifting repayment rates well above individual standards. The occurrence of severe drought, however,

Table 10.6 *Wedza: effects of drought on pooling of draft power, 1981–84*

	1981–82 (following good rains) (% of households)	1983–84 (following severe drought) (% of households)
Borrow draft oxen	53	35[a]
With whom		
'Relatives'	68	72
'Strangers'	32	28
Payment		
Nothing	45	60[b]
Cash	30	16[b]
Labor	15	25[b]

[a]p=<0.001.
[b]p=<0.01.

Table 10.7 *Zimbabwe: seasonal loan recoveries, individual and group lending schemes*

Terms	AFC individual (individual liability)	SH group (voluntary joint liability)	AFC group (mandatory joint liability)
1980–81 (good rains)	53%	72%★★	92%★★★
1981–82 (partial drought)	54%	71%★★	78%★★★
1982–83 (severe drought)	28%★★★	18%★★	9%

Agricultural Finance Corporation, Statements of Loan Account (at Sept. 30) Silveira
House, Loan Repayments Summary (at Sept. 30).
★★★ p=<0.001
★★ p=<0.02

led to a reverse finding. In 1982–83 repayment by individuals (28%) significantly outstripped both group approaches (18%, 9%, respectively). The logic of collective action in this case is for all group members to default, once a natural disaster leads to widespread crop losses. Members who are able to repay have nothing to gain by doing so, either because the group will lose eligibility for future credit anyway, or because their profits will be appropriated to pay the debts of others. At worst, the combination of drought and mandatory joint liability can even lead to a situation of 'internal debt' within groups, where some members owe money to others. This can weaken, and even break up, otherwise cohesive groups. Group lending, therefore, appears to be viable only with a 'normal' range of climatic variability and is counterproductive when farmers are exposed to extreme environmental stress.

The social organization of consumption

Alternative sources of food

Having seen how small farmers organize to produce food during times of drought, let us now examine the organization of food consumption. What institutions are available to which a food-deficit household can turn? Since farmer groups are essentially production organizations, they do not, to my knowledge, become involved in distributing among members supplies of food for home consumption. In Zimbabwe, supplementary supplies of food were available from the other social organizations with a presence in the countryside – the extended family, the market and the state. All of these organizations, however, are based in the urban areas. As reliance upon them increases, the capacity of rural households to guarantee autonomously their own food security is undermined.

The institution usually considered as the first refuge is the *extended family*. Traditionally, Shona and Ndebele societies required that close relatives respond to one another in time of need. Although the glue of reciprocal social obligation still has residual power to bind the behavior of people in the countryside, it is dissolving in the urban areas. Throughout the crisis, political leaders felt it necessary to remind people of the 'generosity and hospitality of Zimbabweans who offered what little they possessed [to] keep up the momentum of the war' [Republic of Zimbabwe Information Service (ZIS), press release 723, 1983]. Urban workers were urged to send maize meal and other basic commodities to their relatives in the communal areas (ZIS press release 496, 1983). A heavy volume of maize meal was, in fact, transported to the countryside on the roof of passenger buses at the end of every month during the drought. Another strategy was for parents to send their children to stay with urban relatives while food supplies were scarce at home. In some cases, the entire family migrated temporarily to town.

On the negative side, the strain on the extended family became hard to bear, with some wage earners complaining of being 'bled' by relatives (Wohler, 1984). The incidence of vagrancy and begging increased in rural and urban centers and some unfortunate individuals found that their family was no longer willing or able to support them.

An alternative structure of support in times of stress is the *market*. Food can be purchased, subject to its availability, by those who can afford it. The network of rural traders in Zimbabwe is extensive, and consumer goods are relatively plentiful by rural African standards. With the exception of a national shortage late in 1983, and an official suspension of supplies at the height of the government's campaign against dissidents in Matabeleland earlier in the same year, maize meal was generally available to consumers throughout the drought. Maize beer was also in good supply. The main drawback was that food prices rose rapidly after 1980, as a consequence of general inflation (over 20% per year) and the partial removal of food subsidies. The price of foodstuffs for lower income urban families more than doubled (up 107%) between 1980 and 1984 (Republic of Zimbabwe, 1985). The increase was even higher in the countryside, as rural traders generally found it easy to evade government price controls.

For people without ready cash, the options were more limited: either to sell a household asset or to seek paid employment. Jobs were hard to come by because of the depressing effect of the drought on the economy as a whole. Commercial farmers cut back on labor recruitment, not only because their crops were damaged by drought, but in response to the introduction of minimum wages in 1980. Often, the only employment available on commercial farms was daily paid seasonal labor for women and children.

As a consequence, many households turned to the sale of their most valued possession – cattle – as a way of raising money. Special inducements to sell were made by the Cold Storage Commission, through periodic rural markets and the relaxation of quality standards. In the first year of drought, peasant farmers were reluctant to sell cattle. They argued that draft power would be needed once the rains came and, anyway, the price of cattle might rise, and they would be unable to buy more. As the drought extended over subsequent seasons and as large numbers of animals died, sales to the national market increased, almost doubling (from 40 000 to 75 000) between 1980 and 1984 (Republic of Zimbabwe, Department of Veterinary Services,

1971–85). There was also an informal trade in livestock with local businessmen in which farmers bartered cattle, goats and chickens in return for bags of maize meal. The businessmen were in a position to dictate terms and farmers were able to realize only a fraction of the true value of the livestock. Cattle thefts rose during the drought in both the commercial and communal areas; the stolen animals were either slaughtered for food or found their way into informal markets.

Where the market fails, as it surely did for the poorer peasants in the Zimbabwe drought, the *state* can provide an institutional alternative. At the outset of the drought, the state had two programs of hunger assistance: a monthly financial aid payment to indigent families (run by the Department of Social Services), and a child supplementary feeding program (under the Ministry of Health). The former was largely urban-oriented, and the latter was aimed exclusively at malnourished children under the age of five. Neither was comprehensive enough to address the demand for food that arose throughout the country among a wide cross-section of the population.

In May 1982, the Zimbabwe government set up a Cabinet Committee on Drought Relief chaired by the Minister of Labor and Social Services. Its first act was to launch a program of food aid for the drought-stricken south. The aim was to coordinate and vastly augment efforts at drought relief begun by disparate charitable and church organizations. The program was based on the delivery of a basic monthly per capita ration of maize meal (one 20 l container), salt (1 kg), groundnuts or beans (2 kg), dried fish (2 kg), and a protein drink mix (5 l). The program also provided water bowsers or borehole sinking equipment 'to discourage people from moving about in search of water' (*The Herald*, 30 May 1982). By August 1982, the program had become so important that the Minister of Water Resources and Development could declare that 'drought relief takes precedence over everything else' (*Sunday Mail*, 22 August 1982).

At the height of the crisis, one economist said of the relief effort that 'the country [had] never faced a problem of this nature and magnitude' and predicted that 5 million people would need assistance (RAL Merchant Bank, 1983). In actuality, the numbers were lower, but still dramatic. After 1 year, the program had fed 888 000 people; after 2 years, more than 2.1 million (*Herald*, 23 December 1983). This constitutes about one-half of the estimated communal land population. There is no doubt that the government's impressive organization effort helped to stem the immediate hunger problem. In April 1984, the Provincial Medical Director, Midlands, could report that there had been no increase in malnutrition cases in the worst year of drought (*The Herald*, 22 March 1984). The Minister of Health could also announce that, due to immunization, oral rehydration and supplementary feeding, the infant mortality rate for Zimbabwe had been halved to 60 per 1 000 in the first 5 years of independence (*Sunday Mail*, 23 June 1985).

This is not to claim that the government overcame the substantial logistical problems that often plague large-scale relief efforts. There was inadequate transport to move food supplies to the field and, even when help was recruited from the private sector, supplies continued to arrive late. In 1983 the author saw hundreds of recipients waiting for several days at a supply point in Gutu for a relief truck to arrive. In Mberengwa-Mwenesi there was said to be a 2 month delay in the monthly food disbursement (*The Herald*, 11 November 1983). The problem had local causes – bad roads in the remotest areas, harassment and intimidation of relief workers by dissidents in Matabeleland North, corruption by officials who sold relief supplies for personal profit – but at root arose from the highly centralized system of grain marketing in Zimbabwe. Legislation on the marketing of 'controlled' products requires that grain be hauled to central depots for sale and storage. In practice, the same grain had to be hauled back to the countryside when food relief was needed. Non-government agencies were quicker than government in organizing shipments directly between contiguous surplus and deficit areas. Indeed, non-governmental organizations, like Save the Children Fund, Christian Care, World Vision International, ORAP, the Zimbabwe Red Cross, and local churches, were in some places the only reliable suppliers.

The most common complaints from peasants were that the government took too long to determine who should get food aid and that supplies were never sufficient. The guideline was that the 'needy' should benefit, defined as those with insufficient grain in the home granary and without a close family member working for a wage. In practice, the distribution pattern was indiscriminate; those who were ineligible received relief food, while those who were truly needy may have gone short (Coenrad Brandt, University of Zimbabwe, Department of Urban and Rural Planning, June 1985, personal communication). Some families, especially those who had to 'pay back' food borrowed from other families at an earlier date, consumed their drought relief allocation long before the next delivery was due.

Peasant consumption choices

How did peasant farmers respond to the institutional alternatives that were available? Where did they turn, when short of food?

Because the drought had differential impacts across the country, households in some communal areas were always able to feed themselves. There were, however, some food-deficit households in every area that had to look for an outside source of maize meal. The proportion of households that consumed their staple grain from an outside source varied greatly across the country. As shown in Table 10.8(a), in Chipuriro (region II) a maximum of 9% of households were in deficit, compared with 92% in Dande (region IV). It is clear, however, that as the drought deepened, more and more households lost the ability to feed themselves. From a national average of 16% in 1980–81, the proportion of food-deficit households rose to 56% in 1982–83.

It is instructive to see where peasant households turned to obtain their supplementary food. The *extended family* did not play a prominent role. At no time in the drought did relatives provide food to more than 9% of food-deficit households, and the proportion actually declined as the drought deepened, as suggested in Table 10.8(b). It is possible that some family contributions, for example, from husbands and sons earning wages in town, are reflected as purchases from the market. But, to all appearances, the family is in decline as an institution of social insurance against hardship in Zimbabwe.

In normal times, peasants in Zimbabwe rely heavily on the *market*, by purchasing extra maize for household consumption from retail vendors. The market remained an important institutional source of food throughout the drought, though the proportion of households making purchases of food gradually declined (from 75% to 40%) with time. This is partly explicable by the fact that, as the drought progressed, farmers were less likely to have disposable income accumulated from crop sales during previous seasons.

The last noteworthy trend is that the market was gradually displaced by the *state* as the main institutional supplier of supplementary food for peasant farmers between 1980 and 1985. For the last 2 years of the crisis, more rural families were receiving supplementary food from the national drought relief program than from any other source. By 1983–84, 55% of households countrywide reported that they were receiving maize meal from the government. This reflects a very heavy level of dependence; in the space of 2 or 3 years the drought had compromised the autonomy of rural dwellers as providers of their own food.

The final inquiry is whether membership in farmer organizations can help to prevent a slide into food dependency during drought. One might expect group households to be particularly vulnerable, because they usually sell more maize than do individuals and might, therefore, retain fewer bags

Table 10.8 (a) Zimbabwe: consumption of maize meal from sources outside the household, 1980–84

	Wedza	Gutu	Chipuriro	Dande	All
1980–81 (good rains)	21%[a]	10%	5%	51%	16%
1981–82 (partial drought)	36%	42%	9%	92%	38%
1982–83 (severe drought)	74%	72%	2%	92%	56%
1983–84 (partial drought)	50%	n.d.	n.d	n.d.	n.d.

[a] % of households

(b) Zimbabwe: sources of maize meal consumed from outside the household, 1980–84

	Given by Relative	Purchased from Market	Drought Relief Program	Other[a]
1980–81 (good rains)	9%[b]	75%	0%[c]	16%
1981–82 (partial drought)	6%	63%	25%	6%
1982–83 (severe drought)	4%	41%	51%	4%
1983–84[d] (partial drought)	3%	40%	55%	2%

[a] Includes government supplementary feeding and non-governmental organization drought relief programs.
[b] Percentage of households.
[c] Government drought relief program not in effect in 1980–81.
[d] Wedza only.

for home consumption. In fact, there is no evidence that group households retain fewer bags of maize per person; in fact, they retain slightly more. In addition, group households are more likely to grow a diversity of crops – four or more – and, therefore, can expect to produce at least some food or cash when the season is poor for maize.

During the drought years in Zimbabwe, fewer group than individual households found themselves in a food-deficit situation (see Table 10.9(a)). The advantages of farmer group membership were felt most significantly in the early years of the drought. In 1981–82, for example, when almost half of all individuals (48%) had to find an outside source of maize meal, a mere one-quarter of group farmers (24%) faced the same quandary. The advantage began to fall away after a couple of years of drought, however, as group members consumed reserve food supplies in household storage. It is evident that grain retained by group households can carry them through a short period of drought, but not a long one.

The benefits of group organization derive not only from higher household retentions, but also from higher agricultural incomes. When group members

Table 10.9 (a) Zimbabwe: consumption of maize meal (%) from sources outside the household, individual and group farmers, 1980–84

	All farmers	Individuals	Group members
1980–81 (good rains)	16%	20%	11%
1981–82 (partial drought)	38%	48%	24%***
1982–83 (severe drought)	56%	60%	42%***
1983–84[a] (partial drought)	48%	54%	46%

[a]Wedza only.

(b) Zimbabwe: sources of maize meal, individual and group farmers, 1982–83

	Purchased from market	Drought relief program
1982–83		
Individuals	27%	59%**
Group Members	57%***	44%

*** p=<0.001.
 ** p=<0.01.

eventually turn to outside maize meal sources, they are more likely to be able to afford cash purchases in the market and are less likely to depend on free 'handouts' from the state. At the height of the drought in 1982–83, when the majority (59%) of individuals had been forced to accept drought relief, the majority of group members (57%) was still able to rely on its own resources by buying bags of maize meal from local merchants (see Table 10.9(b)).

Conclusion

There can be no denying that the social impact of drought is ubiquitous and fundamental. It throws otherwise autonomous households into dependency on outside agencies. It sweeps aside well-established social organizations of peasant farmers. We are led to agree with the argument that drought undermines rather than encourages mutual aid among peasants and the development of rural institutions.

I have argued from the case of Zimbabwe that collective action among small farmers affords a slim measure of protection. Organization for self-help in agriculture cannot prevent the disruptive effects of drought, but it can help to delay and ameliorate them. In Zimbabwe, farmers in groups appear to be buffered from food deficits for approximately one season longer than are individuals. Because they produce, retain and sell more maize, group farmers are better able to guarantee a grain supply for home storage or to purchase grain from the market. They turn to the state for famine relief only after their individual counterparts have long done so.

The full social impact of farmer organization is best observed, however, in the years between droughts. The Zimbabwe case suggests that farmer organization is an essential component in any small-farm development strategy. It serves to plug the gap between small and scattered households and large and centralized institutions like the state and market.

The Zimbabwe case suggests that certain institutional conditions must be met, if farmer organizations are to take root. I will conclude by mentioning just three interrelated conditions as a way of directing the debate toward practical policy issues.

First, the productivity advantage of group organization appears to derive principally from the level of exchange, that is, by connecting farmers to production services and helping them to enter the cash economy. This requires that state and market provide attractive incentives and guarantee that services are delivered efficiently and reliably. Only then will farmer organizations arise at the local level and link with agricultural institutions at the center.

Second, farmer organizations cannot be created from above, especially by government *fiat*. All forms of state-sponsored cooperatives have fared badly in Africa. This is especially true at the level of production. Small farmers in Africa have shown themselves likely to resist policies that require organized resettlement, the pooling of household property, or the sharing of agricultural output. It is far better to allow farmers to elect their own arrangements for pooling household assets with collaborators of their own choosing. Local organizations controlled by farmers are most likely to be socially valued and economically productive.

Third, special attention will always be needed to the very poorest households, many of which in Africa are headed by women. Voluntary collective organization does not break the 'law of big farmer dominance' that attends all forms of rural institution-building. One role for the state is to ensure that selective incentives are provided for self-help organizations of rural women and the poor. This can be done by delivering services on a preferential basis to groups of the underprivileged that can show evidence of collective organization and self-management. Only then will the prospects for food security of the most vulnerable small farm households be improved.

Notes

1. All micro-level data in this chapter are derived from a survey of a random sample of 464 households and 50 farmer groups. The respondents were spread over four communal areas – Chipuriro, Wedza, Gutu and Dande – selected for diversity in terms of natural conditions, population density, proximity to town and development infrastructure.

References

Berg, E. (1984). The Africa Report: an overview. *Rural Africana*, **19–20**, 45.
Bonnevie, H. (1983). The impact of migration in Zimbabwe: Preliminary findings, *Discussion paper*. Harare: University of Zimbabwe, Center for Applied Social Studies.
Bratton, M. (1984). Draft power, draft exchange and farmer organization, *Working Paper 9/84*. Harare: University of Zimbabwe, Department of Land Management.
Bratton, M. (in press). Financing smallholder credit: A comparison of individual and group credit schemes in Zimbabwe. *Public Administration and Development*, forthcoming.
Bratton, M. & Truscott, K. (1985). Fertilizer packages, maize yields and economic returns: An evaluation in Wedza communal land. *Zimbabwe Agricultural Journal*, **82**(1), 1–8.
Callear, D. (1984). Land and food in the Wedza communal area. *Zimbabwe Agricultural Journal*, **81**(4), 163–8.

Chavunduka, G. L. (1982). *Report of the Commission of Inquiry into the Agricultural Industry.* Harare, Zimbabwe: Edinburgh Press.

Collinson, M. (1982). Demonstrations of an interdisciplinary approach to planning adaptive agricultural research projects: Chibi District, Southern Zimbabwe, *Report No. 5.* Nairobi, Kenya: International Maize and Wheat Improvement Center.

Esman, M. & Uphoff, N. (1984). *Local Organizations: Intermediaries in Rural Development.* Ithaca, New York: Cornell University Press.

The Herald (1981–85). Harare, Zimbabwe.

Korten, F. F. (1982). Building national capacity to develop water users' associations: experience from the Philippines, *Staff Working Paper No. 528.* Washington: World Bank.

Lineham, S. (1972). Evidence of a mid-season break in the rains in Rhodesia, *Meteorological Notes, Series A, No. 33.* Salisbury: Government of Rhodesia.

Norman, D., Baker, D. & Siebert, J. (1984). The challenge of developing agriculture in the 400–600 mm rainfall zone within the SADCC countries. *Zimbabwe Agricultural Journal*, 81(6), 205–14.

Peake, J. S. (1952). Normal variation of rainfall. *Notes on Agricultural Meteorology No. 1.* Salisbury: Government of Rhodesia, Department of Meteorological Services.

Posner, R. (1980). A theory of primitive society, with special reference to law. *Journal of Law and Economics*, 23(1), 1–53.

RAL Merchant Bank (1983). *Executive Guide to the Economy*, Harare: RAL Merchant Bank.

Republic of Zimbabwe (1971–85). *Annual Reports.* Harare: Ministry of Agriculture, Department of Veterinary Services.

Republic of Zimbabwe (1981–85). *Press Release* series, Zimbabwe Information Services.

Republic of Zimbabwe/FAO (1983). *Zimbabwe: Food and Nutrition Policies and Programmes.* Harare: Ministry of Finance, Economic Planning and Development, and joint FAO/WHO/OAU Regional Food and Nutrition Commission for Africa.

Republic of Zimbabwe (1984). *Estimates of Expenditure for the Year Ending June 30, 1984*, presented to Parliament by the Minister of Finance, Economic Planning and Development, July 26. Harare.

Republic of Zimbabwe (1985). *Monthly Supplement to the Digest of Statistics.* Harare: Central Statistical Office, February.

Russell, C. S. & Nicholson, N. K. (eds) (1981). *Public Choice and Rural Development.* Washington, DC: Resources for the Future.

Shumba, E. (1984). Reduced tillage in the communal areas. *Zimbabwe Agricultural Journal*, 81(6), 235–9.

Sunday Mail (1981–85). Harare, Zimbabwe.

Uphoff, N. (1985). People's participation in water management: Gal Oya, Sri Lanka. In *Public Participation in Development Planning and Management*, ed. J.-C. Garcia-Zamor, pp. 131–75. Boulder: Westview Press.

Vincent, V. & Thomas, R. G. (1961). *The Agro-Ecological Survey*, Part 1 of Federation of Rhodesia and Nyasaland, *An Agricultural Survey of Southern Rhodesia.* Salisbury: Government Printer.

Wohler, J. (1984). Drought takes a heavy toll in Tsholotsho. *The Herald*, September 19. Harare, Zimbabwe.

11

The social impact of drought in Ethiopia: oxen, households, and some implications for rehabilitation*

JAMES McCANN

African Studies Center, Boston University

Recent famines in Ethiopia have sparked world concern, generated a massive relief effort from donor agencies and captured the imagination of journalists of all descriptions. Remarkably, however, these crises have resulted in only a trickle of empirical social and economic research. In fact our understanding of the local effect on drought-affected farming systems derives almost exclusively from scattered data drawn from those who have come to relief camps, a demonstrably poor cross-section, and a distinct minority of the rural population affected by drought. Since most of the rural population affected by drought remain on their land, the task of researchers is to understand the terms under which they continue to survive and reconstruct their social and economic lives. Only when these social processes are understood can well-planned rehabilitation take place.

My research contributes to the debate on the local effects of drought and famine by offering some pertinent data on micro-level socio-economic institutions and early responses at the level of household production in northern Ethiopia. This chapter bases its argument on two types of data: first, my own oral and archival work on the long-term decline of the Lasta region of Wollo administrative region carried out between 1980 and 1982; and second, recent field work with peasant associations in Ankober and Sela Dengay *weredas* (sub-districts) of Tegulet and Bulga *awraja* (district), drought-affected regions which I believe replicate the conditions that existed in areas of Tigray and Wollo 2 years ago, before the massive migration to relief camps like Korem and Ibenat.

My recent work involved the evaluation of the social impacts of an oxen/ seed distribution project sponsored by Oxfam America and carried out by the Integrated Drought Relief Committee of Tegulet and Bulga *awraja* and the Staff Council of the International Livestock Centre for Africa (ILCA). The project has distributed a capital package, consisting of an ox, seed and a single-ox plow, to drought-stressed farmers who had not yet abandoned their land. The program has distributed over 500 oxen/seed packages within 25 target peasant associations of the Tegulet and Bulga districts of northern Shoa. In addition to oxen and seed, ILCA has developed a training and evaluation component to support the local construction and use of a single-ox plow instead of the pair of oxen required in the traditional plow, a technical innovation in oxen-poor northern Ethiopia (Gryseels *et al.*, 1984).

This chapter, therefore, focuses specifically on the key role of livestock, especially oxen, in determining patterns of response to food-production shortfalls, the distribution of available resources and, in short, social impact. My unit of analysis is the rural household, since it is the primary unit of production and distribution of resources as well as the dominant social unit. In this context I consider inter-household institutions and intra-household

Fig. 11.1. Ethiopia.

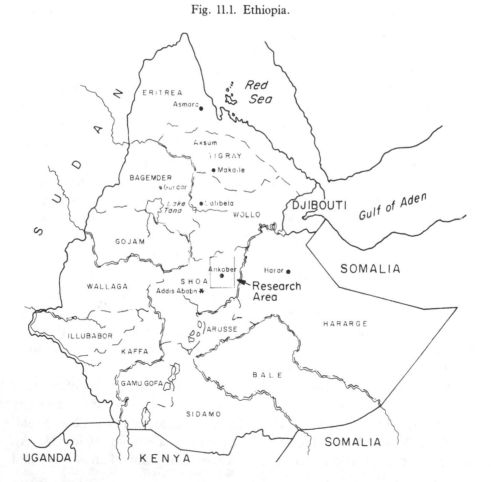

relations, both of which determine local allocation of food and productive resources between and within households. My basic premise, derived directly from my research and field experience, is that the response to drought and subsequent famine is not chaotic, but is a process determined by pre-existing social and economic institutions and that the most important of these practices involve the borrowing and exchange of oxen. Further, the effects of drought are distributed unevenly, in that they include a process of relative enrichment for some households just as they represent tragedy for others.[1]

The geographic setting

The overall effect of the project on farmers, households and regional rehabilitation cannot be assessed without a clear understanding of the ecological and historical circumstances which have conditioned local farming systems. There are two categories of land prone to famine in northern Ethiopia. The first is the highland zone between 2000 and 3500 m elevation in the northeast which stretches from northern Shoa through Wollo and Tigray into Eritrea. Rainfall in this region averages between 600 and 1000 mm per year but a rain shadow effect often restricts moisture from the southwest from reaching areas east of the Blue Nile and Takkaze river valleys. In these areas the effects of drought are exacerbated by pressure on land, erosion and poor conditions for livestock reproduction which result from the long period in which land has been in intensive production. Population pressure, partible inheritance and a farming system based on animal traction have determined recent demographic shifts in the zone, with these effects being further advanced in the north of the zone than in the south. In the last century highland agriculture, which was formerly concentrated in the 2000–3000 m range, has expanded into high elevation areas near the frostline and down into lowland river valleys prone to flooding, malaria and livestock diseases (Messerli & Aerni, 1978; McCann, 1984a).

The other zone is the agro-pastoral lowland area east of the main Addis Ababa–Asmara caravan route. There, the problem of drought stems less from problems of environmental degradation or low *average* levels of rainfall than it does from the high degree of rainfall variability (Degefu, this volume). In these areas, which include some highland zones of Tigray and Wollo, the average annual rainfall is less than 700 mm and the growing season less than 120 days. Despite these factors, lowland areas have in the past 2–3 generations absorbed a large migration of highland agriculturalists who cultivated land previously used as wet season pasturage by Afar pastoralists.

In many of these areas highland farming systems have not yet fully adapted to the drier, lowland environment.

My recent fieldwork took place in both of these ecological zones, one in the lowlands of the Ankober sub-district (*wereda*) and the other in a highland sub-district of Tegulet with its capital at Sela Dengay. Therefore, research results in these areas of northern Shoa will have a wide range of applicability for rehabilitation in other areas of Ethiopia. The ecological and demographic characteristics exhibited in project areas extend north into areas such as the eastern areas of Wollo and Tigray which have experienced the recent mass migration to feeding centers. In fact, I would argue that research results from these two sites will suggest the types of conditions and social responses that prevailed in Wollo and Tigray 2 years ago before mass out-migration took place. Research on social and economic response in those areas is almost non-existent, and current approaches to rehabilitation appear unrelated to the social context of the crisis. Therefore results of research from the ILCA oxen/seed project and further work on household level response have important implications to overall rehabilitation policy and for understanding local impact.

The following sections describe project sites.

Ankober

The research area in Ankober *wereda* generally lies between 1500 and 1800 m elevation. As one descends from the highlands around the capital of the *wereda* to the project area, there is a dramatic shift in vegetation, temperature and soil type. Deciduous trees end at about 2000 m and are replaced by thorny acacia and low-lying bushes. Cultivation takes place primarily in the better-watered areas adjacent to river valleys. Though there is open pasturage, available browse depends to a large degree on grasses which in turn depend on rains. Animals, therefore, must range rather widely to find sufficient food during dry periods.

Such an arrangement is conducive to pastoralism, and the pastoral patterns of dry-season migration to the Awash valley are well adapted to it. For agriculture, however, animals must be grazed closer to the permanent homestead site and the limitations on browse mean that fewer animals can be maintained. In lowland, newly settled areas pasturage is communal and unregulated by either the peasant associations or by individual claims. By contrast, older adjacent highland areas pastureland is allocated to individual households and they are jealously guarded. This ecological zone overall extends from northern Shoa north to Eritrea and generally represents the zone of contact between Christian highlanders and Muslim (often Oromo) lowlanders. It is also a transitional zone between pastoral areas below 1500 m and the margins of cultivation. Given the steady migration of population into this area and others like it, research results have important national implications.

The Denki peasant association (PA), the site of the ILCA evaluation and most of my interviews, is at 1500–1600 m and is typical of the agro-pastoral zone east of the escarpment. Denki's 100 tax-paying households are evenly divided between Muslims and Christians though this configuration is only about a generation old. Twenty years ago most of Denki was occupied by Muslim weavers, many of whom spoke an archaic Semitic language called Argobba. They engaged in agriculture only sporadically and subsisted primarily on income from sales of homespun cotton cloth to the district-level market at Aliu Amba 2 hours away in the foothills of the escarpment. At least two of the Muslim farmers I interviewed, including the current PA chairman, had taken up agriculture after starting life exclusively as weavers. Some still practice weaving to provide supplemental income (about US$50–150 per year) but most have abandoned it, citing better, more secure income as their reason for shifting to agriculture. Though reasons for the shift are not entirely clear, their decisions generally coincide with the penetration of the national market for grain, especially for teff, which generally grows well in the 1500 m zone.

In virtually all cases Christian farmers arrived in Denki as first-generation migrants. Most came from Lalo PA (1 hour's walk north of Denki) and from adjacent areas at higher elevations where population pressure prompted younger sons to leave their families' holdings and take up tenancy in lower areas like Denki where land was readily available. Before their abrogation in 1975, land reform tenancy agreements generally required a payment to a landlord of one-fourth of produce in addition to local taxes. That Denki's Christian residents continue to belong to distant (1–2 hour's walk) parishes in the highlands is further evidence of their recent immigration. Though few attend weekly services, burials, christenings and other rites of passage take place at Lalo and other highland areas. Wives are also brought from these areas, since most farmers arrived as young men and there are few local Christian women of marriageable age. Since most migrants were in a poor position to claim parents' land (*gulema*), they were also in a poor position to accumulate livestock either through inheritance or through marriage contracts (see below).

Thus, most of the farmers in Denki are fairly new to the requirements of agriculture in the area. Highland migrants have had to adapt to new soil and crop types and to different types of cattle (Sanga as opposed to the smaller and less hardy Highland Zebu). Many Muslims, though in residence longer, have only recently begun full-time agriculture. Given the fairly recent beginnings of intensive agriculture, one could argue that the viability of intensive agriculture in the area still needs to be proven in the long term. Certainly, it is worthwhile raising the question of how many additional migrants the area can support, since the PA's open policy of distributing land to new immigrants will increase the population in the next decade (J. McCann, unpublished Oxfam America Social Impact Report).

Tegulet

The Tegulet project area presents a startling contrast to the Ankober lowlands. Whereas Ankober lowland agriculture is fairly recent, farming systems in Tegulet have been in place since at least the thirteenth century, when the region was the primary area from which the Amharic language and political culture expanded to dominate the central highlands. Initially the highland population occupied the ideal middle altitude zones from 2000 to 3000 m but, within the last century, population pressure has forced new households into the lowland valleys of the area's river system (which are hot, malarial and flood-prone). Most farmers receiving oxen were from the lowland zones adjacent to the Mufer Wuha River. According to farmers in the area, the lowland regions are now as crowded as the middle and highland zones.

Land holdings in the Tegulet area are considerably smaller than in Ankober, averaging around 1.3 hectares with each farmer holding 3–4 separate plots, divided when possible between several altitude zones (J. McCann, unpublished Oxfam America report). These fragmented holdings allow farmers to diversify crops and reduce risk of rainfall variability. In the areas above 2500 m, barley and horsebeans predominate; lower areas produce teff, sorghum and pulses. As with Ankober, the major failure of the rains was in 1984 when the main rains and the short spring rains failed; in the previous 3 years farmers reported fairly consistent yields (though I do not have data from the lowest project area which is likely to have had more rainfall variability).

Severe pressure on land has prompted a number of restrictive land policies by peasant associations in the Tegulet project area which contrast with Ankober. Most of the project farmers in the area held land as tenants prior to the 1975 land reform proclamation, a most surprising revelation since most work about Ethiopia assumes land in the area was distributed exclusively by cognatic descent. They were tenants of a former prime minister in the imperial government, and paid one-fourth of their produce (before taxes) as rent. Though this arrangement suppressed farmer incomes, it also allowed farm sizes to be maintained better than in areas where cognatic descent land systems dominated. With land reform, many more people applied for land and reductions in farm size took place. Prior to the decree halting land redivision in 1982, the land had been redistributed four times. Most tenant farmers lost land in that process, not because they were large landholders before, but because of the increase in requests for land locally. Elimination of rents partially offset the effects of the reduction in farm size and gross income. The social effects of increased pressure on land were more significant. Since 1982 no new land applications have been accepted by the PAs. Young men reaching adulthood receive land only if it is provided by their parents or if a PA member dies without heirs. Consequently, the age of first marriage for males has risen steadily and it is not uncommon to find 30 year old men living with their parents for lack of land to establish an independent household. This situation contrasts with the Ankober lowlands which have open land policies.

Small farm size notwithstanding, the primary impact of the land shortage in the Tegulet project area and demographically similar zones is the constraint it places on livestock, especially draft animals. The best evidence for the long-standing nature of this shortage is the fact that Tegulet farmers regularly use cows as draft animals teamed with oxen, a practice unheard of in better-endowed areas. Pasturage in the area is almost non-existent and the forage which exists is apportioned to individual households and is protected jealously. Animals are either stall-fed with straw from the previous harvest or fed through 'lopping browse' on leaves and cactus roots. A donkey load of straw (about 1 month's supply) costs E$12. The 1984 crop failures reduced straw available as fodder in the dry season and reduced the oxen population further, creating severe feed shortages for those which survived. Moreover, failure of the 1985 *belg* (spring) rains and the lack of forage weakened surviving oxen for the key plowing period prior to the main summer planting.

Rural capital and the household economy

The Amhara-Tigrayan mixed-farming complex dominates the economy of the northeast highlands of Ethiopia. Most rural smallholders are sedentary, cereal-growing agriculturalists who supplement their diet and incomes by small-scale animal husbandry and trade. Muslim pastoralists dominate the lowlands to the east and trade livestock and their products for highland grain in a series of markets along the foothills of the escarpment which runs from northern Shoa north to Eritrea. Agriculture is based almost exclusively on the use of a pair of draft animals (usually oxen) pulling the *maresha*, a single-tine, locally made plow. Plow-based farming systems in the highlands are of long standing and have allowed the highland farming complex to expand dramatically over the last millenium.

In most ways, oxen function as rural capital, the possession of which allows the accumulation of labor, food and access to land (McCann, 1984*b*). Recent ecological and demographic trends have increased the importance of oxen in the distribution of rural wealth. Unlike land, the equitable distribution of which was decreed by the 1975 land reform proclamation, patterns of oxen ownership have been affected only by ecological constraints.

The impact of oxen on the local economies of northern regions is a function of relative scarcity and poor distribution among rural households. In general, areas most seriously affected by drought in the last few decades have, over the course of the past century and a half, been net oxen importers. The major reason has been the area's and the individual household's inability to reproduce its own oxen supply, principally because of the lack of sufficient pasturage and the fairly low rate of reproduction of highland breeds. The historical record shows a steady flow of cattle and oxen from southwest to northeast since at least the middle of the nineteenth century (McCann, 1984*b*).

The maldistribution of oxen within rural society, a trend exacerbated by drought conditions, indicates the extent to which oxen-poor households

Table 11.1 *Oxen distribution in selected peasant associations (PAs)[a]*

Number of oxen	Denki (Ankober)	Lalo (Ankober)	Segatna Jer[b] (Tegulet)	Debre Zeit	Debre Berhan	National[c]
(0)	40%	34%	18%	9%	31%	29%
(1)	40%	47%	55%	12%	38%	34%
(2 or more)	20%	19%	22%	79%	31%	37%

Gryseels *et al*., 1984; J. McCann, unpublished Oxfam America report.

[a]The first three figures are based on interviews with PA committees while the latter two are ILCA figures for the areas around their research stations, published in 1984.
[b]Figures of Segatna Jer are higher than others because they reflect a PA in the higher altitude zones. I would expect figures for lowland areas to be significantly lower for the latter two categories.
[c]The figure for the national average derives from a Ministry of Agriculture Study (1980), cited in Gryseels *et al*. (1984).

(having none or only one animal) must depend on wealthier or more fortunate neighbors. Table 11.1 summarizes the distribution of oxen within three peasant associations in the drought-affected area of Tegulet and Bulga as well as comparative statistics for ILCA research station environs at Debre Berhan and Debre Zeit. In the former cases they reflect conditions after failure of the 1984 short (*belg*) rains, the 1984 main season (*kiremt*) rains and, in the case of Segatna Jer, the failure of the 1985 short rains.

As these figures suggest, the distribution of draft animals is highly skewed in the project areas, with over 75% of all households having to seek capital from outside. Capital-poor households cope with this dilemma through a series of schemes which force them into situations of risk which in many cases result in permanent dependence. Institutions of capital borrowing and exchange differ within highland areas in terminology, terms of payment and frequency of application. There are several basic forms with a number of regional variations.

Oxen-sharing (makanajo or mallafagn)

The oxen sharing arrangement is between households that have one animal and agree to pair their animals. Plowing takes place on agreed upon days on each partner's land (either alternate days or blocks of days). Farmers who used this method paired with either neighbors or relatives (sometimes parents), though they expressed no clear preference for the type of relationship with the partner. Proximity was the most important criterion, though one of the interviewed lowland farmers paired with an in-law from a distant PA in 8-day blocks. In this arrangement no goods, labor, or cash changed hands nor were other social obligations incurred. The major productivity drawback to oxen-sharing is the reduced access each household has to oxen, during critical planting periods, when timing of oxen use is critical.

Oxen rental (qolo or melwawel)

In the oxen rental arrangement farmers rent an ox for an entire season in exchange for an agreed-upon amount of grain or, more rarely, cash. This institution is almost always used by farmers with no oxen to allow them to participate in a *makanajo* agreement. Most interviewed farmers rented the *qolo* ox as their primary animal and then paired with neighbors. In past years farmers paid 200 kg of grain (usually a combination of teff and sorghum) at harvest for the rental oxen. *Qolo* rates, however, vary with the market price of oxen (and less so with the price of grain). In June 1985, for example, anticipation of the rains and short supply had driven up the

price of oxen and also *qolo* rates. Farmers estimated 300–350 kg of grain
would be required. Similar rates are found at Debre Zeit, where productivity
is much higher.

In many cases *qolo* oxen are the property of merchants at Aliu Amba, a
local market town, who have bought oxen at depressed prices during the
drought, paid for fodder to keep them alive, and rented them to the same
farmers who had sold them. In other cases the *qolo* ox may be rented from
a local farmer who has an extra ox (called *ganja*) or one unable to use his
animal effectively because of illness or injury. In most cases, however, it is
those persons able to accumulate oxen during crises who in turn rent them
to farmers during the period when prices and demand have recovered.
Obviously, a further failure of the rains or low production not only deprives
poor farmers of their harvest but also places them further in debt. Therefore,
those farmers forced to obtain oxen through *qolo* are at great risk. The
economic effects of *qolo* arrangements are almost completely negative for
local farmers, because of increased risk of debt and the high prices which
accompany anticipation of better rains the next season.

Labor exchange (*balegn* or *menda*)

In the labor exchange system farmers with no oxen exchange their labor
with another farmer who plows their land at the rate of 3 days of human
labor for one day's plowing. More commonly, the farmer borrows only the
oxen and does his own plowing at the rate of 2 days human labor for 1 day
of oxen use. This system is very common in the Sela Dengay area, where
plots on steep slopes require extensive use of the *doma* (hand hoe). It is
less common in flatter, lowland areas. Farmers who depended on labor
exchange usually made such arrangements with relatives, even parents.
Many had subsisted for several years with this system with few prospects
of ever acquiring their own oxen. Obviously such a system reduces flexibility
in the timing of cultivation. Given the restriction in working days in
Christian areas like Sela Dengay, this system of cultivation has already had
serious effects on productivity and restriction of crop choice, since land
preparation before the critical sowing periods is related to both factors.

Charity

In a few cases, indigent farmers had their land plowed by relatives or
neighbors, as a gesture of charity. In these instances, however, the household
usually had only a token plot of land and the arrangement lasted only 1 or
2 years. Households in these circumstances are usually on the verge of
dispersal. Individuals in these households will make up, and already have
further north, the first wave of migrants to relief camps.

The above institutions and the figures on numbers of capital-dependent households suggest the extent to which other economic resources can flow to capital-rich households. In fact, a household with a pair of oxen can obtain labor (through *balegn*), food (through *qolo*), cash (through *qolo*) and access to land. Access to land, through the provision of oxen, was much more common before land reform, but in areas of land surplus, like Denki, families can give seasonal access to their land to those with oxen and labor in exchange for a portion of the produce (between 1/2 and 1/4). Though these forms of dependence have existed in northern farming systems for some time, the dependence they represent has been exacerbated by drought conditions which tend to concentrate oxen in the hands of wealthier, mature households.

The complexity of local arrangements for obtaining oxen, seed and land are demonstrated in one case involving Bedru Ahmed, a 32-year-old farmer with a young wife. Bedru, whose ox died last year, controlled some usually marginal land in a flood plain of the Denki River. With the failure of rains in the spring and summer of 1984, the fertile soil next to the stream was in some demand since it could be irrigated. Bedru himself was not in a position to cultivate it, though his family had some maize seed. Another man from Denki with no ox approached him and made an agreement to use the land during the short rains season in exchange for one-fourth of the produce. The man then rented one ox through a *qolo* agreement and paired it with another man's ox by *makanajo* with the agreement that they would share the harvest. Belg rains were good and the maize crop was the best I saw in the entire area. Though it was not harvested by the time I left the area, Bedru anticipated receiving back his seed plus one-fourth of the remainder in exchange for use of the land, while the other partners who provided labor and oxen expected to share the remainder with the rent of the *qolo* ox subtracted from one of the halves. Such arrangements in Denki appear to be common and acceptable and they would clearly favor those households able to provide seed and draft animals. Though Bedru was able to take advantage of controlling desirable land, he was at a distinct disadvantage because of his lack of oxen. Having obtained an ox, he is now in a position to put the land under cultivation himself (the soil is friable and the plot level; it is, therefore, suitable to the single-ox plow) and obtain 100% of the benefit.

Complex sets of capital-borrowing arrangements pre-date current drought conditions and reflect long-standing patterns of capital shortages that have underpinned rural relations of production in northern Ethiopia. Though individual household production equations shift over time, depending on the relative scarcity of factors of production in the short-term, long-term trends reflect changes in ecological and political conditions. The importance of institutions for capital exchange has developed over time, as the ratio of population to available draft animals has increased. The effects of this process were first felt in northern areas, such as Tigray and northern Wollo, but have reached areas of northern Shoa and newly cultivated lowland areas. Recent droughts have exacerbated the effects.

Distribution of the effects of famine

Available evidence strongly suggests that the distribution of the effects of famine closely parallels processes of stratification and economic vulnerability already present in northern Ethiopia's rural society. The effect is not chaotic, though the absence of rain and food shortages cause special problems. Who has access to what resources determines who migrates, which households fall into a permanent cycle of poverty, and what circumstances allow other households to survive in a strong position to re-enter agriculture. More importantly, a knowledge of the process of economic decline in food shortage periods should directly inform efforts to rehabilitate the rural economy in ways that benefit all sectors of the society.

My own observations in the Tegulet and Bulga area indicate that the effects of famine are distributed unevenly. Households with a pair of oxen, a short rain (*belg*) crop in the fields, and healthy children are found within 500 m of households with no livestock, malnourished children and heavy debts. The range of nutritional and economic well-being is astonishing, especially given the biblical imagery of universal starvation offered by media reports and television cameras. Major contrasts in the effects also appear between zones of different altitude, only 1 or 2 hours' walk apart. In May 1985 the green fields of maturing barley in the zones above 2500 m were in stark contrast to the dry, drought-stressed plantings below 2000 m. In the latter areas farmers, who had access to oxen and got a crop in, received very poor yields because of the low moisture available during maturation of the grain.

The evidence from interviewing farmers and visiting local markets makes it clear that seed, food and livestock still existed in these drought areas. In some cases this reflects the long-standing highland–lowland trade patterns which bring highland grain into interstitial markets. In other cases some farmers still retained food and seed from previous years or from debts owed by their neighbors and collected in the face of the drought crisis. Given this evidence, it is not possible to maintain that famine represents an 'absolute absence of food' or even the destruction of the rural economy.[2] How and why are these resources retained in the face of drought? Who manages to survive? Who needs outside intervention to keep them alive and on their land? In these circumstances the answers to the questions of who starves, who loses their ability to make it economically, and who prospers are a product of relationships between and within households. The impact of regional drought and food shortages on individual households is a result of a number of complex, but definable factors. A number of such factors emerged in the course of my interviews with drought-stressed farmers. They include those described below.

Placement of land

Some land, by virtue of soil type or location, is able to produce, despite rainfall shortages. In some cases land near permanent water supplies can be irrigated or fed by subsoil moisture. In other cases farmers retain land in other ecological zones where more moisture is available. Farmers in Segatna Jer PA, for example, maintained plots at altitudes differing by as much as 600 m (J. McCann, unpublished Oxfam America report).[3]

Phase in the household development cycle

Northern Ethiopian households exhibit a particular cycle of development which reflects the dominance of partible, ambilineal inheritance. There is no transgenerational family or estate. Therefore, each unit has a distinct creation phase (marriage), expansion phase (birth of children and acquiring of extra members as servants, herders, etc.), and dispersal (departure of mature offspring, divorce, or death of head) (Bauer, 1977). Based on my preliminary interviews, it seems that households with a high ratio of consumers to workers (e.g. households in the creation phase or early expansion with young children and no extra laborers) make up the group most seriously affected by drought. Newly established households are especially vulnerable, since they generally have not been able to establish livestock herds or seed stocks. Households entering the dispersal phase, with mature offspring removing their labor, also appear to be in a weak position, since marriage of offspring depletes household livestock supplies, grain supplies and skilled labor. Often, in this stage, parents are reaching the age when they are unable to work effectively. Households in the best position are those in the expansion cycle which had sufficient time to accumulate livestock, obtain labor from adolescent offspring, and acquire well-situated land.

It is interesting to note that peasant associations, when asked to compile a list of oxenless households to receive oxen from the project, chose a very high percentage of young households early in the expansion phase (J. McCann, unpublished Oxfam America report). Based on lists prepared originally by PA committees, young households were preferred over larger households or those close to the dispersal phase.

Access to forage

Given the importance of maintaining healthy, well-nourished oxen, households must have access to pasture and supplemental feed (usually straw from the past season's harvest) for their working animals. Forage in much of drought-prone northeast Ethiopia is not a free good and, often, the first effect of drought is the reduction of available forage. In many areas, such as Lasta and other areas with high population pressure, pasturage is apportioned by households in the same manner as is arable land. Not all households receive such land. Supplies of forage are scarce in wet periods, but in drought periods they are non-existent. Several restrictions on production result: first, oxen are weakened and tire easily; second, weak animals succumb to parasites and disease; finally, household labor is required to hand-feed animals by 'lopping browse' with leaves, young cactus shoots, or (in extreme cases) the roofs of dwellings. Households with cash resources can usually purchase straw or hay from less-affected areas or support a herder to feed their animals, but others face pressure to sell or go into debt to obtain food. Many households are unable to support their animals and fall into capital dependency, which means that they have to borrow or rent oxen.

Access to outside income

Resources in drought-affected areas are available to those with cash. Markets at Aliu Amba and Sasit (for the two project areas) offer grain, seed, fodder and cattle. Regional markets, such as Dase or Debre Berhan, draw grain from the national market. These, however, are well beyond the resources of households without incomes independent of agriculture. Some households have access to cash through craft production, especially weaving in lowland areas (though prices decrease during periods of stress), cash remittances from working relatives, and office-holding. The latter category includes PA officials who receive per diem cash payments when on association business and clergy who receive cash or in-kind payments for ecclesiastical services.[4]

All of these conditions, and others, make up what Sen (1981) might call a 'bundle of endowment and exchange entitlements' which allow particular households to claim scarce resources. No ideal combination exists, but in the course of research it was clear that some households have some or all of these factors and will survive while others which have only some or none may not. Rehabilitation programs which address a range of these needs are likely to maintain themselves within the rural economy.

Risk and enrichment in drought-stressed areas

The ability of households to survive depends largely on their success at minimizing risk, a difficult task during drought. A corollary to this proposition is that the ability of some households to resist the effects of drought and food shortage can in those circumstances transmogrify into a process of enrichment, through the use of existing social institutions. The ownership and labor power of oxen play a major role in this process, where rural capital (i.e. oxen) is concentrated in the hands of particular households which therefore have a strong relative advantage. The process has as many forms as there are households, but relies essentially on their institutions of capital borrowing and transfer described earlier.

On the basis of life histories of farmers who participated in the oxen/seed project, I would suggest the following pattern. The recurrence of drought brings about an overall reduction in the numbers of livestock through disease and sale for the purchase of food, and the consequent drop in prices accelerates the process (Sen, 1981; Cutler, 1984). Though oxen are usually the last resource sold, many of the farmers in the project area had sold their oxen in 1985 to cover debts incurred after the failure of the 1984 *kiremt* rains. Though some animals reach regional markets, my interviews indicate that most oxen were purchased locally by farmers and merchants with access to feed. The younger and healthier of these animals never found their way to district slaughterhouses, remaining within the local economy to allow their new owners to take advantage of the expected rise in the price cycle when anticipation of June rains drove prices to higher levels. In the case of the Ankober area oxen (which were sold in late 1984 and early 1985 at around E\$150, but as low as E\$70 in some areas) sold in June 1985 at E\$350 to farmers willing to risk debt in order to have an animal for summer planting (J. McCann, unpublished Oxfam America report). Spring and early summer oxen sales and rental rates do not reflect grain prices but, rather, indicate the shortage of oxen and farmers' expectations of a return to adequate rains. The oxen supply is short partly because of the earlier sales and deaths but also because farmers are unwilling to sell just prior to planting season. Rental rates also reflect this price rise. Rates of 200 kg of grain per season's use of an ox are common, but such a payment could only be sustained under ideal growing conditions. Farmers who commit themselves to such terms, therefore, take a major risk of default.

The effects of capital concentration go beyond increased economic risk to include a negative impact on crop choice and overall productivity. Drought conditions mean that greater numbers of farmers depend on borrowed or rented oxen (or both!). Most borrowing arrangements (including those who rent their first ox) restrict plowing time during the key plowing days after the first rains and at other critical times in the agricultural cycle, since oxen owners prefer using their animals on their own land at peak season. Effects on productivity can be immediate. Farming systems research in Ethiopia and elsewhere indicates that households which borrow draft animals plant less land than those which own them (Cossins, 1975). In the same vein, restricted access to oxen affects the method of plowing, since to cover more land farmers must restrict the number of passes made on a particular plot. Fewer passes result in either lower productivity or necessitate a different crop choice. Teff, for example, has the highest income potential but requires more land preparation (up to 5 passes). Horsebeans, on the other hand, require only one pass but are less productive in terms of yield and price than any of the cereals. My recent interviews on cropping choice indicate that those who depended on borrowed oxen, especially those practicing labor exchange (*balegn*) planted a significantly higher proportion of pulses over cereals than those who had oxen. The effects of a large population dependent on borrowed oxen is therefore significant, even under favorable climatic conditions.

Increased indebtedness and loss of productivity have a number of effects, including stimulating short-term as well as permanent migration. It should be noted, however, that large-scale migration from the northern highlands is not a new phenomenon. Over the course of the last 4–5 decades, thousands of highland Ethiopians have migrated annually to seek wage labor in Sudan, Italian Eritrea and, more recently, to irrigation schemes in the Awash valley (McCann, 1986). These patterns reflect less the effects of drought than the lack of opportunity in the highland economy and the gradual impoverization of the rural population. More important for the long-term stability of the rural economy of drought-prone areas is the impact on those who return or stay behind. With adequate rains the normative cycle of household growth expected by farmers would allow dependent households gradually to re-acquire livestock and rebuild supplies of food and seed. Though expected, available evidence suggests this trend is unlikely for a number of reasons. First, rainfall in lowland areas will continue to be variable, thus constraining long-term growth of the livestock herds of sedentary populations. In highland areas population pressure has already reduced available pasturage in such a way as to limit the potential for sufficient increases in livestock to allow universal ownership of a pair. Second, my data on the effects of reduced productivity indicate that households dependent on borrowing may never achieve independence in capital. The life histories that I collected in the Tegulet area strongly suggest that farmers who practiced *balegn* (labor/oxen exchange) rarely, if ever, were able to obtain a pair of oxen even in periods of adequate rainfall. The dependence created in the recent drought, therefore, is likely to continue and to make the majority of the rural population increasingly vulnerable to any type of production shortfall.

Given existing institutions and the skewed distribution of capital in the project areas, what is the likely social impact of the distribution of over 500 oxen, seed, and single-ox plows to previously oxenless households? What effects will it have on income, risk, and productivity? Potentially, the single-ox plow itself extends the advantages of draft animal self-sufficiency to the large group of households with only one ox. The effect is two-fold. On the one hand the single-ox household is released from dependence on other households that it incurred through sharing or rental agreements which increase risk or reduce net income. It may also increase productivity, since the timing of planting, especially in highland vertisol areas, is critical to yield. ILCA field tests indicate that single-ox plows can cover 70% of the area covered by a pair (Gryseels, *et al.*, 1984). Even without the use of the new technology, the provision of an ox to an oxenless household will allow it to enter sharing rather than rental agreements thus reducing overall demand and possibly reducing rental rates.

On the other hand, single-ox plow technology should allow a single-ox household to lend its draft capacity to others as a means of obtaining labor or produce (i.e. through labor exchange). While the actual practice has not begun, one could guess that the rental rates for the single ox with plow would be less than for the pair, thus reducing the potential debt of the remaining oxenless households.

The implications of the above argument should affect the types of rehabilitation programs designed to protect populations from drought vulnerability. Afforestation, terracing, and other standard food-for-work programs do not address the issue of capital distribution nor do they provide for regeneration of household productive capacity. Increasing the numbers of oxen and otherwise reducing capital dependency, on the other hand, will immediately affect the rural population's ability to survive the vagaries of nature. The Oxfam America program to introduce the ILCA-designed single-ox plow not only has the potential to increase the total numbers of oxen available locally, thereby driving down opportunity costs for obtaining capital, but will also decrease capital dependency under existing conditions by making the single-ox household self-sufficient in cultivation capacity. Increased options in crop choice and improved productivity should improve significantly the ability of marginal households to resist drought.

Intra-household effects of famine and famine aid

Just as the effects of food crises are not distributed equally among households in Ethiopia, effects differ *within* the household based on such factors as age, social status, and gender. Information on intra-household relations during famine is scarce, consisting primarily of a few sets of interviews with migrants conducted during the 1972–74 famine in the northeast (Wood, 1976; Cutler, 1984). Nevertheless, analysis of intra-household relations is essential to the design of rehabilitation projects as well as to shorter-term relief efforts which attempt to distribute aid at the household level. The evidence for project areas, and probably northern Ethiopia as a whole, is that distribution of resources at the household level is not necessarily an effective means of providing resources to all members of the household.

Northern Ethiopian households, including those in the project areas, are relatively unstable over time. They are rarely extended family units and, since they are not transgenerational estates, they dissolve fairly easily. The laws of property which affect marriage, divorce and inheritance result in unequal distribution of productive resources, an effect which increases in times of economic stress. Moreover, since the rate of divorce is normally high, women average between 3 and 4 marriages and males between 5 and 6 over the course of a lifetime (one farmer in the interview sample had married 10 times in 10 years). The welfare of the household, therefore, does not effectively indicate the welfare of its constituent members over time. The relative vulnerability of household members, especially women and young children (the primary inhabitants of relief camps), can be understood by examining the transfer of capital through institutions of property, particularly marriage and inheritance. Since the oxen/seed project operated on the basis of large capital inputs distributed under the names of male heads of household these questions are of prime importance to overall social impact.

Marriage and divorce

Marriage practices appear to be fairly uniform in the areas considered here.[5] Since property contributed at the time of the first marriage constitutes the 'capitalization' of that household, the circumstances of the first marriage are good indicators of a household's potential. The dominating principle of Christian marriage is the equality of contributions from both sides and the notion that the unit is expected to establish itself as an independent production unit in land, capital and labor. In fact, none of the households I observed were self-sufficient in oxen within the first 2–3 years of marriage. Moreover, only a small percentage of wives brought any livestock to the union. This pattern is especially true in the lowlands, where recent migrants were predominately younger sons from highland households with no claim to land or livestock. In other areas the overall shortage of oxen made it impossible for parents to endow their children, especially female offspring, with any livestock without dissipating their own resources. In the vast majority of all cases interviewed, it was the male who brought livestock to the marriage.

The simple rule of property distribution upon divorce is that spouses retain what they brought to the marriage. Under these conditions, men invariably retained oxen since they either brought them to the union, purchased them through grain sales (production of grain is a male domain), or obtained them as progeny of cows they had contributed (Weissleder, 1965; J. McCann, unpublished Oxfam America report). At the same time, women usually retain custody of young children and are responsible for their welfare. Though women in theory retained rights to half the household land, their actual claim is poor, because they lack the means to cultivate it (since it requires both oxen and mature male labor), and usually return to their parents, remarry quickly, or adopt some form of income generation (prostitution, beer-brewing, cotton spinning). Conventions of marriage call for a woman to join her husband on his homestead and thus women have weaker kinship and political networks in the immediate area. Therefore, divorced and widowed women make up a class of mobile individuals, moving between PAs. One PA chairman named registration of divorced women as his most tedious task.

Women who control livestock, usually those from wealthy backgrounds, are in a stronger position. Their ability to initiate divorce and thus deprive the household of important assets provides some leverage within the relationship. When they divorce, they are able to establish an independent household since they are able, for example, to hire male labor to cultivate their land. In other cases older children take on that role. The percentage of such female-headed households is small and they are generally short-lived but they demonstrate the importance of capital ownership. In the vast majority of cases, however, divorced women lack the necessary assets to manage an independent household. In my sample of PAs there was invariably a certain percentage of indigent, non-tax-paying households headed by widows and divorcees who subsisted on charity and small craft production. In one Ankober region peasant association, 15% of the households registered were indigent, female-headed units. For these reasons Oxfam has recommended to the oxen/seed project that women be incorporated into the project as direct beneficiaries by receiving oxen (or even cows in Tegulet) registered in their names (J. McCann, unpublished Oxfam America report).

Obviously, women's lack of access to oxen and other resources is a major constraint on their welfare and that weakness makes them particularly vulnerable to the effects of extended drought or famine. Preliminary evidence also suggests that divorce increases during economic stress. Even the normal rate of divorce would put women in a tenuous position to remarry or to support themselves in a female-headed household. Since relief-food distribution relies on PA registration lists, women who leave a marriage in one area fail to be registered in their new area and can, therefore, be overlooked during distribution (J. McCann, unpublished Oxfam America report).

Inheritance

Normative patterns of inheritance in the areas discussed here generally conform to the principles of cognatic (ambilineal) descent. In practice, however, capital resources and land are not distributed equally among offspring. Given the overwhelming evidence that women in these areas did not receive property at marriage, it appears that males are the main inheritors. To avoid sub-division of holdings, parents usually pass on land and livestock to a favored son. The evidence from Ankober further suggests that even many male offspring receive no resources, either through inheritance or marriage endowment.

Conclusion

The imagery of tragedy and human suffering which tends to dominate our view of famine conditions in northern Ethiopia conveys the impression of a world gone mad, obscuring the fact that peasant response to prolonged drought and famine is ordered and even predictable. Institutions which functioned to distribute economic resources before food shortages continue to do so but with somewhat different effects. Drought reduces supplies of capital (oxen and seed) which in turn lends increased manipulative powers to households which control the scarce remainder. The results are increased risk for farmers who must borrow oxen and seed for the next planting and a serious debt crisis in the event that rains (or locusts, army worms, etc.) reduce crop yields for the next harvest.

The importance of oxen in this equation has not generally been appreciated, since they are usually seen mainly in terms of trade between grain and livestock (Sen, 1981; Cutler, 1984). In fact, the key institutions affecting household access to resources and overall economic viability are those directly involving the exchange of oxen and their labor. These exchanges are an extension of the types of interaction which take place under any climatic conditions. The relative scarcity of oxen over other factors of production is exacerbated by drought but is not the product of it.

A final conclusion from the findings of this chapter concerns the distribution of the impacts of a food crisis as well as of the benefits of relief and rehabilitation. The reconstruction of the rural economy in northern Ethiopia must somehow link the provision of short-term relief with the rebuilding of productive capacity at the household level. Obviously, the latter requires an understanding of the terms under which production and distribution of resources take place. Rehabilitation of productive capacity must, however, go beyond the short-term effects of drought to address longer-term trends which have reduced the carrying capacity of the land. In the case of northern Ethiopia this is expressed in terms of shortages of oxen at the regional and household levels as well as the more traditional questions of afforestation, terracing, micro-dams, and others emphasized in standard food-for-work programs. In this case the introduction of a new technology, the single-ox plow, and the distribution of draft animals seek to solve short-term and long-term needs in a way that will reduce debt and risk within production units.

Rehabilitation should also be concerned with the overall apportionment of benefits through expanding entitlements. Social institutions such as marriage, inheritance, divorce, and others, which govern the creation and dispersal of rural households, also determine the nature of resource distribution. In most cases they allocate key resources to the control of adult distribution within households. The *de facto* dispersal of household resources, including relief and rehabilitation aid, through these institutions has routinely benefited adult males over all others and created a class of vulnerable individuals with little ability to command food or economic resources. Only when these factors become an integral part of planning will a progam of rehabilitation be successful in allowing rural producers to participate fully in a viable rural economy with increased food security.

Notes

I am grateful to Oxfam America, the Staff Council of the International Livestock Centre for Africa (ILCA), and the Integrated Drought Relief Committee of Tegulet and Bulga awraja for their assistance in carrying out recent fieldwork in Ethiopia, as part of the evaluation of their joint oxen/seed project in northern Shoa administrative region. Work on Lasta was carried out in Ethiopia and Europe from 1980–82 with the support of the Social Science Research Council and the Fulbright-Hays Doctoral Dissertation Abroad Program. The opinions expressed in the paper are my own and not necessarily those of the above agencies.

1.

My agreement is with Sen's notion that the effects of food shortages are weighted by a 'bundle' of entitlement rights which give some persons better access to existing food and resources (Sen, 1981). I, therefore, disagree with Mesfin Wolde Mariam's notion of 'absolute' lack of food (see Wolde Mariam, 1984). Mesfin's empirical contribution addresses questions of spatial distribution of drought, rather than social impact. The areas I worked in clearly conformed to Sen's model. I agree with Cutler (1984), however, in rejecting the validity of Sen's price data from Dase since that town is tied to a national market and is peripheral to the 1972–74 famine areas.

2.

Again, this evidence belies Mesfin Wolde Mariam's notion of 'absolute' absence of food and generally contradicts the impressions created by media coverage of relief camps. The effects of famine are socially and politically distributed (Hancock, 1985, 79–87).

3.

Recent research at ILCA argues for moisture availability to be calculated by precipitation plus moisture stored in the soil and potential evapotranspiration of the crop. Consequently, soil type, depth and location of plots can have an effect on yield, even in areas receiving the same rainfall (see LaRue, 1984, 2–3; Henricksen & Durkin, 1985, 2–9).

4.

This information comes from my interviews with three peasant association chairmen. Ato Abdullah Sewmeweded of Denki PA, for example, received E$4 ($2) per day when on PA business.

5.

There are two basic differences within areas affected by drought. Muslim marriage does not require donations on behalf of the wife (which resembles the actual practice in most Christian marriages). In Tigray a dowry, rather than equal donations from both sides, is the rule. Practices in all areas of Tegulet and Bulga appear similar in practice, if not in theory.

References

Bauer, Dan F. (1977). *Household and Society in Ethiopia*. East Lansing: African Studies Center, Michigan State University.

Cossins, Noel (1975). *The Day of the Poor Man*. Addis Ababa: Drought Relief and Rehabilitation Committee.

Cutler, Peter. (1984) Famine forecasting: prices and peasant behavior in northern Ethiopia. *Disasters* (August 1984). 48–56.

Gryseels, G., Astatke, A., Anderson, F. M. & Assemenew, G. (1984). The use of single oxen for crop cultivation in Ethiopia. *ILCA Bulletin*, **18**, 20–5.

Hancock, Graham (1985). *Ethiopia: The Challenge of Hunger*. London: Victor Gollancz.

Henricksen, B. L. & Durkin, J. W. (1985). Moisture availability, cropping period and the prospects for early warning of famine in Ethiopia. *ILCA Bulletin*, **21**, 2–9.

LaRue, G. Michael (1984). Land and social stratification in Dar Fur, 1785–1875. *Boston University African Studies Center Working Paper*, No. 96.

McCann, J. (1984a). Households, peasants and rural history in Lasta, northern Ethiopia 1900–35. PhD thesis, Michigan State University.

McCann, J. (1984b). Plows, oxen and household managers: a reconsideration of the land paradigm in northern Ethiopia. *Boston University African Studies Working Paper*, No. 95.

McCann, J. (1986). Household, demography and the push factor in northern Ethiopian history. *Review*, **9** (3).

Messerli, B. & Aerni, K. (eds). (1978). Simen Mountains-Ethiopia. Vol. 1: *Cartography and its Application for Geographical and Ecological Problems*. Bern: Geographische Institut der Universtät Bern.

Sen, Amartya (1981). *Poverty and Famines: An Essay on Entitlement and Deprivation*. Oxford: Oxford University Press.

Weissleder, W. (1965). The political ecology of Amhara domination. PhD thesis, University of Chicago.

Wolde Mariam, Mesfin (1984). *Rural Vunerability to Famine in Ethiopia, 1957–77*. New Delhi: Vikas Publishing House.

Wood, A. P. (1976). Farmer's responses to drought in Ethiopia. In *Rehab: Drought and Famine in Ethiopia*, African Environmental Report No. 2. London: International Africa Institute.

12

Role of government in combatting food shortages: lessons from Kenya 1984–85

JOHN M. COHEN* and DAVID B. LEWIS†
*Harvard Institute for International Development, †Cornell University

Up to now, individual nations . . . have dealt with the specter of mass starvation as an unexpected crisis – as something to react to when it occurs rather than as a likelihood to be planned for in advance. Prevention has been the exception rather than the rule.
Jean Mayer (1974, 98)

International agency reports, proceedings of professional conferences, and studies by experts consistently urge nations in drought-prone areas to develop institutional structures to cope with food shortages. Typically, it is recommended that governments prepare a National Food Security Strategy. This usually involves establishment of a national Food Security Agency and a Famine Code to guide its operations. Proponents of permanent government preparedness argue that, despite the emphasis in the recent development literature on the role of the private sector, it is the responsibility of governments to ensure that food shortages are anticipated and responded to effectively and efficiently. The argument is made most strongly for African countries where the majority of the population is rural and poor, the key sector in the economy is agriculture, and the government has assumed primary responsibility for promoting development.

In practice, however, it is neither easy for governments to establish administrative systems to respond to threats that might lead to famine, nor is it certain that it is essential. Most recommendations for drought-prone countries emerge from retrospective analysis of famine situations by technical experts (typically medical doctors, nutrition specialists, or relief workers with non-governmental organizations) who have limited knowledge of the broad economic and political context in which their prescriptions must operate. As a result, recommendations tend to be proposed in an *ad hoc* manner making it difficult to assess how, when taken together, they will impact on a government's administrative and budgeting capacity. Were a government to adopt the full set of recommendations, the financial and administrative costs of doing so would likely be unsustainable during non-crisis periods.

This chapter addresses the issues that government planners must face in formulating an effective strategy for anticipating and responding to the periodic threat of large scale food shortages. First, it reviews the problems involved in seeking to draw on international experience. Then, the essence of the recommendations most commonly advocated is set forth. To test the utility or necessity of following such recommendations, the chapter reviews the actions of the Government of Kenya during and after the 1984 drought. Kenya was selected because it offers an encouraging example of what an African government is capable of achieving. This is particularly important, given the horrifying press reports on famine in Africa at that time, and the images of governmental incompetence they generate (Ross, 1983; Watson, *et al.*, 1984). It is also a case in keeping with methodological trends toward avoiding the stultifying effects of 'negative social science' (Chambers, 1983) by recognizing the utility of analyzing success stories (Paul, 1982). This review reveals several disadvantages inherent in the international recommendations for formal planning and capacity-building. An analysis of these points is undertaken in the concluding section of the chapter.

Finding the lessons of experience

Government planners seeking coherent advice to guide their efforts to prepare for food shortage crises, particularly in Africa, will find it difficult to come by, despite the substantial literature of potential utility to their concerns.

Many reports, articles, and books have been published on the topics of agricultural development strategy, food policy and international food aid in Africa (Eicher & Baker, 1982). In addition, there are many studies on specific African famines (Shepherd, 1975; Hussein, 1976; Dalby, Harrison Church & Bezzaz, 1977) and reports on the administration of particular relief programs (Kenya, 1962; Sheets & Morris, 1974; Bryceson, 1981). There is also a relatively large literature set in the African context that focuses on such topics as the relationship between environmental degradation and famine (Hardin, 1968; Wade, 1974; Picardi & Seifert, 1977), effects of international capitalism on food production (Perelman, 1977; Lappé & Collins, 1979; Franke & Chasin, 1980), and the political causes of starvation (Lofchie, 1975; Ball, 1976). More focused studies are typically confined to health, nutrition and migration topics (Newman, 1975; Seaman, Holt & Rivers, 1978). Behind this literature is an even larger body of materials that cover such topics throughout the world (Ball, 1981; Sen, 1981). However, despite this relative wealth of descriptive and analytical material, governments will have to look long and hard to find specific advice on what steps they should take to anticipate, identify and respond to temporary food shortages.

Planners can begin with the famine handbooks published by the British colonial service for India or other countries deemed to be threatened by drought, or they can turn to a handful of specific papers outlining lessons learned from previous famines. These will be hard to find and incomplete in describing what a government should do, and of limited relevance to the African context. Only a few studies stand out for their attempts to restate basic steps governments can take to deal with food shortages; the most notable is G.B. Masefield's *Famine: Its Prevention and Relief* (Masefield, 1963). It, however, is dated and does not cover the full scope of possible government action. Other papers, such as those by Jean Mayer (Mayer, 1974, 1975), are more recent but too general, perhaps because the authors have, in most cases, worked for international agencies rather than governments. Hence, in the end there is no alternative for planners but to sift through the literature to identify experience-based suggested activities their governments can, with adjustment, carry out (Blix, Hofvander & Valquist, 1971; Aykroyd, 1974; Jodha, 1975; Morris, 1975; Glantz, 1976; Manetsch, 1977; Hay, 1978; Robson, 1981; Cahil, 1982; Garcia & Escudero, 1982).

International recommendations

Food shortages are typically caused by crop failure resulting from combinations of drought, plant disease, pests, natural disaster, civil disturbance, or war. Governments committed to denying famine a future can take steps to address each of these causes. Droughts and natural disasters can be anticipated and planned for, investment in agricultural research on plant diseases and pests can be made, political disputes can be settled before they erupt into widespread fighting that disrupts farming, and policy choices can be made not to use food as a weapon, should war or civil disturbance take place in an area marked by crop failure. However, history suggests that African governments have frequently followed inappropriate food policies, given agricultural research low priority, been unable to settle seperatist movements and, as in Biafra and northern Ethiopia, used crop failure as a military weapon.

Given the possible scope for government action, and the politics that complicate such action, most international agencies and experts have confined their recommendations to strategies focused on planning, administering, and financing major food security and relief/rehabilitation efforts. Larger political issues surrounding development policy and conflict resolution are typically ignored or left to more venturesome academic studies (Bates, 1981). Significantly, the recommendations commonly found in the literature eschew consideration of the administrative and financial capacity of governments to carry them out. These capacities are limited (World Bank, 1981; Hyden, 1983).

The general principles and recommendations drawn from the publications of international agencies and conferences is summarized below. The literature was first reviewed to establish its content. Then, recurrent themes were organized into the four major categories which emerged from this analysis.

I. Establish institutional structures responsible for food security.

   1. Establish a permanent national Food Security Agency (FSA) responsible for detecting crop failures as early as possible, monitoring the emergence of food shortages, planning and administering relief operations, and overseeing rehabilitation efforts.

   2. Establish a National Committee for Food Security (NCFS) composed of senior officers of operating ministries as an advisory body to which the FSA can report on its activities and from which it can receive evaluative review.

   3. Charge the FSA with producing a comprehensive review of all data and experience related to anticipating and responding to food shortages.

   4. Charge the NCFS with submitting to the government a draft Famine Code (to outline for national and local officials the activities to be undertaken by the government). Among the topics it should cover are public security provisions, price and market controls, feeding and health care station operating rules, distribution procedures and entitlement priorities, disease prevention guidelines, water supply and protection, food-for-work activities, and monitoring/evaluation requirements.

   5. Review, revise, issue and publicize the Code as an executive order or as a legislative document.

   6. Charge the FSA with undertaking appropriate activities in support of the Code, including research, training, and simulation activities.

II. Carry out hunger-prevention activities during non-crisis periods.

   7. Using the FSA to coordinate and promote its policy objectives, the governments should charge appropriate ministries and bodies with formulating and publishing policies and plans aimed at increasing food production and nutritional security, including an integrated set of national food, family planning, land use and public works policy documents.

8. While regarding outside aid as supplementary, the government should request responsible ministries and agencies to promote international food security programs and establish relationships based on contingency plans with donors likely to provide food and relief assistance should production shortfalls occur.

9. Charge the FSA with working with specialized ministries to establish and continually improve an early warning system based on improved weather forecasting, satellite surveillance and monthly crop production reports from field officers.

10. Charge the FSA to work with government agencies to establish methodologies and decision rules governing the financial and administrative decisions that must be taken in times of food shortage.

11. Working with the FSA, charge responsible government units with formulating policies, programs, and projects that enhance national and local capacity to respond to food shortfalls, giving particular attention to agricultural research on drought resistant crops, soil and water conservation, on-farm storage, special feeding formulas, transport and off-loading facilities, and emergency medical supply depots.

12. Establish capacity to locate, order, and import food rapidly when needed to respond to national shortfalls in production.

13. Have the FSA send famine assessment team(s) to any area of potential food shortage identified by the early warning system to evaluate needs and prepare a plan of action.

III. Ensure rapid and effective response during crisis periods.

14. Appoint a senior government official to direct relief and rehabilitation operations.

15. Charge the NCFS with ensuring that key government decision makers and officials are immediately aware of the crisis and keep them informed throughout it.

16. Require the FSA to establish a crisis information and monitoring control center.

17. If the crisis involves only aggregate national shortages and no omens of potential famine yet exist, charge the FSA and NCFS with assuring that local food resources are used to maximum extent possible, adequate supply is obtained in international markets, or from aid agencies, on a timely basis, and entitlement programs ensure fair access for lower income groups.

18. If the crisis escalates to involve famine in particular areas of the

country, then charge the FSA and NCFS with carrying out actions in accordance with the Famine Code.

IV. Ensure rapid and effective post-crisis rehabilitation.

    19. Prepare and implement an agriculture rehabilitation strategy for the famine area, based on evaluation of drought's effects, supply of agricultural inputs, ecological repair programs, restoration of basic human services, and re-establishment of commercial market mechanisms.

Case study of Kenya: 1984–5

Alan Berg argues that the Indian government and international community's actions in the 1965–66 Bihar famine 'constitutes the first time in modern history that a government declared war on a large-scale famine – and won' (Berg, 1971). The 1984–85 performance in Kenya went one step further, when the government prevented a potentially catastrophic famine from beginning. Significantly, it did so without the kinds of permanent structures experts typically call for.

Kenya was badly hit by drought in early 1984. It was the worst shortage of rain in the last 100 years. Production of maize, the nation's principal food crop, was approximately 50% below that normally expected for the main rains of March–May. Wheat, the second most important grain, was nearly 70% below normal. Potato production was down by more than 70%. Pastoralists reported losing up to 70% of their stock. The situation had the potential for a famine of major proportions. The disaster, however, never developed. Prompt, effective action by the government ensured that adequate supplies of food were available during the entire drought and its aftermath. While there were acknowledged problems of local distribution that resulted in genuine hardship in some isolated areas, famine was averted.

Overview of Kenya

With a population of nearly 19 million people and a land area of 582 646 square kilometers, slightly larger than France, Kenya has a reputation for having one of the most dynamic economies in sub-Saharan Africa (Kaplan, 1976; Leys, 1975; Hayer, Maitha & Senga, 1976). Gross domestic product is approximately US$390 per capita per year.

Independent since 1963, Kenya has enjoyed a remarkable period of stability and economic growth. The nation's administrative system is based on the concept of a one-party state with a strong presidency and a popularly elected parliament. There are 41 districts organized into seven provinces.

Each district is adminstered by an appointed District Commissioner who reports to a Provincial Commissioner, who in turn reports to the Office of the President. The economy is mixed, with a vigorous private sector operating in parallel with a substantial public sector based on parastatal corporations. Coffee and tea exports and tourism are the leading earners of foreign exchange.

The nation's natural resources are limited. There is no domestic petroleum production, and approximately 80% of the land area is arid or semiarid. Geographically the country is located at the southeastern end of the arid zone that sweeps across the continent from the West African Sahel to Ethiopia. The nation's agricultural food production is roughly equal to its needs; in a good year it can export maize, but in a bad year imports are

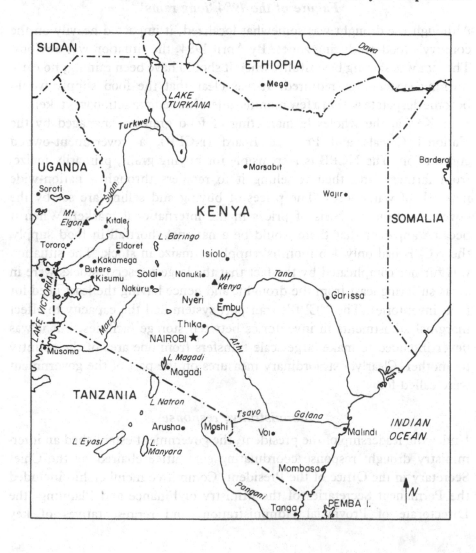

required. Approximately 85% of the population lives in rural areas, primarily on the 20% of the land that has enough rainfall to support agriculture.

The annual agricultural cycle revolves around the normally reliable seasonal rains. The 'long rains', which produce most of the annual precipitation, occur from mid-March through mid-May. The 'short rains' follow from late October to early December. There are local variations but this is the dominant pattern. In food-producing areas, the 'long rains' are typically used to raise maize or wheat, and the 'short rains' are used for other crops with a shorter growing season. The amount of irrigated agricultural land is relatively minor by comparison with the rain-fed crop areas.

Failure of the 1984 'long rains'

Although the drought was somewhat localized, it impacted heavily on the country's food-producing zones. By April 1984, the situation was obvious. The sun was shining beautifully, when it should have been raining; no early warning system was required. It was clear that the food supply would become desperate within a few months unless immediate action was taken.

In Kenya, the wholesale marketing of food grains is managed by the National Cereals and Produce Board (NCPB), a government-owned corporation. The NCPB is responsible for buying grain, primarily maize, from farmers, and then re-selling it to retailers through a nation-wide network of warehouses. The prices of buying and selling are set by the government on the basis of prices in the international market. When it became apparent that there would be a national shortfall in food supply, the NCPB had only 4–6 months' supply of maize in stock. The situation was further complicated by the fact that the largest reserve stocks were in areas suffering least from the drought, and hence having the least need for food inventories. The NCPB's transport system had the capacity to effect marginal adjustments in inventories between storage facilities, but it was never intended to make large-scale transfers from one area of the country to another. Clearly, extraordinary measures on the part of the government were called for.

Administrative response

Under the leadership of the president, the government established an inter-ministry drought response coordinating committee chaired by the Chief Secretary in the Office of the President. Committee membership included the Permanent Secretaries of the Ministry of Finance and Planning, the Directorate of Provincial Administration, and representatives of key operating ministries. Once created, the committee moved swiftly to assess the situation, establish the government response policy, begin commercial imports of food, negotiate with the donor community for food assistance, and establish a task force to manage the food import and distribution program.

A highly productive working relationship emerged between the Planning Department of the Ministry of Finance and Planning and the Office of the President. The Planning Department had strong analytical capabilities and access to much of the background data required for assessing food requirements. It also had the capacity to develop the management information systems needed by the task force. The drought response effort was given top priority, and Planning Department staff who had had no previous experience with this type of work suddenly found themselves putting in 12-hour days. Analytical capability tended to preempt official formality. Junior clerks who operated microprocessors found themselves with immediate access to the Director of Planning. Productivity for those involved reached unprecedented levels.

In response to the flood of requests for information, the Planning Department initiated a program of high-level weekly briefings for the international donor organizations. These served to keep the donors up-to-date on progress with the drought-response program, thus reducing the problems resulting from action taken on misinformation. It also demonstrated to the donors that the Kenyan government was well informed and firmly in control of the coordination of all the drought-response activities.

Establishment of drought-response policy

The government decided early on that the drought would be dealt with as a serious problem requiring immediate and high-level attention, but it would not be considered a crisis. The normal administrative systems would be used, and if additional logistic capacity were needed, the private sector would be employed on a regular commercial contract basis. The military would be used for logistic support only as a measure of last resort. (It was never needed.) The drought-response program was to be managed by Kenyans; there would be no call for expatriate crisis-management advisors.

The government's low profile policy served many purposes. It contributed to keeping the level of public concern in check, and thus helped avoid most of the potential hoarding and public security problems. By using existing administrative systems, it helped to minimize the total cost of the drought response. Most importantly, however, it provided a conceptual frame of reference within which everyone could work. There were to be no crises-motivated *ad hoc* programs.

Importation of food

The decision to import food was inevitable; the shortfall in production was too great to be bridged by any available domestic resources. As a matter of national policy, the NCPB seeks to maintain sufficient grain buffer stocks to ensure adequate supply even in years when there is a shortfall in production, but the shortfall in 1984 far exceeded those reserves. The critical question was not whether to import food, but how much to import and how to import it in an efficient manageable manner.

It was clear from the early assessments of need that the cost of the required imports would be enormous, more than Kenya could reasonably afford without substantial foreign donor assistance. It was also clear that, although the primary need was for maize, there was a substantial need for wheat for the bread commonly consumed in urban areas, and also a need for beans and milk powder for school feeding programs. Once the international donor community fully appreciated the need, its response was dramatic. The government found itself in the peculiar position of welcoming aid while discouraging the general public and the donors from characterizing the situation as a crisis.

Food requirements and timing

Among the most critical tasks the government faced were the determination of the amount of imported food required, and of the schedule on which it should be imported. The first task involved estimating aggregate need; the second required estimating port and inland-transport capacity.

The Ministry of Finance and Planning approached the task of assessing import requirements from two perspectives. One approach involved estimating the number of people who would experience a food deficit and multiplying that by an estimate of their average individual nutritional requirements. The other approach focused on estimating the national production shortfall, then using this as a measure of the need to import. In the final analysis, the second approach was used, because its relative simplicity facilitated operationalization, and because it would reproduce market conditions experienced in a normal year. Import requirements were agreed upon for 1984 and 1985. This estimate was revised, as new crop production data became available. In the end, a total (commercial, aid grant and aid loan) of more than 850 000 metric tons of grain was imported.

Developing a time schedule for imports proved to be particularly challenging for the Planning Department analysts. The amount of food in the NCPB storage depots was known from inventory reports. By comparing these reports over time, an assessment was made of draw-down rates. Using extrapolation, a rough estimate was made of when the stocks might be expected to be exhausted. This, then, established the outside date by which the initial shipment of imported food must have arrived and have been transported inland for distribution. Making the optimistic assumption that the short rains of 1984 would not fail and that the situation would be back to normal with the harvest from the long rains of 1985, the last imports would have to arrive by the end of June 1985, in order to be available for use during the last few weeks before the harvest. With the time frame thus bounded on the beginning and end, the import requirement was distributed in as uniform an arrival schedule as possible.

Commercial imports

Early in the negotiations with the donors it became evident that, even with the most efficient effort on the part of all parties involved, foreign assistance would not be available until well after the domestic reserve stocks were depleted. It was further evident that virtually all assistance would be in the form of food shipments rather than money that could be used to purchase food. It was clear that Kenya would have to begin buying and importing grain commercially with its own resources in order to bridge the period between the exhaustion of domestic food stocks and the arrival of donor-supplied food, and to ensure adequate stocks in the event of delivery delays for the donor-supplied grain. This would have a serious impact on the national budget and on the nation's already strained foreign exchange reserves.

Choice of yellow maize

The choice of type of maize to import was critical. Although there is little nutritional difference, Kenyans strongly prefer to eat white maize as opposed to the yellow varieties. Unfortunately, there is very little white maize grown commercially for human consumption outside of East Africa. Given the supply/demand situation in the world market when the 'long rains' failed in Kenya, white maize was commanding a premium of some 30%. Since the government was already facing major unanticipated expenses for food imports, it was felt that this premium for aesthetic preferences could not be justified. Therefore, it was decided to order yellow maize. It was already expected that all of the donor-supplied maize would be of the yellow variety.

Aside from the major cost saving, the importation of yellow maize had the secondary positive effect of encouraging informal rationing by those consumers who had access to alternative food supplies. While this group would have consumed normal amounts of white maize, faced with the prospect of eating yellow maize, they reduced their consumption by substituting other foods (until exhausted). This reduced the demand on the amount of food the government had to import, without denying nutrition to those who could not afford to exercise taste preference.

The commercial import of yellow maize, however, carried with it an element of risk. Being the food of second choice, the demand for it could be expected to drop substantially as soon as the rains resumed and the first harvest of white maize became available in the market. If the government imported too much yellow maize, it would be left with an inventory of valuable, but unmarketable, grain. It was, therefore, not only important to import enough maize, but to avoid importing too much. Furthermore, distribution for sales purposes was important. If some areas of the country continued to enjoy white maize from the hold-over NCPB stocks while other areas were being supplied only the imported yellow varieties, problems could be anticipated.

Donor-supplied food imports

One of the more challenging aspects of managing the government's response to the drought was the process of negotiating with donors for food assistance. The major problem was the uncertainty about how much would be supplied, and when it could be expectecd to arrive. The arrival date was critical both because the food had to be in the country when it was needed, and because the limited port and interior-transport facilities necessitated the careful coordination of ship arrivals with domestic truck and rail availability.

Donor negotiations were complicated by the need to work out a program not only for the immediate period, but also for the post-drought rehabilitation period. There were conditions to be agreed upon and approvals to be obtained from the donor governments in their respective capitals. Among the large donors, the World Food Programme (WFP) was particularly efficient in dealing with Kenya's needs in an expeditious manner. The WFP guidelines were clear, agreement was quickly reached, and the food began to flow. Encumbering conditions were minimal. Among the remaining donors, the small countries were the most efficient in clarifying what they could do to help, and then proceeding to do it.

In the final analysis, despite all the problems, the donor assistance was immensely helpful. Donor-supplied food grants totaled 108 000 metric tons of maize and 177 400 metric tons of wheat. This accounted for nearly 34% of the total grain imports.

Food-distribution strategy

In keeping with its policy that the drought was a serious problem but not a crisis, the government chose to use the well-established commercial distribution channels as the primary means of getting food to people. The objective was to keep food available through normal outlets at normal prices. This would minimize the potential problems of black marketing, and would help maintain an atmosphere of stability and confidence. The government was fully aware that keeping prices at their normal level would require a substantial subsidy of imported grain and associated transport.

Realizing that there would be many people who could not afford to buy maize through the regular channels, the government also undertook a joint program for employment generation and relief. The option of food-for-work was seriously considered, and a system developed for payment vouchers redeemable in food at local merchants. The approach was dropped, however, in favor of generating employment through government-sponsored rural-development activities, paying workers in cash, and then letting them have the discretion to spend it for food or other necessities. This strategy complemented the policy of supplying food at normal prices through the normal commercial channels. Since workers could earn more on a daily basis than they needed for food, it allowed people to build a small reserve to carry them through to the period when they would have to leave government employment to return to their farms for the next planting.

The necessary employment was generated through a major expansion in the number of labor-intensive local development projects funded under the Rural Development Fund. Districts were encouraged to submit proposals for labor-intensive projects and the Ministry of Finance and Planning processed the requests as rapidly as possible. The program, however, was still frustratingly slow in gathering momentum.

Early on, the government realized that despite all the effort to maintain the operation of the normal market system, there would be some households that would need direct relief. While it was hoped that the size of this group could be kept to a minimum, District Commissioners were authorized to make food available free wherever needed to prevent hunger. This took mainly the form of distribution through the local administrative chiefs. These men knew the needs of their people, and by most reports did an effective, equitable job of distributing the government-supplied grain. There was also a modest amount of relief food distributed through an expansion of the school feeding program.

Role of the non-governmental organizations

A large number of non-governmental organizations (NGOs) were involved in the drought-response effort. Their individual programs varied both in nature and in scale. Some groups had their own food-importation programs. To facilitate these efforts, the government established a system by which import orders could be consigned for shipment directly to the NCPB, and the NCPB would give the NGO credit for an equivalent amount of grain to be drawn immediately from stocks already in Kenya. From the NGO's point of view this greatly reduced the amount of time required between placement of an order and receipt of grain. By consolidating the large number of relatively small NGO orders into a few economically sized shipments, the NCPB was able to increase the efficiency of transport.

Although the government, through the NCPB, undertook to transport all of the imported grain inland to the NCPB wholesale depots, some NGOs had difficulty transporting food from the depots to the relief-distribution sites. Mounting an unusual program effort, CARE and UNICEF made a point of assisting with these transport problems. CARE paid for certain transport costs directly; UNICEF paid for the rehabilitation and operation of critically needed trucks. Most of the NGOs concentrated on food-for-work programs and on preventing hunger in particularly hard-hit areas. Some also provided seeds for the next crop cycle. Coordination of effort among NGOs, with the exception of food transport, proved difficult. While virtually everyone involved wanted to be helpful, each organization had its own localized program and its own mode of operation. These had been established before the drought, and the expectation was that they would continue afterward. There was a reluctance to surrender autonomy to coordination of the larger effort. The government had originally hoped that, with appropriate support, the NGOs would have been willing to expand their operations and take on a major share of the responsibility for the distribution of food to the people who could not afford to buy it. As the situation developed, however, the government readjusted its expectations, and accepted the NGOs for what they were willing to do.

Mobilization of information

People working on the drought response effort struggled continuously to obtain adequate information to make sound planning and management decisions. At the early stages donors did not know how much they might be able to supply; the railway did not know how much sustainable capacity it could mobilize; the Department of Agriculture did not know how much of the crop from the 1984 'long rains' season might be salvaged; and no one knew the extent of famine relief required.

The Planning Department of the Ministry of Finance and Planning moved quickly to assemble all the pertinent available information, identify the missing data to be collected, and organize a series of analytical reports that were made available to the wider community working on the drought-response effort. A weekly report was prepared on the inventory status of each NCPB depot, indicating the rate of draw-down on stocks and the date by which they would have to be replenished to avoid exhaustion. A weekly report was also prepared on the arrival and off-loading of incoming grain ships. An up-dated monthly report on the overall import schedule was initiated, and periodic crop-forecast reports were prepared as the data became available. With these reports it was possible to monitor the overall situation.

Going beyond this basic monitoring function, the Planning Department collected all the information available on the school feeding programs, commodity prices in various local markets around the country, domestic commodity transport costs, rainfall, food-relief distribution, crop-production estimates, livestock condition, nutritional status of selected low-income communities, NGO program activities, and the status of railway rolling stock. All of this was analyzed and made available to the task force and other key personnel involved in the drought-response effort.

Coordination of transport

The task force, operating out of the Office of the President, was responsible for ensuring that the food import and distribution program ran smoothly and efficiently. Responsibility for the purchase of grain from abroad remained with the government committee that normally oversees this task; transport and storage of grain was the responsibility of the National Cereals and Produce Board (NCPB). In normal times, the NCPB system operates smoothly in making many modest volume purchases over an extended period of time and over a widely dispersed network, and transporting them in relatively small consignments over short distances to nearby storage depots. However, the massive volume of imported maize, arriving in an unrelenting torrent at one point, and requiring immediate movement halfway across the country over a single transport axis, grossly exceeded the NCPB's normal capacity.

The task force represented the government in helping the NCPB get the grain cleared through the port, put into bags, and transported to appropriate depots inland. Membership included representatives of key units of ministries involved with the drought-response program. Food-movement problems identified by the task force received top priority in each of these Government bodies.

The Port of Mombasa, the nation's only facility for receiving large ocean vessels, dedicated its best berths and all of its bulk grain-handling equipment to the effort of meeting the import schedule. Even with this total dedication of resources, there was just enough capacity, under ideal conditions (no rain, no arrival delays, no equipment problems) to accommodate the need. There was no margin for safety, and the situation looked bleak. A few days before the first ship arrived, however, the dockworkers' union, an organization not known for its cooperation with management, spontaneously announced that it would take on, as its national duty, responsibility for ensuring that the incoming grain ships were unloaded expeditiously and that there were no delays in the port. True to this commitment, the off-loading process ran smoothly. The dock workers frequently sustained a pace that enabled the port to operate at more than 100% of design-rated capacity.

While there were few problems associated with unloading of the berthed ships, the task force faced a myriad of other concerns ranging from ships that drew too much water to cross the bar at the entrance of the harbor, to arriving grain that had been spoiled by dampness, to customs complications regarding the import of the spare parts to repair the locomotives needed to pull the grain trains to the NCPB depots. In each case the problem was addressed and solved with dispatch, and the food moved steadily inland. When the railway could not cope with the volume of grain to be transferred, the task force organized private-sector trucking services to meet the need. When the truckers pointed out that the government ban on night movement was severely limiting the amount of grain being moved, the task force arranged to have the ban temporarily suspended. When grain inventories started building up in the port area, additional trucks were mobilized. At one point nearly 1000 trucks were involved in the effort. The drivers, like the dockworkers, went to extraordinary lengths to keep the food moving.

In addition to overseeing the grain movement through the port and up-country, the task force monitored relief operations. In this connection it followed school feeding programs, NGO programs, and Government-food distribution. Problems were solved as they came up. When District

Commissioners were uncertain as to their responsibilities, the task force arranged with the Office of the President to issue a clarifying circular. When the NGOs were having difficulty obtaining relief supplies, the task force negotiated with the World Food Program to provide the required grain.

Post-drought rehabilitation

In the intensity of the effort to cope with the immediate impact of the drought, there may be a tendency to overlook the longer-term considerations. The need, however, for planning post-drought rehabilitation is critical. The problems do not automatically end with the return of the rains. In mid-1985, the rehabilitation program was still gathering momentum. Importantly, it was being carried out by the same Government officers and administrative systems that successfully responded to the food shortage. No hand-over from special agencies to established ministries was required.

Reassessment of conventional strategy

Analysis of the Kenyan drought response experience provides more than a rare example of a government's successful effort to prevent famine. It offers insights of potential use in challenging and improving current conventional prescriptions for responding to food shortages, particularly in Africa.

Most of the recommendations of experts are based on experience or analyses of the worst of famines. These are cases where, by definition, existing structures and systems failed to cope with the needs thrust upon them. Understandably, the problems are often ascribed to a vacuum of specialized institutional capability. The conventionally recommended solution to this problem is to create institutional structures specifically designed to respond to food shortages.

By focusing on the study of significant food shortage crises, there has been a tendency to overlook the quietly effective action taken by some governments in averting such catastrophes. The documented list of these successes is short. Such cases by virtue of their non-crisis nature do not draw attention, and so they often go unrecognized and their significance unappreciated. Their importance, however, lies in the fact that potential famines are prevented from growing to the point of attracting international attention.

Given that the conventional package of recommendations is not being adopted in Africa, it is particularly important to examine the lessons of indigenous success cases such as the Kenyan experience. They suggest that there may be two major approaches to addressing food shortage problems: (*a*) the permanent structural strategy typically recommended by the international community, and (*b*) the functional standby strategy followed by Kenya. The Kenyan government, without establishing the purpose-specific structural apparatus usually advocated by experts, drew upon the management and operational resources of existing systems to obtain the required capabilities. While it had taken some of the steps deemed essential to food security, such as issuing a food policy paper (see number 7 of the recommendations drawn from the publications of international agencies and conferences listed earlier in this chapter, henceforth referred to as R-7), these were not part of an integrated structural effort aimed at creating crisis-response capability. They were part of a much more extensive national development strategy. This points to the possibility that the building of functional capacity to cope with occasional food shortages might be part of broader on-going development programs.

Lessons drawn from the Kenyan experience

This chapter was written in late 1985 before the full impact of the 1984 drought had run its course. Much of the rehabilitation still lies ahead. There are, however, a number of lessons about the role of governments that can be drawn from the drought response effort to date.

Financial and managerial cost

Had the Kenyans adopted the full range of international recommendations described earlier in this chapter, they would have been over-prepared for the food shortages they faced, and they would have incurred significant additional costs. Despite the fact that the 1984–85 crisis was the result of a so-called 100-year drought, Kenya did not have to undertake the expensive emergency actions that comprise much of the collected wisdom about food shortage relief and rehabilitation (particularly R-13 and R-19).

Carrying out the range of international recommendations would be impossible for most African states; they are too costly and administratively complex. Kenya has more administrative and technical capacity than many governments, yet the case study suggests that even the limited demands of the 1984 drought severely stretched the capacity of the nation's planners and managers and the financial resources of the treasury. Rather than dismiss the set of international recommendations as utopian and unrealistic, it is better to view them as an argument for more forward planning, so that scarce human and institutional resources can be used more effectively and efficiently in times of food shortages.

Need for flexible response capability

The Kenyan experience suggests that creating flexible adaptive problem-solving capacity in existing governmental organizations is more appropriate and realistic than building a permanent, institutionalized response capacity. When the crisis threatened, the Kenyan government assigned some of its best administrators and professionals to work on it. The policies and actions they adopted are quite similar to those outlined in the consolidated recommendations (R-17). Yet, they were worked out with little awareness of the accumulated international experience. Does this mean a country with good problem-solving capability can forego the institutionalized food-security and famine-relief systems that expert reports call for? The Kenyan cases suggests that while leaders of drought-prone countries need to formulate a conscious strategy for building functional capacity, they may not need to create the complete structural system usually recommended.

The Kenyan experience identifies some problems that might have been anticipated and prepared for, had the objectives, if not the structure, of international recommendations been followed prior to the drought. For example, it is clear that some forethought could have been put into sorting out NGO roles and coordination, NCPB transport capacity, management information systems functions, District Commissioner responsibilities, and public works projects availability. To be sure, these were dealt with as they arose, but how much better might they have been met had they been considered ahead of time? For example, had R-7 been followed, there might have been well-designed local development projects that could have been more rapidly started up, and would have had a higher probability of being completed and promoting long-term development objectives.

Establishing the basic objective

In responding to a national shortage of food, it is critically important to clarify the basic objective. Kenya was able to do this, without having specialized organizations staffed with food crisis experts. Attempting to provide people with enough food to remain healthy is quite different from importing enough food to make up the production shortfall. Analysis of the Kenyan data suggests that, if the government had pursued the first objective, it would have imported far more maize and wheat than would have been indicated by the second. Given that cases of serious malnutrition are not common in Kenya the difference between the import requirements of the two objectives was explained by the fact that people are normally quite resourceful in informally supplementing their cereal intake with other foods. Had the government imported enough maize and wheat to cover the requirements (to meet an international standard) of everyone in need, there would have been more food in Kenya than ever before in its history. This would have created unsustainable expectations and a loss of self-reliance.

Realizing the existence of genuine hardship cases, the government sought to ensure that the normal amount of food would be available in the country, and then concentrated on targeting special programs for those segments of society with unmet needs. This created a situation that would provide for those people who would suffer seriously without special assistance, while simultaneously facilitating the return to normal food supply conditions after the drought.

Importance of using existing systems

Kenya's response to the drought taught everyone involved how demanding of good management this type of activity can be. Even modest enterprises, such as expanding the school feeding and rural works programs, were frustrated by the shortage of management resources. It soon became evident that major new administrative initiatives were not feasible. It would be necessary to use existing administrative structures to meet the new requirements of the drought response.

Kenya's existing institutions and systems had the latent capacity to take on functions called for in the international recommendations. The national administrative system is hierarchically organized with primary decision-making centralized in the Office of the President. The senior civil servant in the Office of the President became the 'famine czar', issuing directives and resolving problems throughout the government (R-14). This made it possible to achieve effective implementation of decisions made by the inter-ministry coordinating committee and the task force.

The task force served the functions of the National Committee for Food Security (R-2). Once the decision had been made on how much grain was to be imported and the schedule on which it had to be received, it was possible to move quickly into international markets, solicit bids, and execute all the procedures associated with ordering and shipment. Briefings were established to keep the public and donors well informed and confident of public actions (R-15 and R-16). NGO activities were coordinated (R-7 and R-8). District Commissioners were informed of their relief responsibilities (R-7). Grassroots organizations and the private sector were used (R-4). Had this latent capacity been more consciously developed prior to the drought, the Kenyan performance might have been stronger. That it came to the fore when needed supports the pragmatism of and potential for the functional standby strategy.

The amount of new institutional capability that had to be created was minimized by the use of existing systems. For example, once the grain had been delivered inland to the NCPB depots, the commercial retail traders collected and distributed it just as they would normally do. The only distribution the government had to manage was the grain going to the relief programs, and even a portion of this was taken care of by NGOs.

It is clear in retrospect that it would not have been feasible to assemble and coordinate enough managerial talent to operate the drought response program on an *ad hoc* basis. Even if it had been possible, using external assistance, the cost of assembling such a large-scale program would far exceed the costs of using existing systems. Finally, the creation of new systems would have frustrated the process of returning to regular operations after the drought. Normal systems, such as retail commercial distribution of grain, would have atrophied, and new, hard-to-break dependencies, such as relief camps, would have been established. While the approach 'business as usual', using existing systems, may lack the appeal of more drastic action, in the long run it is likely to be less costly, more efficient, and more conducive to a smooth transition back to a sustainable situation after the drought.

Had the drought continued for several seasons, the purchasing power of the general public, particularly in rural areas, would have declined, and greater emphasis would have had to be put on distribution of free food for famine relief. This might have put a greater burden on the system than it was able to sustain without specialized structures. It should be noted, however, that much of the food that passes through the formal marketing system moves from rural to urban areas, and the income of urban dwellers is relatively independent of drought. Hence, even in a prolonged drought, commercial channels would remain a vital distribution system.

The Government was not tested to the extreme on its capacity to respond to hunger, health, nutrition, and shortages of drinking water. Nor has its success in rehabilitating drought-stricken areas yet been established. The case study demonstrates that the government faced increasingly difficult management problems as it moved from macro planning (R-17 type actions) to famine-relief actions (R-18 type actions). It is likely that the more famine-relief actions required, the more useful a permanent structural capacity to respond is. Forward planning, established agencies, and famine-code provisions are likely to be more important at the field level when relief and rehabilitation are the concern rather than in regard to import and logistics actions.

Importance of management information system

The centrality of good information to effective government response to food-shortage crises is repeatedly recognized by the international recommendations and confirmed by the Kenyan case. This suggests that building knowledge relative to famine response and crises identification and tracking information systems (R-3, R-4, R-9, R-15 and R16) are essential, whether a country accepts the international community's permanent structural recommendations or relies on its capacity to build a functional standby system based on enhanced problem-solving skills and conscious preparation for a range of possible problems.

In times of food shortage, it is essential that the government monitor the situation carefully, in order to respond with enough, but only just enough, intervention to address the problems. Efficient targeting, through measured response, allows maximally effective use of scarce resources and minimizes the problems of phasing out the special measures after the crisis is over.

Good information is also critical for gaining sufficient lead time to arrange the necessary resources. Kenya found the international donor community willing to help, but a 2- to 6-month lead time was required between ordering and receiving grain from abroad. After arriving in Kenya, it took another 2 weeks to unload it and move it inland for distribution.

Planning and decision-making during the early phases of the drought were hampered by a lack of critical information such as railway capacity, on-farm food reserves, and grain-production estimates. Relevant data were collected from any accessible source, including such disparate origins as remote sensing, out-patient records in rural clinics and files in locomotive-repair shops. The data were analyzed by the Planning Department and assembled into various report formats, as needed by the different parties involved in the drought-response effort. Microcomputers proved invaluable in facilitating the rapid processing of data.

As the trickle of management information grew into a steady flow, it quickly became the integrating element of the entire effort. Information networks were established to link the different parties working on different parts of the program. Projected crop-yield data, for example, would flow into the Planning Department from several different sources in the field. These would be analyzed and the discrepancies between different sources resolved. The information would then be combined with data on current grain inventories in the NCPB depots and imported grain on order. This would give an overall view of the food-supply situation. By combining this with a projection of demand (based on recent trends and anticipated harvests), an estimate would be made of the amount of food still to be imported. From this, the outstanding donor commitments would be subtracted to obtain an estimate of net commercial imports required. This information would then be passed to the grain-import committee for placing the necessary orders, and to the Central Bank for use in estimating foreign exchange requirements to be negotiated with the International Monetary Fund. It was critically important to keep information flowing in a timely fashion to those who needed it.

Difficulty of targeting relief assistance

After planning the food imports, ordering the grain, and having it shipped halfway around the world, off-loading, bagging and transporting it inland, one of the most difficult tasks proved to be the final process of delivering relief supplies to those who needed them. There were two dimensions to the problem: first was identifying those people with serious need; second, guiding the system to deliver food to them in an expeditious manner. The functional standby system proved capable of dealing with these problems, although not perfectly.

The first issue was the definition of need. An on-going nutrition survey with baseline data from before the drought indicated that, for some people in the study area, nutritional status had declined seriously by mid-1984. For the small sample involved, it was clear who was not getting enough food and was in need of relief assistance. Unfortunately, it was only a modest study in a limited area. It would have been prohibitively expensive and administratively infeasible to monitor the entire population at this level of detail. Surrogate measures had to be used. It was found that when local chiefs were asked to list all the people in their area in need of relief assistance, the names on their lists corresponded closely to those found to be needing assistance through the nutrition survey. It was, therefore, decided to use the existing administrative system and make the chiefs responsible for the distribution of relief-food supplies.

This system was reasonably effective in general, but it missed some sub-groups that needed special attention. As experts would expect, these included pregnant women, small children and nursing mothers. A school feeding program was a partial solution, but it missed the pre-school children, the school-age children who were not enrolled in school, and all the women in need. The system of relying on chiefs also missed those households which still had food supplies, but were rationing their use so severely they were experiencing malnutrition.

Even when the population in need is clearly known, it is difficult to effect relief efficiently. District commissioners may be reluctant to call attention to food shortages in their area and, hence, may not push as hard as necessary to get adequate relief supplies to the individual chiefs. Non-governmental organizations may have effective delivery systems in their respective areas, but leave adjacent areas completely without service. The shortage of managerial capacity can constrain the expansion of school feeding programs, and valuable time can be lost while officers wait for administrative directives. All of these are solvable problems, but the solutions require enormous amounts of managerial attention. This resource was in critically short supply.

Need for strong political commitments

The Kenyan case reveals the importance of political commitment. Kenya's drought-response effort had the critical ingredient of a strong, unwaivering political commitment from the highest level of government. While the initial mobilization of effort might have been quicker and more efficient had there been a well-developed plan to follow for addressing each type of problem that arose, any problems that resulted from lack of planning were addressed with whatever resources were required. The drought was treated as a matter of national urgency, and the response received the political support required for success. The active continuous involvement of senior government officials contributed to an increase in efficiency over what normally would have been expected from middle-level bureaucrats following the administrative instructions of a pre-approved plan. The Kenyan case suggests that politically stable, development-oriented bureaucracies are probably essential to effective drought response. Repressive, coup-prone political systems are unlikely to make such progress converting existing capacity into effective famine-response capability.

Matching response systems to economic and administrative realities

The experience in Kenya suggests that the conventional recommendations for famine response need to be examined. Kenya was able to cope effectively with the worst drought in 100 years, without most of the prescribed permanent structural mechanisms. This impressive accomplishment was achieved by mobilizing existing functional capabilities and focusing them on the drought-response effort. In effect, Kenya accomplished much of the intent of the structural recommendations by using management systems and resources that were already in place and that were only marginally different in functional character from the requirements of the drought-response effort. The Directorate of Planning deals with resource mobilization and allocation on a daily basis. It has the analytical and computational capacity to absorb and use large quantities of data in making allocational decisions. Because of the breadth of material it covers in the normal course of its work, it has access to an extensive information network and the flexibility to take on new analysis and planning tasks with minimum organizational adjustment. Similarly, with the National Cereals and Produce Board normally monitoring grain inventories at each storage facility on a weekly basis, it was relatively easy to obtain accurate up-to-date data on the food supply situation in virtually every area of the country. Perhaps most critical, Kenya's well-developed administrative system, managed by the Office of the President, facilitated efficient communication and coordination of effort between the various parties working on the drought response. The government gave the effort high priority and, as a result, the system functioned efficiently.

The strategy of developing fundamental capacity that normally serves other purposes, but that can be called into service in time of food crisis, has much to recommend it over building standby structural systems that are expensive to maintain and stand virtually idle most of the time. Most African countries lack the administrative capacity required to mount and sustain the structural systems of famine response typically recommended by the international community. The pressing day-to-day demands of administration and development virtually guarantee that most governments will not be able to set aside substantial resources to wait idly for the possibility of a food shortage. In economic terms, the effective discount rate is too high to make the possibility of a future benefit sufficiently attractive to offset the current cost. In the common situation where near-term political survival is a leading objective, governments cannot afford to withdraw resources from administration and development activities. Viewed from the standpoint of risk analysis, governments prefer to be 'self-insured' by gambling on the relatively low probability of incurring a substantial loss at some indefinite time in the future rather than sustaining the current costs associated with building and maintaining an idle structural capability for food-shortage response.

Many of the functional capabilities required to mount and manage an effective food-shortage response program are generically similar in character to those required for normal administration and development activities. If consciously planned and incorporated into the overall administrative system, potentially powerful crisis-response capability can be developed at very modest cost. The key to success is a strong political will with a conscious commitment to preparedness. The international recommendations, although structural in nature, provide insight into the capabilities that must be developed. The task is to spread the cost of building and maintaining these capabilities over the broad base of administrative and development activities, rather than concentrating it on a separate system for drought response.

This chapter represents only a small step in what must be a continuing quest for better understanding of how to avert famine. In suggesting that the structural nature of most recommendations is inappropriate to many developing countries, it raises the questions of how the structural approach emerged and how, in the interest of improving the utility of recommendations, this conventional wisdom might be revised. There are at least two dimensions to these questions. First, most recommendations have been made in response to specific aspects of famines. The literature tends to be fragmented. Individual recommendations are developed in detail, without a comprehensive overview of all the other things a government would need to do to prepare adequately for responding to a major food shortage. If a country were to implement all of the recommended actions, the burden would be overwhelming. It is clear why the governments of drought- or famine-prone countries have not gone further to adopt the recommendations that, when considered individually, appear to be so appropriate. Second, most analyses and subsequent recommendations are made after disaster has struck and the potential response systems have collapsed. Structural recommendations emerge naturally from this context. The relative success of the Kenyan experience suggests, however, that if structural collapse can be prevented through the use of existing functional capability, new structural institutions may not be required. Thus, in the effort to deny famine a future it is important to look at the entire set of problems from the point of view of the decisionmakers within the affected developing country, and to look at cases of success for clues as to what can be done with existing capabilities to prevent deterioration to the point where they can no longer cope.

References

Akroyd, W. R. (1974). *The Conquest of Famine*. London: Chatto & Windus.

Ball, N. (1976). Understanding the causes of African famine. *Journal of Modern African Studies*, **14**(3), 517–22.

Ball, N. (1981). *World Hunger, A Guide to the Economic and Political Dimensions*. Oxford: Clio Press.

Bates, H. (1981). Food policy in Africa: political causes and social effects. *Food Policy*, **6**(3), 147–57.

Berg, A. (1971). Famine contained: Notes and lessons from the Bihar experience. In *Famine: A Symposium Dealing with Nutrition and Relief Operations in Times of Disaster*, ed. G. Blix, Y. Hofvander & B. Valquist, pp. 133–9. Uppsala: Almquist & Wiksells.

Blix, G., Hofvander, Y. & Valquist, B. (eds.) (1971). *Famine: A Symposium Dealing with Nutrition and Relief Operations in Times of Disaster.* Uppsala: Almquist & Wiksells.

Bryceson, D. F. (1981). Colonial famine responses: the Bagamoyo district of Tanganyika, 1920–61. *Food Policy,* 6(2), 91–104.

Cahil, K. M., ed. (1982). *Famine.* New York: Orbis Books.

Chambers, R. (1983). *Rural Development: Putting the Last First,* pp. 28–46. London: Longmans.

Dalby, D., Harrison Church, R. J. & Bezzaz, F. (1977). *Drought in Africa.* London: International African Institute.

Eicher, C. K. & Baker, C. (1982). *Research on Agricultural Development in Sub-Saharan Africa: A Critical Survey.* East Lansing: Michigan State University, International Development Paper No. 1.

Franke, R. W. & Chasin, B. H. (1980). *Seeds of Famine: Ecological Destruction and the Developmental Dilemma in the West African Sahel.* Totawa, NJ: Allanheld Osmun & Co.

Garcia, R. V. & Escudero, J. C. (1982). *The Constant Catastrophe: Malnutrition, Famines & Drought,* 2 vols. Oxford: Pergamon Press.

Glantz, M. H. (ed.) (1976). *The Politics of Natural Disaster: The Case of the Sahel Drought.* New York: Praeger.

Hardin, G. (1968). The tragedy of the commons. *Science,* 162, 1243–48.

Hay, R. W. (1978). The concept of food supply system with special reference to the management of famine. *Ecology of Food and Nutrition,* 8, 65–72.

Hayer, J., Maitha, J. K. & Senga, W. M. (1976). *Agricultural Development in Kenya: An Economic Assessment.* Nairobi: Oxford University Press.

Hussein, A. M. (1976). *Drought and Famine in Ethiopia.* London: International African Institute.

Hyden, G. (1983). *No Shortcuts to Progress: African Development Management in Perspective.* Berkeley: University of California Press.

Jodha, N. S. (1975). Famine and famine policies: some empirical evidence. *Economic and Political Weekly,* October, 1609–23.

Kaplan, I. (1976). *Area Handbook for Kenya,* 2nd edn. Washington, DC: Government Printing Office.

Kenya, Government of (1962). *Inquiry into the Drought Relief.* Nairobi: Government Printer.

Lappé, F. M. & Collins, J. (1979). *Food First: Beyond the Myth of Scarcity.* New York: Ballantine Books.

Leys, C. (1976). *Underdevelopment in Kenya.* Berkeley: University of California Press.

Lofchie, M. F. (1975). Political and economic origins of African hunger. *Journal of Modern African Studies,* 13(4), 551–67.

Manetsch, T. J. (1977). On the role of systems analysis in aiding countries facing acute food shortages. *IEEE Man and Cybernetics,* 7, 264–73.

Masefield, G. B. (1963). *Famine: Its Prevention and Relief.* Oxford: Oxford University Press.

Mayer, J. (1974). Coping with famine. *Foreign Affairs,* 42, 98–120.

Mayer, J. (1975). Management of famine relief. *Science,* 188, 571–7.

Morris, M. D. (1975). Needed – a new famine policy. *Economic and Political Weekly,* February, 284–94.

Newman, J. L. (1975). *Drought, Famine and Population Movements in Africa.* Syracuse: Syracuse University Press.

Paul, S. (1982). *Managing Development Programs: The Lessons of Success.* Boulder: Westview Press.

Perelman, M. (1977). *Famine for Profit in a Hungry World.* Totawa, NJ: Allanheld, Osmun & Co.

Picardi, A. C. & Seifert, W. W. (1977). A tragedy of the commons in the Sahel. *Ekistics*, **43**(258), 297–304.

Robson, J. R. K. (ed.) (1981). *Famine: Its Causes, Effects and Management*, 2 vols. London: Gordon & Breach Science Publishers.

Ross, J. (1983). Africa: the politics of hunger (4 part series). *Washington Post*, 26–29 June.

Seaman, J., Holt, J. & Rivers, J. (1978). The effects of drought on human nutrition in an Ethiopian province. *International Journal of Epidemiology*, **7**, 31–40.

Sen, A. (1981). *Poverty and Famines: An Essay on Entitlement and Deprivation*, pp. 217–49. Oxford: Oxford University Press.

Shepherd, J. (1975). *The Politics of Starvation*. Washington, DC: Carnegie Endowment for International Peace.

Sheets, H. & Morris, R. (1974). *Disaster in the Desert: Failures of International Relief in West Africa*. Washington, DC: Carnegie Endowment for International Peace, Humanitarian Policy Studies. Also published in M. H. Glantz, ed. (1976). *Politics of Natural Disaster: Case of the Sahel Drought*. New York: Praeger.

Wade, N. (1974). Sahelian drought: no victory for western aid. *Science*, **185**, 234–7.

Watson, R., Wilkinson, R., Harrison, C., Callotta, J. A. & Whitmore, J. (1984). An African nightmare. *Newsweek*, 26 November, pp. 59–64.

World Bank (1981). *Accelerated Development in Sub-Saharan Africa: An Agenda for Action*, pp. 24–90. Washington, DC: The World Bank.

13

The social impacts of planned settlement in Burkina Faso

DELLA E. McMILLAN

Center for African Studies, University of Florida

Introduction

It is increasingly clear that any long-term solution to famine in sub-Saharan Africa will involve the spontaneous or organized relocation of massive numbers of rural people away from the more degraded, drought-prone agricultural regions. National planners are often attracted to planned settlement schemes as a means of achieving this sort of population relocation. There is a critical need, therefore, for policy planners concerned with short-term problems of drought recovery through the planned relocation of impoverished groups to study the local social and economic processes that are involved in the design of successful planned settlements. The question remains, however, whether the existing models for planned settlement schemes are in fact the best means for achieving government goals of famine relief and drought recovery. Indeed, there is a large literature that shows that past experiences with planned settlements in Africa have not been successful (Hilton, 1959; Lewis, 1964; Christodoulou, 1965; Mellor, 1966; de Wilde *et al.*, 1976*a*, *b*; Moris, 1968; Chambers, 1969; Takes, 1975; World Bank, 1978; Dettwyler, 1979; Scudder, 1981, 1984; Hansen & Oliver-Smith, 1982; Reining, 1982).

This chapter examines the early history of one of the largest planned settlement schemes that was started in the wake of the 1968–73 drought and evaluates the project's performance in achieving its goals for population transfer, food production and economic development. The evaluation process, however, raises the question of *how* and *when* to measure project success. Scudder (1981, 1984) suggests that it is not realistic to measure a planned settlement scheme's achievements after only 1 or 2 years. Instead, a longer time frame is required, because most successful projects pass through a series of 3–5 year stages of settler adjustment which, in turn, affects the project's economic performance and regional impact (Chambers, 1969; Colson, 1971; Nelson, 1973; Brokensha, Horowitz & Scudder, 1977;

Moran, 1979; Scudder, 1981, 1984; Scudder & Colson, 1982; USAID, 1985*a*, *b*).

Scudder describes four stages, each with a relatively distinct set of problems and opportunities. The first stage focuses on the design of new settlement areas and the physical transfer of the settlers. The second 'transition' phase refers to the first 3–4 years during which the settlers are in the initial stages of adaptation to scheme policies and to the physical dislocation associated with the move. During this time, they typically

> adopt a conservative stance, their first priority being to meet
> their subsistence needs. They favor continuity over change; and
> where change is necessary, they favor incremental change over
> transformational change. (*Scudder, 1984, 16*)

This second stage ends when enough settlers 'shift from a conservative stance to a dynamic open-ended one, hence initiating a third stage of economic and social development' (Scudder, 1981, 2). This usually occurs only after settler security has increased, as a result of the production of sufficient food and the increasing tendency of the settlers to feel more at ease in their new environment. It is in stage three that one can anticipate the settlers' shift from their earlier emphasis on extensive agriculture, to a more diverse range of investment strategies that are designed to achieve higher levels of labor productivity. A fourth stage which Scudder refers to as 'handing over and incorporation' is intended to emphasize the fact that no successful project is complete, until it has gone through the difficult task of handing over the day-to-day problems of scheme administration and land tenure to a second generation and the direction of many project-specific activities has been taken over by local, regional and national authorities. Unfortunately, most evaluations of planned settlement schemes have taken place during the first 3–5 years of a settlement's existence, when it is unreasonable to expect rapid increases in food or cash crop production.

Policy planners need to realize that these stages in the development of planned settlement schemes exist and influence the social impact of a project. Both to test Scudder's ideas and to evaluate the impact of a large-scale planned settlement effort as a famine-relief solution, this chapter evaluates the performance of the *Aménagement des Vallées des Volta* (Volta Valley Authority, henceforth AVV) of Burkina Faso.

Since 1974, the AVV has coordinated a capital intensive program of planned settlement and agricultural extension to work with settlers moving into the country's river basins. This represents an area of some 30 000 km², about one-tenth of the total land area of the country that has remained sparsely inhabited because of the high incidence of disease (Fig. 13.1). With the advent of a World Health Organization (WHO) program to control the disease onchocerciasis (or river blindness) in a seven-country area of West Africa in 1973, one of the main obstacles to a more intensive settlement of the river valleys was presumably removed.[1]

One of the original goals of and major justifications for international funding for the regional Onchocerciasis Control Programme (OCP) and the AVV was to provide new economic opportunities for rural farmers from the densely populated areas most severely affected by the 1968–73 drought. The project was also expected to have a range of economic benefits in terms of drought recovery at the national, regional and local levels. While the AVV planned settlement program was not expected to provide an all-purpose panacea for hunger and economic development in Burkina Faso, it was considered to provide a temporary relief from some of the most urgent pressures and to give the country

a vitally needed breathing spell, time to address the fundamental problems: improvement and eventual transformation of farming

Fig. 13.1.

LAND ASSOCIATED WITH THE VOLTA VALLEY AUTHORITY

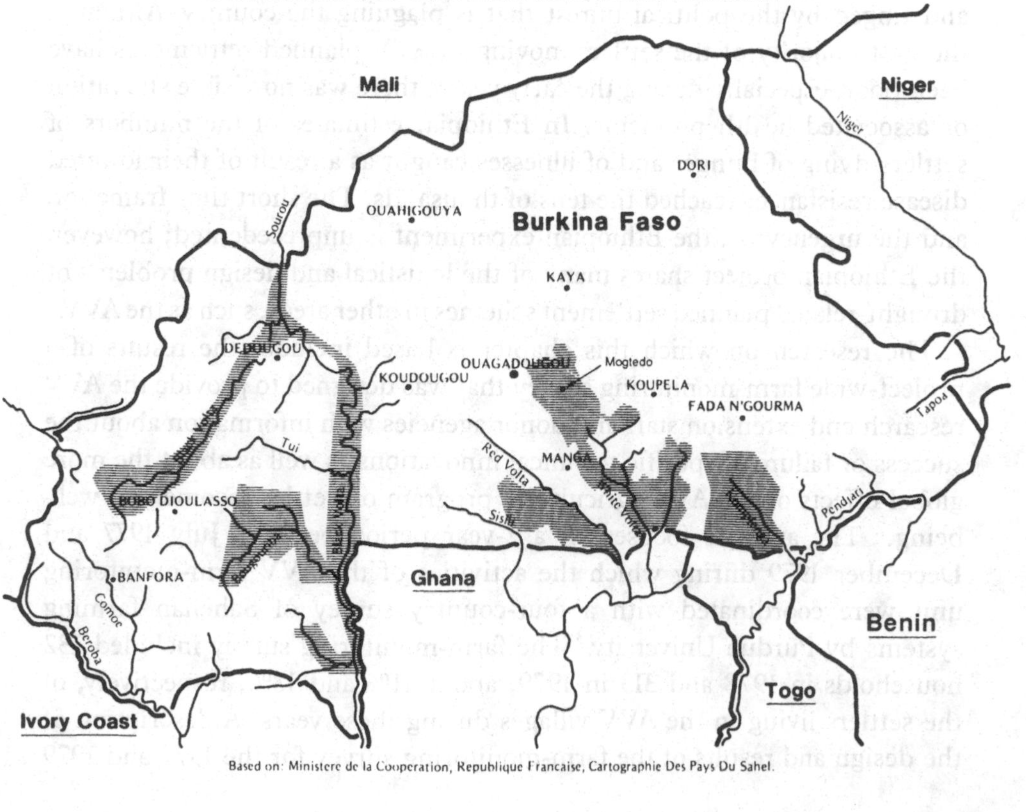

Based on: Ministere de la Cooperation, Republique Francaise, Cartographie Des Pays Du Sahel.

practices in the direction of permanent intensive cultivation. (*Berg et al., 1978, iv*)

In any case, the gains in settled and cultivated areas in the valley since the eradication program began were estimated to amount to more than 10% of the total land area under cultivation in Burkina (about 2 million hectares), and the cultivation of these new lands undoubtedly accounts for the fact that there was only a slight drop (or possibly even an increase) in cereal production in 1983, in spite of the deplorable weather conditions (Rochette, 1982, and Albergel *et al.*, 1984, quoted in Hervouet *et al.*, 1984).

As a voluntary planned settlement program, the AVV is different from contemporary cases of involuntary planned settlement as a result of drought and/or civil war. The most prominent example today of this type of involuntary planned settlement is in Ethiopia where an estimated 1.5 million farmers from the country's drought- and famine-stricken northern provinces are being moved to newly created villages in the south and southwest. In Burkina Faso, although the settlers' move to AVV planned villages was government-assisted, the final decision to move has been voluntary. In contrast, in the case of Ethiopia, the decision to move is often seen as forced. Moreover, the motivations of the government have been suspect and tinged by the political unrest that is plaguing the country. Although the vast majority of the settlers moving to AVV planned settlements have been poor, especially during the early years, there was no visible starvation or associated health problems. In Ethiopia, estimates of the numbers of settlers dying of hunger and of illnesses caught as a result of their lowered disease resistance, reached the tens of thousands. The short time frame for, and the urgency of, the Ethiopian experiment is unprecedented; however, the Ethiopian project shares many of the logistical and design problems of drought-related planned settlement schemes in other areas, such as the AVV.

The research on which this chapter is based includes the results of a project-wide farm-monitoring survey that was designed to provide the AVV research and extension staff and donor agencies with information about the success or failure of specific technical innovations as well as about the more global effects of the AVV agricultural program on settler income and well-being.[2] The analysis focuses on a 2-year period between July 1977 and December 1979 during which the activities of the AVV farm-monitoring unit were coordinated with a four-country survey of Sahelian farming systems by Purdue University.[3] The farm-monitoring survey included 132 households in 1978 and 313 in 1979, about 11% and 18%, respectively, of the settlers living in the AVV villages during these years. A description of the design and results of the farm-monitoring survey for the 1978 and 1979 crop years appears in Murphy & Sprey (1980). One of the unusual aspects of the cooperative agreement between Purdue and the AVV was the decision to fund two intensive case studies to complement the unit's survey research.

The case study presented in this chapter focuses on the economic and social consequences of the AVV for a group of voluntary settlers from one of the main recruitment zones. The study compares a single group of settlers from the same home village who are living in the same project village and have related households who had remained in the settlers' home area.[4] The baseline research for the case study was conducted over two agricultural seasons from April 1978 to April 1980. A short restudy of the same group of settlers was conducted during the summer of 1983 (McMillan, 1984).

The chapter is divided into four sections. Section one describes the background, goals, and design of the AVV project. This is followed by a discussion of the results of the AVV in terms of population transfer, technological innovation, increased income and food production, and the costs of achieving these results during the first five years, based on an analysis of project records and the AVV farm monitoring survey. The third section compares this macro-level assessment of project results with the longitudinal case study. The final section focuses on some of the implications of these different levels of analysis for an assessment of project achievements and the design of future programs.

Background and goals of the Volta Valley Authority

The concept of a four-stage planned settlement process can be usefully applied to the AVV. At the local level the AVV project was based on the progressive installation of groups of six to seven villages known as blocs. The project was responsible for the selection of village, field and house sites; installation of basic infrastructure (wells, roads, bridges, extension worker housing); and coordination of economic and social services. Settlers were recruited from the overpopulated areas most severely affected by the 1968–73 drought and assisted with their move. Since the majority of the settlers were from impoverished households with little or no reserve food stores, they were provided with a monthly ration of grain, oil and fish, until they harvested their first crops. Each settler household was entitled to one or, in the case of an extremely large labor force, two 10-hectare farms that consisted of 6 1.5 hectare bush fields and a 1-hectare plot on which to construct a house. The project was also responsible for the design, testing, extension and evaluation of a new intensive dryland cropping package. Basic elements of the package included cultivation with animal traction;

the use of high levels of fertilizer and pesticides on certain crops; a new system of land allocation; new production techniques; cultivation of the cash crop cotton; and a system of mandatory (i.e. extension supervised) crop rotation.[5] During the first 5 years that a project village was in existence, this intensive cultivation package was supervised by a dense network of extension services, including one male extension agent per 25 settler households and one female extension agent for every 50.

The project plan foresaw the evolution of the AVV program in terms of an annual increase in the number of planned villages, the type and level of infrastructure, and the farmers' income. Although the proposed agricultural program required a greater outlay of cash and labor than the settlers were accustomed to, it was considered that these costs would be offset by higher yields. Moreover, it was assumed that the settlers' per capita food production and cash income would increase every year. During the first 3 years, this increase would derive from the annual addition of a new field. After the third agricultural season, any subsequent increase would come from the greater use of fertilizer and labor on the existing crop area and the expansion of the settlers' livestock activities, rather than the addition of new fields.

The decision of the Burkina government to adopt (and foreign donors to support) this type of capital-intensive development plan for the country's land covered by the Onchocerciasis Control Programme can be attributed to the large areas of the country covered (84% of the total land area) and the dramatic decline in crop and livestock production caused by 2 years of severe rainfall shortages in 1972–73 and 1973–74. It was estimated that between 1972 and 1973, the agricultural production of Burkina fell by 16% (Berg, 1975). Moreover, there was every reason to expect that the associated decrease in export production and the increased need for food imports would have a crippling effect on the country's trade balances (Berg, 1975; Caldwell, 1975; Christensen *et al.*, 1981).

Previous efforts to develop Burkina's agricultural resources and to reduce the impact of 1-year and 2-year droughts that affect the country have met with very limited success. This negative record was attributed to a combination of poor soils, distant markets, limited infrastructure and the extremely low level of existing agricultural technology. Another frequently cited problem was the country's high rural population density estimated at 17.9 persons per km^2 in 1975, the third highest in West Africa (Jeune Afrique, 1975). Even more important is the fact that this population density was mostly concentrated in the central plateau region which covers less than one-quarter of the country's total land area.[6] Population densities on the Mossi Plateau (50–100 persons per km^2) contrasted sharply with the relatively low densities (below 10 persons per km^2) in the far north and south (Figure 13.2). What was especially striking was the virtual absence of population in the major river basins.

The vast majority of farmers on the densely populated Mossi Plateau were and continue to be dependent on subsistence agriculture. There are limited possibilities for irrigation and government programs to introduce animal traction have not met with widespread success (de Wilde *et al.*, 1967*b*; Barrett *et al.*, 1981). Because of high population densities most plateau farmers have been forced to shorten the fallow cycle necessary to restore soil nutrients lost in cultivation. This has had a negative effect on soil fertility and yields (de Wilde, *et al.*, 1967*a*, *b*; Broekhuyse, 1974; Rey, 1980) and increased the farmers' vulnerability to periodic and lengthy drought. One of the results has been a high rate of emigration from the region.[7]

Despite low population densities, relatively high-quality soil, and high population pressure in the neighboring plateau, almost none of the emigration from the Mossi Plateau has been directed toward the valleys of the Red, White and Black Volta Rivers. This has long been attributed to

Fig. 13.2

POPULATION DENSITY IN THE MOST DENSELY SETTLED RURAL AREAS OF UPPER VOLTA

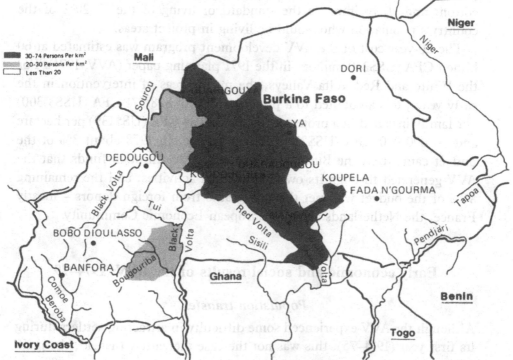

Based on: Ministere de la Cooperation, Republique Francaise, Cartographie Des Pays Du Sahel.

the high incidence of disease, especially onchocerciasis (Hervouet, 1977, 1978; Marchal, 1978). Therefore, it was anticipated that the Onchocerciasis Control Programme could have a positive effect on food production and the long-term development of the country's resources. First, it would provide new settlement opportunities on virgin soils in less drought-prone ecological zones for a considerable number of limited-resource farmers from the areas most severely affected by drought in the early 1970s. Second, it would provide the Burkina government with a vast fund of underpopulated valley land where infrastructure development could take place and increase the GDP of the nation as a whole.

Some of the projected benefits over a 20-year period were as follows (AVV, 1974; Ouedraogo, 1976; Nikyema, 1977): (*a*) to settle an estimated 650 000 persons in planned villages – 290 000 persons, or 38 000 families, in rainfed agriculture and another 360 000 in irrigated agriculture associated with project plans for dam construction; (*b*) to reduce population pressure in some of the more densely settled areas of the plateau; (*c*) to control settlement and development in the valleys in order to minimize the negative consequences of higher population densities, in large part through the introduction of a system of intensive cultivation; (*d*) to provide a regional grain surplus which would allow the country to offset a large part of its projected food deficit; (*e*) to triple Burkina's production of the export-crop cotton; and (*f*) to improve the standard of living of the 15–20% of the country's population who would be living in project areas.

The 20-year cost of the AVV development program was estimated at 60 billion CFA (US$240 million) in the 1971 planning paper (AVV, 1974). For the White and Red Volta Valleys, the major areas of intervention in the early years, this amounted to a projected cost of 828 571 CFA (US$3 300) per family installed in a project farm and 35 024 CFA (US$139) per hectare and 56 000 000 CFA (US$221 306) per village. In 1978 about 3% of the budget came from the Burkina government, 12% through funds that the AVV generated through its own commercial activities, and the remaining 85% of the budget was met through grants from foreign donors – mostly France, the Netherlands and the European Economic Community.

Early economic and social results of the AVV: 1974–79

Population transfer

Although the AVV experienced some difficulty in attracting settlers during its first year (1974–75), this was not the case thereafter. Instead, the main problem the agency faced was in providing a sufficient number of farm sites for the candidates who wished to immigrate. It soon became apparent that the original projection of moving 3 500 families between 1973 and 1977 and 9 000–13 000 by 1979 was too ambitious and would have to be modified. By 1979, 1 800 families had been transported to and installed in AVV planned villages. The project was not always successful in meeting even this modified schedule, because of administrative delays in road construction, tree clearance and mechanical plowing. In spite of difficulties at the start, however, most villages did show a stable rate of population growth during the first 5 years. However, even if planned settlement had been carried out at the anticipated rates, it would have absorbed only a fraction of the annual population increase in the most densely settled areas. Moreover, the planned settlement to the AVV attracted only a small part of the internal migration to less-populated agricultural areas within the country (Murphy & Sprey, 1980).

Infrastructure and extension services

The AVV did achieve a successful record in the organization of support programs to provide credit, extension services and basic infrastructure to the new settlements. The project completed extensive aerial and ground hydrogeologic surveys for planning purposes as a preliminary to planned settlement. The AVV also had a successful record in the recruitment, training and placement of male and female extension workers and other support staff. There was a reliable system of rural warehouses providing farmers credit, fertilizer; animal traction equipment, spare parts and supplementary feed for draft animals. In addition, the payment performance on short- and long-term credit through AVV was much higher than for similar types of credit programs in one of the neighboring regional development authorities during the same time period (Barrett *et al.*, 1981).

Technological innovation

An analysis of the results of the farm monitoring survey for the 1978 and 1979 cropping years indicated that the AVV had failed in its major objective to encourage the settlers' adoption of intensive cultivation methods. Specifically, Murphy and Sprey (1980) found: (*a*) settlers who had been at the project for shorter periods of time tended to follow the extension package more closely than those who had been there longer; and (*b*) cotton was the only crop on which the recommended package of cultivation techniques including monocropping, pesticides, planting in rows, fertilizer, thinning and timely weeding with animal traction was consistently applied. This differential acceptance of agricultural innovations on particular crops and according to length of residence in the scheme was attributed to the higher level of supervision by the extension service when the settlers first arrived, and the extension service's emphasis on cotton which was sold to reimburse settler credits for new agricultural equipment.

Food and cash production

The project was more successful in its attempts to raise crop yields and settler income. In 1978 the AVV settlers marketed about 1800 metric tons of cotton through the national marketing board. This represented about 3% of the total national production in that year and was expected to rise as the number of AVV farmers increased (Murphy & Sprey, 1980). Moreover, the average yields and quantity sold per hectare was substantially higher in the AVV than in other parts of the country.

The picture was less positive for food production. Although the farm-monitoring survey showed that the average yields for sorghum were below the expectations of the project planners, they were still 2–3 times higher than the case study recorded for the home village farmers (700–900 kg per ha as opposed to 200–350 kg per ha). The study shows, however, that these increased yields were primarily a result of the natural fertility of the new soils and not necessarily due to the successful introduction of the recommended package of technical innovations. Based on an analysis of the results of the farm-monitoring survey (which only measured production on the official bush fields and relied on an estimate of production on the house fields), it was possible to conclude that there was little 'surplus' grain production in the years 1977–79 and, based on these trends, the situation would seem unlikely to improve.

To evaluate the impact of the AVV on agricultural income, the farm-monitoring survey used two calculated figures of net and gross 'farm results' based on the recorded kilogram production on the four official bush fields and an estimate of production in the house garden.[8] On the basis of this analysis, it was possible to conclude that, although the settlers' average income was well within the projected levels, the AVV's projections for a steady increase in total farm income seemed unrealistic. The reason for this was that as the settlers increased the number of official fields they were authorized to farm (from two to four) during the first 3 years, they also increased the size of their families.

Cost effectiveness

By far the major criticism against the AVV was the cost effectiveness of the program in directing emigration into the OCP lands. A report prepared for the donor countries in 1978 estimated the cost of settling a single family at approximately US$12 500 (this figure included the costs of basic infrastructure and the first 5 years of operating expenses for the extension service) (BEI-agrer, 1978). Even allowing for inflation, this represents a significant increase over the US$3 300 (82 857 CFA) cost per family anticipated in the 1974 project plan (AVV, 1974). This figure was also high compared to the average cost of non-irrigated settlement programs funded by the World Bank from 1962 to 1975 (US$6 460 per beneficiary family) (World Bank, 1978). While a large part of the high unit costs could be attributed to the smaller number of households that had settled in the project villages, there were some significant cost over-runs, particularly in the areas of road construction and mechanical field preparation.

An additional concern was the fact that by the late 1970s there had been a tremendous increase in the spontaneous settlement, clearance and cultivation of the AVV project lands (Nana & Kattenburg, 1979). Although many of the spontaneous settlers were temporary residents who lived there only during the rainy season, there was a growing number who considered their move to be permanent. In most areas of the AVV, the number of spontaneous settlers was estimated to be at least equal to the number of legal settlers in planned villages. In areas along the Black Volta, the ratio was much higher. At the existing rate of planned settlement, many of the areas where the AVV expected to install blocs of planned villages would be occupied before development planning could begin.[9]

Impact on participating settlers: the case study

Where the case study distinguished itself was in its ability to gather information on areas outside the proposed agricultural innovations and settlement figures that were being monitored by the project (McMillan, 1985). The case study revealed important differences over time in the background and original incentives for emigration between the settlers who left their home village for the project in different years. There were also important differences in family organization, production pattern, sources and levels of household income, as well as changes in intrahousehold patterns of production and distribution between the project model and the settlers' activities.

Changing patterns of settler recruitment and background

It was originally planned that all the settler households scheduled to move to the project village would be installed in 1975. This was frustrated, however, by recurrent delays in almost every aspect of bloc development, including the construction of access roads and wells. As a result, the project

village was settled in three progressive installations, one year apart, from
1975 to 1977. There were some important differences in the background
and motivation of the settlers who emigrated from the home village to the
AVV in different years. All nine of the settlers who went in 1975 belonged
to clans with limited access to inherited fields (an average of 75–100% of
the land area planted was borrowed in most households). Since the fields
that a farmer lends tend to be on poor-quality soils or fields that he is about
to leave fallow, a high percentage of borrowed fields is usually associated
with lower food and cash crop production and fewer livestock.[10] It is this
category of farmers that is most severely affected by drought. There is little
doubt that many of the settlers who came in 1975 were attracted by the
project's promise of supplementary food, until the first harvest. It is also
doubtful that many of these first-year settlers would have had the means
to relocate their families to the AVV or to one of the more traditional sites
of agricultural immigration in the country's west on their own.[11] Although
the settlers who came in 1976 and 1977 tended to come from more prosperous
clan and extended family groups, they still reported cultivating a high
percentage of borrowed land and, from their reports and my familiarity
with the family members who remained behind, would have had very limited
resources with which to finance their family's emigration on their own.

By 1979 the emigration from the home village to the AVV had become
so large that the settlers' decisions to join the project were being influenced
by their desire to join other family members. This accounts for the fact
that by 1979 some of the wealthiest households from the home village and
the home-village-founded frontier towns to the north of it were moving into
a new neighboring bloc of AVV villages. By 1980 there were an additional
15 settler households in other planned settlements in the area and another
10 settler households of those who had either been born in the home village
or had male household heads who were descended from fathers who had
been born there. Almost all the officially recognized household heads from
the settlers' home village who signed up after 1975 had visited the project
at least once, before making their decision.

When the settlers emigrated to the AVV, the inherited fields they left
behind continued to be farmed by any members of the nuclear family who
remained in the home village. If no one was left, the land was reabsorbed
into the clan fields of the patrilineal extended family and a close male
relative would be given the right of determining who should farm the land.
Any redistribution of the inherited land outside the immediate family
members was through borrowing, with the borrower having rights of usage
and the absent settler retaining his permanent claims to the land for purposes

of inheritance. Any borrowed fields reverted to the family of the individual who originally made the loan.

Changing patterns of technological innovation

The results of the case study agree with the general conclusions of the AVV farm-monitoring program; the AVV failed in its major objective to encourage the settlers to adopt intensive cultivation practices. The case study illustrates, however, that the exclusive focus of the farm-monitoring survey on the settlers' participation in and income derived from the recommended agricultural package may have overlooked other areas of positive change. These include new areas of income growth from food and cash crop production, investment in livestock, community development and modifications by the settlers of the recommended practices in order to satisfy specific needs, such as the need for women and married sons and brothers for independent sources of income.

Changing patterns of agricultural production and distribution

In 1979, the settlers' cash income from the sale of cotton and other agricultural products was 172 000 CFA per household (US$750) and 38 000 CFA (US$160) per worker using the AVV system of labor and consumption equivalents.[12] This is more than seven times the recorded figures for the settlers' home village (38 000 CFA per unit labor versus 5000) and represents an increase in sales from the equivalent of 16–39% of the recorded production of all crops.

During the early years at the project, the settlers were involved in the heavy work of clearing new fields, families were small, and there was little time or money for non-crop activities like livestock raising and trade. By the fifth year, the Kaya settlers had accumulated stores of reserve grain and had paid off all or most of their initial debts to the project. Moreover, most households had substantially increased in size, as a result of the immigration of additional family members. This increase in household size on a fixed 10 or 20-hectare land base was an incentive for the settler households to move away from the recommended program through investment in non-crop activities like livestock maintenance, trade and the expansion of fields into areas outside the 1 hectare home site and the four bush fields in the prescribed crop rotation system. A second very powerful incentive for the settlers to increase the time spent on non-crop production activities was the desire to move away from a singular dependence on agriculture in the face of a high level of uncertainty about rainfall and the project's future. There was also, by the fifth year in the AVV village, a greater number of opportunities for commercial endeavors and for specialised trades such as masonry and mechanics.

In contrast to the results of the AVV farm-monitoring survey which concluded that there had been little 'surplus' grain production in the project during the years 1977–79, the case study showed that the average production per worker for the settlers in 1979 was three times the average quantity produced in the home village. This represents an averge of 515 kg per family above the minimum food standards established by the FAO.[13] The substantial differences in results between the two studies can be attributed to the fact that, in 1979, the Kaya settlers of the case study were in their third to fifth year at the project, whereas the farm-monitoring survey figures included all settlers – those who had been there only 1 year as well as those who had been there for 5. Moreover, the case study measured production on all fields, while the survey focused only on the official fields in the extension program.

Although the case study showed a slight increase in the sale of grain both in absolute terms and as a percent of total production between 1979 and 1983, the increase in sales did not appear to reflect either the overall increase in area planted or that settlers now had large cumulative stores. One of the important factors affecting this recorded figure was a substantial increase in the quantity of grain that was given as gifts or in exchange for livestock and hired labor. The largest category of gift exchange in terms of actual quantities was the food given to new settlers. This typically involved an established settler giving gifts of 100 kilogram sacks of grain to supplement the food rations the new settlers were receiving from the project. In most cases the new settlers could claim some sort of pre-existing lineage or affinal tie with their sponsor. By 1983, a growing number of families were involved in the direct sponsorship of new settlers. We estimated that the latter type of food aid was, in many households, the equivalent of 30–50% of the recorded harvest for 1982. Other grain was exchanged for livestock. The case study showed that the majority of the settlers sold the original oxen they had purchased from the AVV during their fifth year at the project and used the money to pay off their remaining debts. In most cases the replacement oxen were purchased from the local Fulani in exchange for grain. The cost of the animals did not appear in any of the data on marketing but was discovered during the research on purchase and resale of livestock.

Changing patterns of intrahousehold organization and production

There were also a number of important changes over time in the internal organization of the settler households. In the settlers' home village, a household's land was divided into fields that were cooperatively and privately worked. Women have traditionally farmed from 20–25% of the total area planted as private fields from which they alone control the harvest. These fields were usually positioned alongside the edge of the cooperatively worked fields in order to reduce travelling time between sites. Women have also had an active role in livestock and trade. No consideration was given to these personal activities in the original design of the AVV agricultural program.

By the fifth year, however, most women had reinstated a small area of private fields (6% of total land area planted in 1979), without any recognition from the AVV extension staff. Thus, they suffered a loss of income caused by a reduction of their private crop (from 16% of the total area planted to 6% and from an average of 22% of total grain production to 2%) and trade, as well as the much higher labor demands of the new technical package (from 622 to 1256 weighted hours per unit of labor). In recognition of this reduction in the women's semi-independent sources of income, the male household heads began to make cash gifts to family members, after the sale of the cooperatively produced cotton. They also gave gifts of 20–30 dried ears of corn to the wives of close friends and allies in the village. Over the course of the harvest, a woman could receive 100 to 200 kg of 'gift' corn in this manner.

The restudy in 1983 showed the reintroduction of semi-autonomous sources of income for women (8% of the total area planted in 1983), including the widespread introduction of small private fields for grain (from the equivalent of 1.2% of the recorded kilogram production of the crop to 4%). In contrast to what we discovered in the settlers' home area, almost all the women's grain was sold rather than used for family consumption. The male household head was then responsible for purchases of sauce condiments, school materials and clothing. The cash that the AVV women earned from cash payments and the sale of food products was usually used for personal needs such as jewelry, travel, clothes and gifts, or to purchase trade goods and livestock. In 1979, very few of the Kaya settlers' wives had animals; by 1983, several of the older wives had large herds of 20 to 30 goats. Two of the senior wives owned their own cattle.

Changing patterns of community organization

The settlers also passed through a series of stages in terms of their development as independent communities. The first 2 years were characterized by the settlers banding together, irrespective of region of origin or previous social status. This period was also characterized by a high level of dependence on the extension agents and few interactions with the indigenous inhabitants of the valleys.

One of the most visible signs of the settlers moving into a new phase of community development occurred when the male settlers' evening talk sessions shifted from the door of the extension agent to the homes of some of the emerging village leaders. This period was also characterized by active participation in evening catechism and Koranic schools. It was in this second stage that the villages began to divide themselves into different groups based on religious affiliation, economic status and extended kin. There was also a growing number of new conflict situations whose origins could be traced to the social impacts of the move, such as increased economic competition among the settlers and the greater importance of religious affiliation. In this same transition period there was a reinstatement of certain types of ancestral celebrations, religious tithing and harvest gift exchange. Greater participation in the ceremonies and rituals of the inhabitants of the surrounding indigenous villages, and greater independence from the extension agent in the resolution of internal problems (including the development of local markets and the administration of a cooperative mill) was also seen. There was also a greater ease in and frequency of dealing with the civil authorities of the new region for birth certificates, identity cards and taxes.

Interrelationships between economic and community change

A comparison of the different stages of community development with the economic results that the settlers were obtaining at different points in the project cycle suggests that the two were interrelated in important ways. It was during the second transition stage (after the first stage of project pre-planning and the actual move), when the settlers were just beginning to adjust to their new social and production environment and during which they showed the greatest dependence on the extension agent, that they were most willing to follow the recommended agricultural program. In the third stage, when the settlers were more at home in their new community and were able to adequately feed their families, they became more willing to experiment. These experiments included the modification of different aspects of the technical package (e.g. the creation of private fields, requests to cultivate corn instead of red sorghum as part of the official rotation, and cultivation of a larger total area than they were authorized to farm); and investment in non-crop production activities such as livestock and trade. Moreover, the major motivation that the settlers had for their sponsorship of new settlers was the desire to solidify their new communities and to insure their long-term survival in the region, which was not a given, due to resentment of the project by the original inhabitants of the valleys.

Conclusion

There is little doubt that the social impacts of planned settlement schemes as a strategy for long-term drought and famine relief are considerable. First, they provide policy planners with an opportunity to integrate their short-term concerns providing impoverished groups food with an alternative and presumably higher yielding and more drought-resistant living environment. Second, they provide the sponsoring governments with an unprecedented opportunity to coordinate the development of infrastructure and extension services in the new settlements to encourage the long-term development of the region.

What I have shown here is that the local level impact of these projects does not reflect either a passive or a static response to the imposition of specific scheme policies. Instead, it is necessary to view the economic responses of the AVV settlers and the aggregate results of these responses only after a period of time sufficient for the planned settlement scheme to undergo Scudder's stages of settler adjustment. An analysis of the case study and farm-management research on the AVV settlers shows that the settlers passed through a series of stages in terms of economic development and their responses to a proposed cropping package. Moreover, these stages of economic response were related to the social adaptation of the settlers to the project. Although it is unlikely that the duration and nature of the different stages of economic response would be the same in another land settlement scheme, or even another planned village within the same project, there is substantial evidence from this study and others (Chambers, 1969; Nelson, 1973; Scudder, 1981, 1984; Hansen & Oliver-Smith, 1982; Scudder & Colson, 1982; USAID, 1985*a*, *b*) that a stage-like pattern of social adaptation and economic performance of the settlers should be anticipated.

Moreover, the role of the AVV as a strategy for drought recovery also changed. Research in the settlers' home village supports the general thesis of McCann (this volume) that the effects of drought are not evenly dispersed but tend to concentrate in social units with differential access to high quality land and other agricultural resources. In the settlers' home village it was the poorer households with limited access to inherited land, few livestock and little (if any) reserve food who emigrated to the project in 1975. Over time, the very success of the AVV in providing an improved economic setting for these impoverished groups was a factor in attracting a wealthier group of candidates.

The implications of these different stages of settler recruitment and social impact are considerable. First, they suggest that the design of agricultural research and extension programs in areas of planned settlement must be viewed as a stage process. In the first year, when the settlers have just arrived, they are dependent on food aid, and involved in the difficult work of clearing fields and constructing housing. Their primary concern is to develop a stable subsistence base. In this period their response to a proposed package of technological innovations will tend to be less innovative and more a direct response to proposed programs than in later stages. Moreover, settlers in the early stages will be less willing to adopt voluntarily and of their own initiative proposed innovations that require a radical reorganization of production activities. In the case of the AVV project, it was only after the Kaya settlers had accumulated food stores and had paid off all or most of their initial debts that they were able to explore actively new production possibilities of the area. This is not to say that no attempt should be made to introduce new production techniques at the beginning of a land settlement scheme, but simply to suggest that project goals for large-scale technological transformation should be tempered during the initial stages.

Specific policy suggestions include the need for project administrators and donor agencies to monitor the changing patterns of settler opportunities, constraints, and goals and to incorporate this information into the modification of project research and extension programs. This includes the need to plan for and assess the impact of technological innovation on the intrahousehold organization of production. In addition, more careful attention needs to be paid to the delineation of evaluation goals. For example, the AVV was slow to realize that its emphasis on the cultivation of the four bush fields in the crop rotation system and the attainment of higher yields per hectare through the use of intensive cultivation practices was not necessarily the goal of the participating settlers. Admittedly, there are problems with the design of a farm-monitoring unit within a project. However, incorporating a monitoring system within the project structure increases the possibility that monitoring and evaluation results will feed back into the design of scheme policies and programs. Whenever possible, this type of farm-monitoring program should include the integration of conventional survey research with longitudinal case studies that examine the more specific adaptation of the affected settler and non-settler groups.

Consideration should also be given to having the case studies funded by an agency outside the land settlement scheme, so that the case studies can respect the anonymity of informants. Care should also be taken to distinguish between different project policies within the same category of spontaneous or planned settlement, for example, planned settlement based on different patterns of recruitment or infrastructure development.

Second, the autonomous actions by settlers to create a supportive economic and social environment are an essential ingredient to their long-term survival as independent communities, a theme discussed by Bratton (this volume) for rural farmers in Zimbabwe. The AVV study shows the important role of the settlers' independent dealings with the project administration in terms of modifying specific aspects of proposed cropping packages and support services. This includes the introduction of small areas of private fields for women and the settlers' refusal to sell their cotton to the marketing board on two different dates.

Supplementary food aid that older settlers gave to new settlers played an important role in buffering the severity of the move. Moreover, given the settlers isolation in the region, the social linkages created by the exchange of food aid played an important role in the development of permanent communities. Stemming from this observed pattern of stage emigration, there is a need for project planners to allow space for successive years of settlement from the same home region in the same settlement zone.

Moreover, it is suggested that there is a chronic need for comparative analysis of the impact of different models of village layout and recruitment on the evolution of community patterns and the settlers' economic results. Careful attention should be paid to the delineation of production units. For example, a recent report on settlement in the OCP zones in Burkina (WHO, 1984) compares the economic results of one type of settlement in units of 'undertakings' with the economic results of another type of settlement in terms of 'families' and a third in terms of 'individuals'. This is not the fault of the WHO document which was based on available data. It does, however, emphasize the need for coordination in the design of comparative research on different types of planned settlement schemes.

Third, scheme policies toward recruitment, food aid and credit must be directly linked to the macro-economic, political, and social goals that a project hopes to accomplish. The study of the AVV settlers agrees with Scudder's finding (1984) that settlers attracted to planned settlement schemes tend to be the poorer farmers in an area. It is also this group of farmers that has the least opportunity for emigrating to a more fertile and presumably less drought-prone agricultural zone, without some sort of government assistance. Therefore, if one of the main goals of a resettlement scheme is to provide new economic alternatives for drought-impoverished farmers, it is important to consider how different aspects of a resettlement and crop program can influence the willingness and ability of the poorest farmers to participate.

Fourth, if agricultural research and extension programs are to work with settlers who have been at the project longer, they will have to deal with a growing amount of diversity within the settler population itself. One potential source of variation includes the major shifts in settler background and the resources that the settlers bring to a project over time. Another source of variation includes the emerging differences between settlers in terms of management skills, investment capital, labor force and entrepreneurship.

To conclude, there is little doubt that many African governments will continue to support the planned settlement of rural farmers from areas that are drought-prone to the less drought-prone regions and from areas of persistent food deficits to areas of potential food surplus. An essential element of the long-term success of these development and relief programs will be the extent to which the implementing agencies are able to build on the settlers' initial responses to scheme policies in terms of production, income and community development. Failure to incorporate information on different levels of social impact can lead policymakers to underestimate the effects of planned settlement as a strategy to realize various goals for short-term drought recovery and the more long-term development of a country's agricultural resources. It can also lead to the design of agricultural research and extension programs that are inappropriate to the shifting needs, constraints and goals of the target farmers.

To the extent that a settlement scheme is able to plan successfully the relocation of a group of rural farmers at pre-drought (or higher) levels of per capita income and food production, the project administration has accomplished a difficult task. Once the settlers are able to move from this initial stage of economic and social adjustment to participate in a wide range of new economic activities, the administration is faced with a categorically different situation with a distinct set of policy concerns. If a land settlement scheme is successful in the early stages, it creates a group of farmers with a very different set of opportunities and constraints in later years. The challenge to researchers and policymakers alike is to examine the influence of these different levels and evolving patterns of social impact on the attainment of long-term macro-level project goals for drought recovery, economic growth and development.

Notes

1. Onchocerciasis, or river blindness, is a disease transmitted to humans by the female fly, *Simuleum damnosum*. The fly carries the larva of a parasitic worm, *Onchocerca volvulus*, which spreads into the epidermal tissues of the skin eventually reaching the anterior chambers of the eye. Clinical indication of the disease appears only after repeated bites from infected flies. Effects include skin discoloration, itching, subcutaneous nodules and, in the later stages, eye lesions that may result in blindness. It has long been recognized that the Volta Basin of West Africa is one of the worst endemic onchocerciasis areas in the world. A United Nations survey in the early 1970s estimated that nearly 700 000 km^2 with a population of 10 million were affected. Of this number, an estimated 1 million people were infected and 70 000–100 000 were either blind or suffering serious eye impairments (WHO, 1980, 1). Since 1974 the infected river valleys have been sprayed repeatedly with abate, a biodegradable organophosphate which destroys the fly's larvae. By 1978 the Onchocerciasis Control Programme covered some 654 000 km^2 of a seven-country area including parts of Togo, Benin, Ivory Coast, Niger, Mali, Ghana and Burkina Faso. The spraying is scheduled to continue for a 20-year period to cover the length of time an infected person remains contagious.

2. The primary mechanism for the collection of data was an economic survey of a random sample of households in all of the major AVV village clusters. The unit of research in the AVV was the '*exploitation*', defined as the social unit that cultivates one of the 10 or 20 hectare AVV farms. It is the *exploitation*, so defined, that receives access to a registered landholding. Moreover, it was assumed that crops planted on this landholding would be cultivated cooperatively under the supervision of an adult male who is recognized as the official *chef d'exploitation* or household head. It is this official household head who represents the group in extension programs and in contractual dealings with the AVV for insurance, equipment purchases, credit and sales. Each sample household in the survey was visited once a week by an enumerator with a packet of questionnaires. During the interview, the male head was asked questions concerning any labor and non-labor inputs (e.g. fertilizer, manure and pesticides) on the bush fields that the household was authorized to farm under the crop rotation system. Other questions in the interview focused on the cash income, expense and loss associated with non-crop production (e.g. trade, crafts, livestock) and the sale and non-market distribution of food and cash crops (Murphy & Sprey, 1980). Enumerators were also required to measure the total area planted and harvests for each of the cultivated bush and house fields. Enumerators received the same initial training as extension agents and were supervised by the central office of the AVV Statistical Service. The Statistical Service was also responsible for the collection and analysis of the survey questionnaire.

3. The baseline research was funded through a grant from USAID to the Department of Agricultural Economics, Purdue University (AFR-C-1257 and AFR-C-1258) that was referred to as the Purdue West Africa Project.

4. The case study was one of two funded through the same grant from USAID to the Department of Agricultural Economics at Purdue University that provided supplementary support for the AVV farm monitoring program. A second case study, conducted by anthropologist Mehir Saul (1980, 1983), looked at the effects of AVV planned settlement on an indigenous village near Kaibo. A Technical Assistance Grant from SECID's (South-East Consortium for International Development), Center for Women in Development supported the restudy in 1983. The home village is located in the area outside the regional capital of Kaya, while the project village is in the third village (Village V3) in a six village cluster of AVV settlements outside the town of

Mogtedo. The two sites are separated by 120 km (200 km by the main road). In the first year, the principal investigator (the author) lived in the settlers' home village and in the second year in the AVV project village. She supervised trained enumerators who gathered information on crop and non-crop production activities and income in the two villages [See McMillan (1983) for a more detailed explanation of procedures.] In contrast to the AVV project plan, there was no clearly defined and terminologically distinct unit that was referred to as a household in the settlers' home village. If one used a working definition of household as 'the social unit that works together and eats together', then most of the home village households could be described as members of kin-based residential groups that worked certain fields collectively and/or relied on the harvest of these cooperatively worked fields for the basic food and cash needs of the group.

5. Under the crop rotation system, each of the six fields to which a settler household has access was planted in a cycle of crops that was supposed to preserve soil fertility, when used in combination with the recommended levels of fertilizer and cultivation methods. Since two of the six fields must always lie fallow, no household was supposed to have more than four fields under cultivation in a given year.

6. Various reasons have been given for this uneven distribution of population, including the cultural cohesion of the Mossi who constitute the dominant ethnic group. In the precolonial period the Mossi were grouped into a band of related kingdoms and subkingdoms that were a major cultural, economic and political force in the region.

7. The best-known example of this population movement is the massive emigration of Mossi farmers to work in the more developed coastal countries (Deniel, 1967; Conde, 1978; Coulibaly, Gregory & Piche, 1980; Finnegan, 1980). A second population movement of extreme importance is that of Mossi farmers emigrating toward the less-populated agricultural areas in the country's south (Remy, 1973, 1975).

8. The 'gross result' for each sample household was considered to be the cash value of the recorded kilogram production of each crop at local market prices; minus the cash costs of seed, fertilizer and insecticide; plus the cash value of the recorded kilogram production of the house garden. In the next stage of analysis the cost of tool purchases, depreciation of the animal traction equipment and credit were subtracted to obtain the 'net result'. The analysis of these calculated figures was broken down by village cluster and length of residence in the scheme (Murphy & Sprey, 1980).

9. By 1980, it was increasingly clear to project administrators and donor agencies that it would be necessary for the AVV to transform itself into a more flexible, less costly program capable of working with the indigenous inhabitants as well as with spontaneous and planned settlers. In 1982, the project underwent a complete reorganization from a single centralized planning and development agency into a series of regional planning units (Unité de Planifications or UPs) and separately planned and funded settlement zones (Unités de Développement or UDs) (AVV 1981a, b, c, 1983). Although the central AVV continued to play a role in project planning, coordination, and the provision of special research and development services, the direct responsibility for development of a prescribed package of agricultural innovations as well as settler installation, extension worker recruitment and training, and agricultural credit was administered by the UPs in coordination with the existing structures for agriculture, health, education, forestry and water development in a region.

10. The main exception to this were farmers who managed to supplement their crop production with cash earned from trade.

11. Loading bills for the vehicles that transported the settlers from the home to the project village show several of the settlers arriving with only a few sacks of foodstuffs.

12. The AVV uses a system of labor equivalents to determine the amount of land a

household receives and a similar system to determine the distribution of supplementary food during the first year. This potential for labor is measured by a labor index which assigns weights to persons according to sex and age. Since an adult male is considered to have the work capacity most readily transferred to a variety of tasks, this is the standard unit and is assigned a value of one. Women and children are assigned lesser values (0.75 for adult women, 0.50 for teenage boys, 0.25 for a female over 55, etc.). The use of labor and consumption equivalents to standardize the units of comparative analysis (so that one does not calculate the 'average' household income based on units that may range from 3 to 35 residents and 1 to 12 workers in size) is a standard and hotly debated topic in farm management research. For purposes of comparison with projected income and production figures of the AVV, I have used the AVV system.

13. Figures are based on the 1975 Project Identification Report of the Dutch Government for the AVV which estimated a minimum daily food requirement of 2 230 calories per person per year including losses during storage (Murphy & Sprey, 1980, 22). Figures represent the difference between the recorded grain per resident and the recommended 240 kilogram minimum of cereals.

References

Autorité des Aménagements des Vallées des Volta (AVV) (1974). *La mise en valeur des Vallées des Volta: Principes d'aménagement et perspectives.* Ouagadougou: AVV.
Autorité des Aménagements des Vallées des Volta (AVV). (1981a). *Nouvelles méthodes d'intervention de l'AVV. Tome 1: Principes Généraux.* Ouagadougou: AVV.
Autorité des Aménagements des Vallées des Volta (AVV). (1981b). *Nouvelles méthodes d'intervention de l'AVV. Tome 2: Annexes.* Ouagadougou: AVV.
Autorité des Aménagements des Vallées des Volta (AVV). (1981c). *Nouvelles méthodes d'intervention de l'AVV. Tome 2: (suite) Annexes.* Ouagadougou: AVV.
Autorité des Aménagements des Vallées des Volta (AVV). (1983). *Programme de Développement, 1983–1986.* Ouagadougou: AVV.
Barrett, V., Lassiter, G., Wilcock, D., Baker, D. & Crawford, E. (1981). *Animal Traction in Eastern Upper Volta: A Technical, Economic and Institutional Analysis.* East Lansing: Department of Agricultural Economics, Michigan State University.
BEI-agrer. (1978). Report on the AVV Program, 1978–1982.
Berg, E. (1975). *The Recent Economic Evolution of the Sahel.* Ann Arbor: University of Michigan, Center for Research on Economic Development.
Berg, E., Bisilliat, J., Burer, M., Graetz, H., Melville, R., Volyvan, V., Park, J., Sawadogo, R., Sederlof, H. & van der Meer, K. (1978). *Onchocerciasis control program.* OCP Economic Review Mission (mimeo). Ouagadougou, Burkina Faso: Onchocerciasis Control Program.
Broekhuyse, J. (1974). *Développement du Nord du Plateau Mossi*, 4 vols. Amsterdam: Département de recherches sociales, Institut Royal des Tropiques.
Brokensha, D., Horowitz, M. & Scudder, T. (1977). *The Anthropology of Rural Development in the Sahel: Proposals for Research.* Binghamton, NY: Institute for Development Anthropology.
Caldwell, J. (1975). *The Sahelian Drought and Its Demographic Implications.* Overseas Liaison Committee, Paper No. 8. Washington, DC: American Council on Education.
Chambers, R. (1969). *Settlement Schemes in Tropical Africa: A Study of Organizations and Development.* London: Routledge & Kegan Paul.
Christensen, C., A. Dommen, A., Horenstein, N., Pryor, S., Riley, P., Shapouri, S. & Steiner, H. (1981). *Food Problems and Prospects in Sub-Saharan Africa: The Decade of the 1980s.* Foreign Agricultural Economic Report No. 166. Washington, DC: Africa

and Middle East Branch, International Economics Division, Economic Research Service, US Department of Agriculture.

Christodoulou, D. (1965). Land settlement: some oft-neglected issues. *Monthly Bulletin of Agricultural Economics and Statistics*, **14** (10), 1–6.

Colson, E. (1971). *The Social Consequences of Resettlement: The Impact of the Kariba Resettlement upon the Gwembe Tonga.* Kariba Studies, IV. Manchester: University of Manchester Press.

Conde, J. (1978). *Migration in Upper Volta.* Washington, DC: World Bank, Development Economics Department.

Coulibaly, S., Gregory, J. & Piche, V. (1980). *Importance et ambivalence de la migration voltaïgue. Les migrations voltaïgues*, Tome I. Ouagadougou: Institut National de la Statistique et de la Demographie.

Deniel, R. (1967). *De la savanne à la ville: Essai sur la migration des Mossi vers Abidjan et sa region.* Aix-en-Provence: Centre Africain des sciences humaines appliquées.

Dettwyler, S. (1979). Khashm El Girba irrigation scheme: an examination of agricultural development in the Sudan. In *Changing Agricultural Systems in Africa*, ed. Emilio Moran, pp. 15–36. Williamsburg, Virginia: Department of Anthropology, College of William and Mary.

De Wilde, J., McLoughlin, P., Guinard, A., Scudder, T. & Maubouche, R. (1967a). *Experiences with Agricultural Development in Tropical Africa: The Synthesis.* Baltimore: The Johns Hopkins University Press for the International Bank for Reconstruction and Development.

De Wilde, J., McLoughlin, P., Guinard, A., Scudder, T. & Maubouche, R. (1967b). *Experiences with Agricultural Development in Tropical Africa: The Case Studies.* Baltimore: The Johns Hopkins University Press for the International Bank for Reconstruction and Development.

Finnegan, G. (1980). Employment opportunity and migration among the Mossi of Upper Volta. In *Research in Economic Anthropology*, vol. 3, ed. G. Dalton, pp. 291–322. Greenwich, Connecticut: JAI Press, Inc.

Hansen, A. & Oliver-Smith, A. (eds.) (1982). *Involuntary Migration and Resettlement: The Problems and Responses of Dislocated Peoples.* Boulder, CO: Westview Press.

Hervouet, J. (1977). *Peuplement et mouvements de population dans les vallées des Volta Blanche et Rouge.* Ouagadougou: ORSTOM.

Hervouet, J. (1978). La mise en valeur des vallées des Volta Blanche et Rouge: un accident historique. *Cahiers ORSTOM: Serie sciences humaines*, **XV**, 1, 81–97. Ouagadougou, Burkina Faso: Onchocerciasis Control Program.

Hervouet, J., Clanet, J., Paris, F. & Some, H. (1984). *Settlement of the Valleys Protected from Onchocerciasis After Ten Years of Vector Control in Burkina Faso.* OCP/GVA/84.5 (mimeo).

Hilton, T. (1959). Land planning and resettlement in Northern Ghana. *Geography*, **XLIV**, 227–40.

Jeune Afrique (1975). *Atlas de la Haute-Volta.* Les Atlas Jeune Afrique. Paris: Editions Jeune Afrique.

Lewis, W. (1964). Thoughts on land settlement. In *Agriculture in Economic Development*, ed. C. K. Eicher & L. W. Witt, pp. 299–310. New York: McGraw Hill.

McMillan, D. (1983). A resettlement scheme in Upper Volta. PhD. thesis, Northwestern University. Evanston, Illinois.

McMillan, D. (1984). *Changing Patterns of Grain Production in a Resettlement Scheme in Upper Volta.* Washington, DC: Center for Women in Development, South-East Consortium for International Development.

McMillan, D. (1985). Monitoring the evolution of household economic systems over time in farming systems research. Revised version of paper presented at the *Workshop on Conceptualizing the Household: Issues of Theory, Method and Application*. Cambridge, Massachusetts: Harvard Institute for International Development (HIID).

Marchal, J. (1978). L'Onchocercose et les faits de peuplement dans le bassin de Volta. *Journal des africanistes*, **48**(2)., 9–30.

Mellor, J. (1966). *The Economics of Agricultural Development*. Ithaca, NY: Cornell University Press.

Moran, E. (1979). Criteria for choosing successful homesteaders in Brazil. In *Research on Economic Anthropology*, vol. 1, ed. G. Dalton, pp. 339–59. Greenwich, Connecticut: JAI Press Inc.

Moris, J. (1968). The evaluation of settlement schemes performance: a sociological appraisal. In *Land Settlement and Rural Development in Eastern Africa*, ed., R. Apthorpe, pp. 79–102, Kampala: Nakanga Editions, 3.

Murphy, J. & Sprey, L. (1980). *The Volta Valley Authority: Socio-Economic Evaluation of a Resettlement Project in Upper Volta*. West Lafayette, Indiana: Purdue University, Department of Agricultural Economics.

Nana, J. & Kattenberg, D. (1979). *Etude Préliminaire de la Question des Migrants Spontanes*. Ouagadougou: DEPE, section sociologie, AVV.

Nelson, M. (1973). *The Development of Tropical Lands: Policy Issues in Latin America*. Baltimore: Johns Hopkins Press.

Nikyema, J. (1977). *Mémoire de Fin d'Études: Migration Organisée de Population (AVV)*. Ouagadougou: RHV, Direction de l'Authorité des Aménagements des Vallées des Volta.

Ouedraogo, F. (1976). *L'aménagement du Bloc de Mogtedo dans le Cadre de la Mise en Valeur des Vallées des Volta*. Bordeaux: Université de Bordeaux III, Institut de géographie tropicale et d'études régionales.

Reining, C. (1982). Resettlement in the Zande Development Scheme. In *Involuntary Migration and Resettlement*, ed. A. Hansen & A. Oliver-Smith, pp. 201–24. Boulder, CO: Westview Press.

Remy, G. (1973). *Les migrations de travail et les mouvements de colonisation Mossi*. Paris: ORSTOM.

Remy, G. (1975). Les migrations vers les 'Terres Neuves': un nouveau courant migratoire. In *Enquête sur les Mouvements de Population a Partir du Pays Mossi*, dossier 1, fascicule 2. Ouagadougou: ORSTOM.

Rey, C. (1980). *Analyse de la Situation Agro-Pastorale dans l'ORD du Centre-Nord, Kaya (Janvier 1980)*. Kaya, Burkina Faso: RHV: Service Departémental de Planification du Departement du Centre-Nord.

Saul, M. (1980). Beer, sorghum and women: production for the market in rural Upper Volta. *Africa*, **51**(3), 746–64.

Saul, M. (1983). Work parties, wages and accumulation in a Voltaic village. *American Ethnologist*, **10**(1), 77–96.

Scudder, T. (1981). *The Development Potential of Agricultural Settlement in New Lands*. Third Six-Month Progress Report on United States Agency for International Development Grant No. DSAN-G-0140. Washington, DC: USAID.

Scudder, T. (1984). *The Development Potential of New Lands Settlement in the Tropics and Subtropics: A Global State of the Art Evaluation with Specific Emphasis on Policy Implications*. AID Program Evaluation Discussion Paper No. 21. September. Washington, DC: USAID.

Scudder, T. & Colson, E. (1982). From welfare to development: a conceptual framework for the analysis of dislocated people. In *Involuntary Migration and Resettlement: The*

Problems and Responses of Dislocated People, ed. A. Hansen & A. Oliver-Smith, pp. 267–87. Boulder, Colorado: Westview.

Takes, C. (1975). *Land Settlement and Resettlement Projects: Some Guidelines for their Planning and Implementation*. Wageningen, Netherlands: International Institute for Land Reclamation and Improvement.

United States Agency for International Development (USAID). (1985*a*). Problems and issues in African river basin planning. Washington, DC: Africa Bureau/S/T. For presentation at *Conference on River Basin Development*, 18–19 April 1985. Submitted by Settlement and Resource Analysis Cooperative Agreement (USAID), Clark University and Institute for Development Anthropology. Washington DC: USAID.

United States Agency for International Development (USAID). (1985*b*). Summary. *Conference on River Basin Development, Irrigation and Land Tenure in Africa*, 18–19 April 1985. AID/AFR/DP/PPE.

World Bank. (1978). *Agricultural Land Settlement: A World Bank Issues Paper*. January. Washington, DC: The World Bank. (T. J. Goering, coordinating author).

World Health Organization (WHO) (1980). *Onchocerciasis Control in the Volta River Basin Area: Information Paper*. OCP/74.1 Rev.4, May.

World Health Organization (WHO). (1984). *Socioeconomic Impact of the Onchocerciasis Control Programme in Upper Volta 1973–1983*. Presented to the JPC during the meeting of the Fifth Session, December 1984. OCP/GVA/84.2.

14

Evolution of food rationing systems with reference to African group farms in the context of drought

WILLIAM I. TORRY

Department of Sociology/Anthropology, West Virginia University

Introduction

Populations which fall critically short of food must ration it. Emergency rationing, accordingly, makes up an essential component in the food systems of many societies. Researchers commonly regard the disintegration of traditional food-rationing systems in Africa and elsewhere as a factor in the frequency of recent famine occurrences in drought-prone regions of developing countries. Interventions planned by the governments of these countries have supplemented or replaced the rationing devices in force in precolonial society, but often with dubious results. Ingrained corruption, logistical handicaps, and underdeveloped famine-diagnostic procedures on the part of Third World governments are among the factors recognized as limiting the effectiveness of official relief measures.

This chapter explores the validity of the assumptions that (*a*) 'primitive rationing'[1] affords broad-based insurance against food emergencies, within the productive constraints of peasant and tribal economies and (*b*) the modern state lacks this capacity.

Very little attention is paid *per se* in economics and anthropology to the issue of emergency food rationing in stateless societies.[2] Indeed, just a single family of ethnographic models proves to have an explicit bearing on this subject. These 'social storage' theories are discussed and then critiqued on both factual and logical grounds. With this analysis it is established that stateless societies are not only unfit structurally for rationing food on a sustained basis, but actually defeat a basic emergency rationing goal. That objective involves forestalling famine at its terminal stages, marked by abnormally high mortality rates and mass migration. Other evidence of tribal rationing is adduced to substantiate this claim. Next, discussion turns to emergency food rationing (EFR) systems evolved by the modern state. Their principles of operation are contrasted with 'primitive rationing' modes to highlight what are in theory the advantages that they offer over traditional systems of emergency relief.

The final section discusses transformations in progress in traditional systems of farming and herding with special reference to African societies. Group farming/herding schemes have evolved fairly recently in sub-Saharan countries, partly as an adjustment to persistent food shortages. In this section, possible weaknesses and benefits of these schemes in connection with food emergencies brought on by drought are explored. Particular attention is given to the prospects of integrating EFR measures within the structure of group farming systems so as to combine short-term emergency assistance and long-term development planning.

Social storage theories

Only one area of social theory treats emergency food management in stateless societies as a primary object of analysis. It covers a range of scarcity situations interpreted by a small family of exchange models I label 'social storage' theories.[3] These theories share several characteristics.

1) The societies to which they apply experience abrupt and drastic changes in weather, causing severe reductions in local food supplies.
2) The effect of these perturbations is variable. Some households suffer greater losses of subsistence than others.
3) To avert starvation when weather conditions decline, households will establish long-term exchange partnerships extending to each a line of credit, some of which stays in reserve for emergencies. An institutionalized conversion mechanism redeems stored credit for food or food tokens (items exchangeable with food).
4) Conversion mechanisms equalize uneven food distribution during times of dearth, promoting individual and population survival.

As an institution, social storage is not anything more than an ethnographically ordinary exchange partnership, founded on principles of delayed reciprocity, which processes credit into shares of food and other items. What theorists make of these partnerships in terms of an emergency security function is significant. I shall consider two social storage models, including Wiessner's (1977; 1982) account of the !Kung San *hxaro* institution and Epstein's (1967) characterization of South Indian *jajmani* exchanges.

The patent incongruities between San and South Indian social systems should not distract us from important functional similarities *hxaro* and *jajmani* reciprocities share. This social storage function narrows inter-household differences in food income and awards deficit households a margin of food security, satisfying minimal requirements for survival. I shall construe social storage as a rationing procedure. By this line of reasoning the critique of social storage functions in food emergencies can be extended to include distributive limits of other forms of exchange found in stateless societies. I shall understand all forms, including social storage, as different expressions of a common set of rationing principles, and then show why such primitive rationing principles are similarly wanting as strategies against famine. This argument creates a foil for the claim – developed later – that only EFR systems (at least potentially) furnish effective emergency safeguards.

Hxaro

Wiessner (1977, 1982) has described in admirable detail the *hxaro* system of gift exchange practiced by the Kalahari !Kung San who inhabit northwest Botswana, northeast Namibia, and southeast Angola. *Hxaro* exchanges, she stresses, reduce subsistence risks, especially from droughts. Localized rainfall and biotic conditions in this semiarid territory vary resource productivity even within areas no larger than 50 square miles, checkering San country with pockets of dearth and sufficiency in relatively dry years. *Hxaro* arrangements, made by males and females alike, even out these variations, creating for each family 'ties which distribute its risk over the population and thereby assure that losses will be covered in bad years' (Wiessner, 1982, 66). *Hxaro* describes balanced, delayed exchanges conducted primarily among consanguineal kin (real or fictive blood kin), acquiring maximum strength within cores of perpetually coresident consanguines. People hesitate at 'hxaroing' kin related by marriage (affines) directly, preferring to relay gifts through their spouses.

Hxaro transactions dominate economic life. Non-food items enter *hxaro* networks, and families acquire about 75% of their belonging *via hxaro* channels. People respect the habit of soliciting gifts from *hxaro* partners only. *Hxaro* gives one access to the bush products found within his partner's area of land rights (*n!ori*), to the partner's possessions, and to the hospitality of the partner's bandmates. In this way, *A* of camp 1 may give *hxaro* gifts to *B* of camp 2 who passes them on to *C*, *D*, and *E* of *B*'s camp. When *A* visits *B* he expects hospitality from *C*, *D*, and *E*, who benefit from his (*A*'s) partnership with *B*.

Individual *hxaro* networks cross many camps. A married couple with small children will average 26 partners located in about 7 different areas. *Hxaro* partnerships almost double for couples with mature children and span 8 areas. Partnership density and frequencies of exchanges fall off with distance. Approximately 70% of one's partners live within his/her home area and in adjacent areas up to 50 km away, while about 30% reside at distances of 50–200 km. Wiessner thinks that the number and spatial distribution of *hxaro* partners 'should be sufficient to cover all critical risks' (1982, 74).

Wiessner's analysis is convincing within the perimeters of routine scarcity. When she takes her model, however, beyond the realm of the ordinary and projects it onto situations of acute scarcity, it begins to churn up unsettled questions. She notes 'in times of hardship a person's losses can be absorbed by others in a population, if risk is well distributed' (1982, 65), and states that 'each family creates ties which distribute its risk over the population and thereby assure that losses will be covered in bad years' (1982, 66). These passages signal something askew in Wiessner's model. It postulates a state and imposes a hypothetical condition for inducing that state. Yet, without verification it assumes that the condition will hold good and then concludes that the state eventuates. The hypothetical condition is optimal distribution and the postulated state is food security. But, does the condition come about and must the state inevitably result from it? In fact, Wiessner's ethnography does not tally completely with this optimal distribution postulate. Even if it did, the claim that *hxaro* mutuality 'assures that losses will be covered in bad years' would be far from proven.

Productivity limitations of San technology and some features of *hxaro* exchange challenge Wiessner's thesis. *Hxaro* dependency limits food procurement options and critically so, perhaps, during food emergencies. Only *hxaro* partners share food resources obligatorily, and yet a person's network of *hxaro* associates comprises but a fraction of his total social universe. Practical constraints impinge upon the size of *hxaro* grids. These bonds are demanding, subject to cheating, and are liable to deteriorate after a few years, so that 'partners are chosen with great care' (Wiessner, 1977, 100). Then too, an individual must avoid acquiring more *hxaro* partners than he can handle, without overextending his productive capacity. The set of persons with whom one 'hxaros', accordingly, 'cannot be said to be the maximally capable group for reducing risk . . . ' (Wiessner, 1977, 100–1). Can victims of scarcity find refuge in *n!ori* of people with whom they do not enjoy *hxaro* partnerships? San ethnography provides no outright answer. Considering that sharing among coresident consanguines can be problematic, even in times of sufficiency, we may doubt that such succor would come easily. Even in camps composed of consanguines not all related by nuclear family ties, 'sharing does not always go smoothly', and 'sharing obligations are ambiguous' (Wiessner, 1977, 311–12).

Weak dispute settlement procedures and *n!ori* land-use conventions can also curtail emergency survival options. Long droughts uproot family units from preferred food hubs, as they first assemble around receding pools, then mass near permanent water sources in their *n!ori*. In this phase of aggregation camps grow large. As bush resources near water holes dwindle, the family treks further afield in search of food and tensions within large camps mount. Conflict is difficult to arrest, bands break up and people disperse to *n!ori* that are controlled by more distant *hxaro* partners. The *hxaro-n!ori* system, however, does not provide assured food sanctuaries for drought refugees. In fact, it 'can be troublesome in years where environmental failures are localized' (Wiessner, 1977, 65), as the norm of movement only to *n!ori* of *hxaro* partners superimposes on the drought an additional layer of restrictiveness on land use. In effect, *n!ori* territoriality 'accentuates the consequences of the variation in rainfall and vegetable foods' in lean years (Wiessner, 1977, 60).

A San family on the move might 'on paper' call up two or three dozen *n!ori* options. But intense and/or prolonged droughts rapidly shrink these options. One factor limiting choice is the San custom of avoiding *n!ori* that start crowding up. The placement of approximately 75% of a household's *hxaro* partnerships within 50 km of its *n!ori*, moreover, would preclude for it quite a few migration prospects, assuming the drought fans across a broad stretch of territory. Should that happen, any pockets of land spared by the drought would absorb, for reasons just discussed, only small concentrations of San migrants.

If a general drought sets in, there is no predicting from Wiessner's published records how many *hxaro* partners a household can settle with for the duration of the scarcity. Bonds with more distant partners may be 'ambiguous', she states, unless reinforced by occasional exchanges of gifts. Regrettably, Wiessner's otherwise rich dossier allows her no way of ascertaining !Kung unremitting attention to nurturing far-flung relationships or of gauging the reliability with which these *hxaro* associations satisfy the minimum requirements of scarcity-stricken families. Since only about one-third of a person's *hxaro* relations are ever very active, and the most active are ones established with neighboring households, questions of dependability of more distant partnerships understandably arise.

Wiessner offers scarcely any data of the kind she would need for standing her subsistence security argument up to the test of a food crisis. So, her few accounts of the mettle of *hxaro* associations, during the drought and flood she observed, invite close attention. Wiessner noticed widespread hoarding of food, the denial of aid to the sick and disabled by all but their nuclear families, and the disaggregation of camps. She describes, for example, a case of a married couple too ill to gather food. Their children lived somewhere else, and the more distant relatives with whom they settled refused any responsibility for attending to their welfare, preoccupied with pressing needs of their own close kin. The wife died and her husband got a ride on a truck to his daughter's camp:

> At the burial of the old widow, there were open accusations aimed at her relatives of having starved her to death, but she had no strong ties left in the areas and the accusations fell on deaf ears, bringing little response and no consequences. (*Wiessner, 1977, 133–4*)

Hunter's became more secretive about their kills. One group Wiessner reported on consumed 'as much meat as possible' at a kill site, distributing the rest surreptitiously back at camp in the evening. Wiessner's exposition suggests a pervasiveness of such behavior under the circumstances. Though !Kung would not condone it, food hoarding was rampant.

Inter-camp sharing dissolved fairly rapidly. Sharing between camps 'works well when each person, except for a few disabled, is bringing in food on a regular basis and not too many have needs' (Wiessner, 1977, 157). The system apparently founders, however, when scarcity prevails. Eventually camps break up, and those who move away go only to areas occupied by *hxaro* partners. Wiessner (1977, 164) notes that

> a relationship must be created well prior to calling up on it in times of need. A person cannot expect to and will not go to a new area in the case of environmental failure, to begin a new relationship and expect to be integrated into the group on that basis.

Weissner's data do not actually pinpoint either the particular locations to which displaced households moved or the kinds of accommodations provided by their hosts. Presumably, most relocatees managed satisfactorily and if they did, it might have been because of the location of ranches, clinics and other non-aboriginal facilities stationed at the sites to which they migrated. Perhaps these facilities were sustaining an unusually large and stable population of camps controlling more resources than remote bands could command. If so, this leaves open the question of migratory advantage under aboriginal conditions.

A Jajmani *system of South India*[4]

This system of social storage differs from the hunting/gathering case discussed above. Environmental disturbances in this situation are accessory to land ownership rules, as a cause of household differences in access to food. Further, *jajmani* insurance derives from exchanges between coresident households, while the most potent insurance dividends from San exchanges purportedly stem from inequalities in access to food, although they supposedly lead to short-term equalities. The *hxaro*, as we find it reported, keeps food access fairly uniform in perpetuity.

Ownership and use rights over arable land generally concentrated in the hands of upper-caste residents of early colonial and precolonial Indian villages. Landlords disposed of a share of the grain produced by their estates by exchanging it for labor supplied by dependent service-caste households. Within this structure of intercaste exchanges, landless and smallholding families could earn a fixed minimum annual wage in grain, paid out at harvest time and augmented, perhaps, with occasional gifts for special services. For the lower-caste households and untouchables, *jajmani* arrangements paid off primarily as a hedge against starvation during droughts and other crises, while the landlord secured a cadre of available servants he would need, especially for labor-intensive years of plenty. So, artisan and laborer wages poorly reflected the bounty awarded during bumper years, as patron–client associations put the landlord under no obligation whatsoever to share in the windfall with his clients. But crop failure, often induced by drought, moved the system of reciprocities toward a more egalitarian mode, compelling landlords to share harvests and reserves with dependent families. In this manner starvation risks diffused through the village population.

Epstein (1967) is one of the many students of rural Indian society accepting this characterization of *jajmani* reciprocities without challenge, although she goes further than anyone to examine just how much obligatory reciprocity conventional models of *jajmani* transactions imply. Before irrigation came to Wangala, the principal food of this Mysore village was millet. Epstein works out millet requirements per annum for each Wangala household, then compares this figure with an estimate of the village-wide millet crop for an average poor growing year. Based on yield per acre estimates and a count of landholdings in acres per caste, she calculates average household shortfalls and surpluses for landlord and artisan/laboring-caste families. Relating these figures to her schedule of annual fixed payments in grain, she concludes that lower-caste landowners would be able to make up the deficit in poor years by obtaining grain remunerations from about seven to eight patron households, which apparently they managed to do. Landlord families would consume no more and no less than their dependents. Equal distribution was more than an ideal. It was the norm wherever *jajmani* systems developed.

Epstein's deductive approach makes her conclusions suspect. Variations in important factors that must be quantified before the testing of her formulations are unreported. For example, she would need surveys of intrafamily patterns of food distribution, household size, and the amount, quality and productivity of agricultural holdings per household. She would have to know the extent of crop loss and the size of stocks of each production unit in her sample. She would also want evidence of the propensity of individual landowners to abide by their *jajmani* obligations in a difficult situation. One can now consider fragmentary documentation from other village studies which casts considerable doubt on the validity of Epstein's conclusions.

Epstein relates that landowners denied village newcomers a share of millet, during a bad growing season, effectively forcing them out of the community. Harper (1968, 52) describes a situation during the 1876–78 Mysore famine in which wealthier landowners were 'hard pressed', while many of their bonded laborers starved and died. Mencher (1982, 41) remarks that when Kerala society was still feudal, 'it was always the poor who had less to eat' when 'times were hard'. The wages paid to slaves of feudal Trichinoply, Tanjore and Malabar (Kerala and Tamil Nadu) 'were either cut or stopped altogether when there was no employment', as would be the case during crop failure episodes (Hjejle, 1967, 96). Also, Harijan (*hali*) laborers 'were dismissed everywhere', during the Gujarat famine of 1890 (Breman, 1974, 116). Officials, chronicling nineteenth century famines, repeatedly describe sharply drawn differentials in wealth persisting among caste groups throughout a period of drought (e.g. Government of India, 1901). Landlords from relatively contemporary villages with functioning *jajmani* structures are reported to lower their annual grain payment rate for service- and artisan-caste dependents or to refuse to pay them altogether during a scarcity (Lewis, 1958, 63; Rao, 1974, 304).

Further, Hindu customary law does not enjoin egalitarian conduct of a *jajmani* (upper, landowning-caste) toward a *kamin* (dependent, servicing-caste), under any circumstances. What is especially difficult to accept is the proposition of upper-caste patrons voluntarily putting themselves on the same plane of destitution as their bonded laborers and slaves as a matter of principle, considering the servitude to which masters customarily relegated dependents (Dumont, 1980). More generally, the very idea of patron beneficence toward clients, resulting in a subsistence guarantee during food crises affecting pre- and early colonial southern Asian villages (Scott, 1976), has been persuasively repudiated (Popkin, 1979).

Primitive rationing and reciprocal exchange

Social storage partnerships incorporate definite rationing components. Rationing applies when a plan takes effect for allocating foods and services for which demand widely exceeds supply. A rationing schedule limits every authorized recipient to portions of a scarce resource and/or limits its availability to certain groups. It regulates supplies in a manner that permits every beneficiary a minimum share, yet restricts overall consumption. Equitable allocation would fail totally if price were allowed to dictate access. Suppliers control for this by placing rations under a subsidy which evens out differences in consumer purchasing power. Rations can be apportioned according to random selection, regular users served, or any of a multitude of other criteria.

Partners in stateless societies operate under social storage exchanges as combined suppliers and takers of rations. Rations here consist of stored credit and items obtained with it. When providing rations, an individual will adjust supply and demand levels of items in his possession through a crude form of quota restriction: he parcels out the supplies he controls and limits the number of partners he supplies. Some items are apportioned in fixed quantities, as the *jajman*'s payments of grain. Others, an individual divides in accordance with the amount of goods he owns as well as the role relationship and burden of debt he has with ration takers. When too many demands come down on a person at once, and if time permits for making choices, he may abide by a preference ordering, reserving allotments, or larger allotments, for customarily large or reliable creditors.[5]

Social storage exchanges engender minimum share distribution. *Jajman* patrons see to it, says Epstein, that their clients get no less than the minimum set by agreement. San honor at least part way, within their means, every partner's requests, just to keep the relationship alive. In addition, we may interpret customary creditor generosity as akin to a subsidy, adhering to the principle that credit should be affordable so that the partnership will continue. This subsidy is a concomitant of delayed reciprocity as an exchange type (Sahlins, 1972) and sets it apart from barter and market transactions. *Jajman* patrons routinely help out *kamin* clients with paying fines and debts and present them with occasional gifts of grain, while their clients may do extra work for them and stand by them in a dispute. In sum, a partnership reservoir creates for an individual, as a ration provider, a pool of persons especially entitled to his reserves. In a general scarcity, he allocates his wealth very restrictively and favors exchange partners over most other individuals counted as friends and kinsmen.

Playing the ration receiver turns one's calculations to drafts he can make on the rations of his partners. As a ration receiver, he faces quota restrictions. As a ration giver, he will levy them. The *maximum* expected entitlement of a ration receiver, correspondingly, may hinge on the debt the giver owes him.

Give-and-take, patterned on quotas, offers definite advantages to people who participate in social storage exchanges. In a manner of speaking, partnership allegiances trim down an individual's welfare caseload. Partnership commitments give one legitimate grounds for limiting claims on his stores when food is generally scarce and, in principle, they entitle his family to a minimum share of stores in the possession of other households. This second benefit represents the subsistence safety net that has been given emphasis by social storage theorists.

Yet, social storage is not an ideal system for rationing food, when food is not in abundant supply. At the foundation of the problem of organizing even distribution is the reciprocal structure of social storage exchanges. A reciprocal relationship forms when participating parties come to an understanding regarding obligations each begets when the other provides a good or a service. The partners (as single individuals or groups of individuals) can be unequal in status, exchange goods and services of disproportionate value, and requite some of their obligations promptly, or resign their debts gradually. The social storage cases discussed above involve relationships of reciprocity. This brings us to the next point.

A social storage system of reciprocities cannot guarantee an effective defense against a general food emergency. For several reasons, reviewed below, it is constitutionally unfit for this role.

1) While groups participate in some forms of reciprocal exchange, social storage reciprocities join self-interested *individuals*.

2) Exchange partners are counted on when problems occur. This fiduciary element binding the partnership solidifies with the slow passage of time.

3) The carrying costs of maintaining each partnership may be steep. Sizable investments of time, energy, gifts and trust help explain why exchange partnerships are non-substitutable. Hence, one cannot seek temporary replacements for an exchange partner from that partner's kin group, nor is a kin group collectively liable for defaulted obligations of its members to their partners, unless, of course, the partnership coincides with other exchange compacts for whose solvency entire groups share responsibility.

4) The size of an ego-centered partnership reservoir is no larger than the household is productive. Low overall productivity of the societies that theories of social storage describe, and the relatively high maintenance costs weighing down on individual partnerships, limit a typical reservoir to a mere fraction of a household's total network of relationships.

5) Economic mutualism, facilitated by face-to-face interaction, favors the spatial propinquity of partners. Many partners will live within walking distances of one another for at least part of the year.

6) Partnerships are not welfare instruments. A partner can claim food and food tokens but will not acquire an entitlement to subsistence.

Summing up, these points present a good case for skepticism about emergency-insurance functions that social storage theorists impute to reciprocal exchanges. An affordable partnership reservoir on average neither grows very large nor, because of the non-substitutability factor, expands instantly. And if many partners live nearby, the household's interests can actually be countered, since a fair proportion of these households would presumably be exposed to the same environmental stressor. Thus when a drought develops, most or all of the households in this group will be wanting, if the reference population is hunters/collectors or poorer client households of peasant societies. Prosperous patrons with grain laid away might be benevolent to a degree, but not so generous as to put themselves on or below the level of an impoverished partner's want. Such action is doing more than is expected and is inimical to household survival.[6] Rich behave like poor; they hoard. Thus, there are good reasons for doubting that the food supplied by ego-centered partnership reservoirs would necessarily pull many families through a food emergency. Social storage reciprocities might stave off temporary setbacks threatening the fortunes of single households, but would not produce that salutary effect in a general scarcity persisting for many months – and that is what a drought-induced famine involves. This conclusion fits the facts presented by social storage theorists.

Primitive rationing in a broader perspective

In social storage systems, one finds elements of rationing that may be global for stateless societies, *and* which discourage the aims of food emergency management. One can go to ethnographies and notice many institutions with distributive procedures which accord with principles of rationing. To list a few, rationing can allocate water from deep wells and irrigation canals (Gray, 1963; Helland, 1982), induct individuals into secret societies (Ottenberg, 1971), and designate brides for wife-receiving units in preferential marriage systems (Levi-Strauss, 1949, 39–40, cited in Douglas, 1982, 65; Dickemann, 1979). A comparative analysis of pre-state rationing principles would also include some aspects of status allocation identified by Douglas in her seminal research on primitive rationing (Douglas, 1963,

1982). We will discuss Douglas's investigation briefly, since it should serve
as a touchstone for confirming certain generalizations, which we will come
to presently, about rationing in stateless societies.[7]

In most stateless societies, Douglas explains, so-called primitive money
– the cowrie shells, wampum and metal rods often passing for currency of
market commerce – functions more like coupons issued by a moderate
wartime rationing agency. Money is an exchange medium establishing
standard rates of conversion by which any salable good and service can be
bought or sold with a single currency. The pricing system of a money
economy integrates exchange spheres by converting commodities and
services into equivalencies – one unit of this has the same dollar or pound
value as ten units of that. Coupons as a rule do the opposite. They segregate
exchange spheres and limit important transactions to certain ranges of
persons.

Douglas makes out primitive rationing to be an analog of the group
method of rationing scarce commodities used by European and American
governments during the first and second world wars (de Scitovszky, 1942;
Tobin, 1952). Items transfer under the group system only within their own
classes. In this way, meat stamps, hypothetically, buy beef, poultry and
fish, but not any commodities classified as clothing (e.g. hats, shirts and
pants). Commodities and not coupon currencies fall into ration groupings,
since one currency cannot substitute for another. Primitive rationing systems
make commodities coupons and group them together: 'any of a specified
group of valuables can serve as a coupon for attaining a specific status'
(Douglas, 1982, 66).

Complex rationing economies like the precolonial Tiv, discussed by
Douglas, and the Lele too, might elaborate several ration currencies, each
restricted chiefly to specific items and persons. Tiv placed chickens, baskets
and food crops at the bottom (or subsistence) rung of their three-tiered
hierarchy of exchanges. They exchanged these objects, one for another, but
could not trade them 'up' for guns or slaves, much as the latter could not
sell for wives at the top level of prestige. Trading down the hierarchy only
defeated the object of rationing currency within isolated spheres of exchange,
namely acquiring and maintaining status for senior males. This was done
only in cases of domestic emergencies.

Coupons redeemable for status had to be regulated, so they were kept in
relatively short supply and monopolized by a minority, operating as a kind
of rationing authority. This privileged group controlled the flow of coupons
and consumer commodities. The gate keeping role of influential persons
(usually elders) and the sheer expense of producing or acquiring costly

coupons, like raffia cloth, prized shells, or cattle, limited the production of coupon currency. This control was marshalled other than by the fiat of an agency, and actors did not perceive it in the light of a systematic transfer of resources from one sector of society to another. Rather, it was levied through countless transactions engrossing individual patrons and clients. Douglas (1982, 79) describes rationing controls in these terms:

> In such transactions there is an appearance of centrally imposed control, but it is deceptive. No central governing body imposes the rates of exchange. The exchange control emerges by the decisions of individuals striving to hold to their position of advantage in a particular social structure.

Over the developmental cycle of the domestic group, the monopoly of rationed currency and objects of exchange rotated, as younger clients succeeded their older patrons. Hence, in Douglas's view, primitive rationing, over the long haul, provided equity of distribution. Douglas regards primitive rationing as a system of delayed reciprocity, with one generation taking from the other, then eventually giving back: 'today's giving will be recompensed by tomorrow's taking' (Polanyi, 1944, 54, cited in Dalton, 1981).

We can now return to the project of explaining why primitive rationing is not a satisfactory adaptive approach for a society menaced by acute scarcity of food. As a big drawback, ration givers are also ration receivers. This fusion of roles readily fosters distributive biases favoring duty over need. Taking status rationing as an example, duty arises partly from a senior's obligations to repay his juniors who, while complying with norms of status denial thrust onto the line of their own aspirations, subsidize his privilege. In most societies these subordinates are primarily close junior kin who must bide their time with proportionately fewer wives, cattle, pieces of ornate textile, and so forth, until their turns come and they ascend the status ladder. Meanwhile, senior males monopolize stores of status and wealth. Institutionalized hoarding, which this is, gave Douglas pause to wonder, when she first hit upon the expository value of rationing, if 'the rationing system is a bad analogy, since the allocation of several [Lele] girls to older men in fact created the scarcity which it also regulated' (1963, 66). Later, rethinking the subject, the privilege of hoarding, she detected, inexorably revolves through generational divisions and, therefore, viewed over a community's life cycle, distributes brides and status equitably: 'mere seniority, relatively greater age, in the due course of time is expected to make up to each deprived junior the privileges foregone in his youth' (1982, 64). Thus sharing makes hoarding possible and hoarding makes sharing necessary. Given this insight, she could now state with conviction that 'rationing is an appropriate model for interpreting institutions which seek to ensure an equal distribution of high status' (1982, 64).

The apparent paradox Douglas encountered delivers us back to the theme of emergency rationing. The hoarding *qua* rationing paradox sorts itself out on the axis of time. Seniors very gradually release hoarded goods and power through the social system. But over time, it would appear, everything hoarded is eventually redistributed *and* shared equally. I suggest that any such system does not give ground for effective emergency rationing, because redistribution is long-phased and poorly serves the interests of equality based on need.

Food emergencies are transient, requiring short-order rationing dependent upon extensive pooling and need-based policies of egalitarian redistribution. Status rationing is a drawn-out process, extending over years. Even if groups with corporate functions, such as age-sets, allocate status, they still lack the capacity for amassing productive property owned by their members or for appropriating food as a collective good, except on special ceremonial occasions. The operational unit of redistribution is the household or extended family and not a class of persons. Fathers take from and give back to sons and other close relatives so that reward is governed, to a considerable extent, by the role relationship of ration giver and receiver. Need determines when and how much to give.

An effective emergency rationing system would assume a central rationing agency endowed with authority for compelling senior males to share their rations with one another and with juniors, whoever they may be. Conversely, any system which biases redistribution in favor of role and status will indubitably leave large allocative gaps that food emergencies automatically widen. Slaves, criminals, patrilineal widows, members of despised caste groups and other structurally inferior persons who enjoy few or no exchange relationships with a senior male, or hold on to exchange relationships that are fragile, will in a crisis almost certainly receive relatively small rations, if any at all. Historical documents amply bear out this sad fact.[8]

Social storage exchanges, and indeed all forms of primitive rationing, share the same discriminatory attributes of status rationing. They keep the public at large away from scarce resources and restrict their access to people related by marriage alliance, stock associateship, consanguinity, or in other specific ways. Rather than generating public goods, they privatize availability of needed items. A stateless society thus fragments into hoarding enclaves. Moreover, hoarding within enclaves, while generally denounced, is virtually impossible to keep under control during an emergency. In short, primitive rationing is not a way of providing the greatest good to the greatest number, during a food emergency.

Emergency food rationing systems

Food rationing systems develop at the state level of societal evolution and extend back historically as early as dynastic Egypt. Here, our concern is with modern rationing schemes, which have come into their own within the past century.[9] Industrializing nation-states equipped with strong central governments, extensive road and rail networks, public distribution systems in foodgrains, and a sophisticated agricultural infrastructure, operate the most successful emergency programs. Every developing country severely affected by drought has had practice providing emergency relief. Most tried approaches are variants of the basic methods discussed below.

Food-rationing systems, regardless of structure, adjust food access of target consumers by controlling quantities, prices and routes of distributed supplies. While food-rationing programs cater specifically to nutritional needs of hungry consumers, often they also provide seeds, fodder and credit for the restoration of agricultural production. Ration currency and identification cards and books will validate the consumer's rationing privileges, although some programs do not make ration eligibility this formal. They issue rations on a first-come-first-served basis to anyone in need. In most cases central governments run modern rationing systems.

Common to every food-rationing system are the compound functions of procurement and redistribution. Obviously, supplies must be isolated and aggregated, before being allocated at critical points of need. Procurement ends and redistribution begins, once grain from suppliers reaches state-controlled warehouses. A state amasses this supply by purchasing directly from producers using statutory levies or through commissioned purchasing agents. It might also import stocks from foreign growers at prevailing market prices or accept grain cargo as a charitable donation or at concessional rates packaged into a broad assistance agreement. To increase its share of current harvest offtakes and carryover stocks, the state can impose limits on private trade in grains, even pulling millers and merchants from the market, and it can requisition from various sources rice, wheat and other food commodities. Furthermore, by ordering embargoes on interregional movements of grain, authorities can concentrate grain commerce in food-deficit zones. No matter how stringent the procurement policies enacted by governing agencies are, they leave some grain in the market for consumers and merchants who can afford it.

On the distribution side of a rationing system, practically every government will set some food aside for free distribution through the venerable technique of public feeding. This is one way of bringing food within the reach of destitute children and elderly persons, the disabled, and refugees, whose physical, economic and legal handicaps make the subsistence quest for them especially difficult. Another method, popular with many governments, involves payment-in-kind to the able-bodied for work performed on public relief projects. Food-for-work programs ration food by regulating the quantities allotted and restricting work entitlements in accordance with age, sex, location, need and other criteria. Ration shops constitute another pillar of redistribution. These stores hold licenses to sell procured food grains and other necessities at officially subsidized prices. Money may be rationed too. It is doled out to people afflicted with the handicaps mentioned already or is locked into wages exchanged for work on construction projects.

What do emergency food rationing (EFR) systems have to offer drought victims that primitive rationing cannot? To handle this question it will be convenient to consider a series of contrasts between primitive and modern emergency rationing operations.

1. EFR systems generate public goods collected largely by the involuntary pooling of scarce resources. Stateless societies cannot effect involuntary pooling on any scale. Clans and other groups with corporate functions, by voluntary action of their members, will round up food stocks and other resources for special feasts and in payment of fines, but ethnography affords no evidence of community pooling of food over the better part of a drought emergency, even when many people have some food to eat. If anything, they hoard (Firth, 1959; Turton, 1977; Turnbull, 1978; Dahl, 1979).

2. Some EFR programs make ration entitlements and ration income independent of a household's normal productive capacity and its ability to recompense a rations supplier. In stateless societies this situation could not be more different. Relatively productive households develop a larger reservoir of exchange partnerships than other family units. Very much the same can be said for the higher-status against lower-status households in the status-rationing systems discussed by Douglas (1982). Productive households are not this advantaged under EFR controls. In fact, they may even be compelled to make short-term sacrifices from which poorer households are exempt. Compliance with procurement levies is one such instance. But the sacrifices enjoined on grain-giving households do not reward them with extra rations.

Playing by the rules, the hungry rich receive no more from the rationing system than the hungry poor.[10]

Exchanges directed by norms of reciprocity, which primitive rationing transactions represent, exhibit little tolerance of free-riders. Strings attach to transactions. Upon receiving, you accept a debt, repaying it when you can or when you must. Claims to receive are earned claims. Status rationing may look like an exception, as roles by ascription lay down rights, but it is not. A male can be entitled to share a senior agnate's status dividends, in principle, but for that right to be ratified, the junior must pay dues over time. Accession to a senior status, down the road, signifies that debts or dues are paid off (Stewart, 1977; Baxter & Almagor, 1979). All the while, the rations received from the senior help a junior pay his dues and, additionally, make him acknowledge the measure of his indebtedness. Unearned claims, however, do have a place under EFR provisions. Food and money doles are free. Works projects and ration commodity sales, while requiring the return of one benefit by another, demand payments (in cash or work) that hardly reimburse the government for its expenses.

3. Welfarist in principle, EFR programs run at a loss, aiming at securing for needy households a minimum subsistence income for the duration of an emergency. Distributive schemes discussed earlier are anything but consumer-welfare programs.

4. To the extent that EFR schemes achieve their welfarist agenda, they supplant important community work routines made impracticable by the scarcity. For scarcity victims, primitive rationing has nothing even remotely comparable to offer.

5. Emergency rationing currencies are not meant for surplus accumulation. They cannot be saved for future consumption, becoming invalid after a set date, and are subject to revocation without notice by the rationing authority. Primitive rationing perpetuates the life span of a ration currency and allows it to be stored and even multiplied. Thus, in the short run, status-rationing systems bestow ration entitlements on the (male) minority, but over the life cycle of the community, a (male) majority will acquire rationing rights. Under emergency rationing, the majority obtains rations in the short run, but no one is entitled over an extended interval, because the rationing system is immobilized in a matter of months.[11] EFR systems thus have the adaptive advantage of serving the majority of ration-seeking persons rapidly, while the cycle of redistribution in a tribal society may last a generation and is not compressible into narrow slots of time.

Group-farming systems[12]

Sub-Saharan countries have been experimenting for over four decades with group-farming designs for overcoming environmental and economic problems facing traditional family farms. Some countries have gone far toward putting these designs into action. Cooperative programs in Tanzania, Ethiopia and Mozambique reach, or are planned to include, most agricultural households. Group or collective experiments are planned most often by governments. They range from independent households cooperating in one or two tasks, such as the marketing of a single crop, to the collective organization of most major enterprises, the pooling of all assets, and the sharing of all risks and profits. Tanzanian *ujamaa* cooperatives, Ethiopian *wereda*, and the large Sudanese state farms of Gezira and Khasm el Girba conform in some respects to this more comprehensive model of group farming.

Group schemes, in principle, offer governments several attractions. They supply a framework for efficient delivery and supervision of agricultural inputs. This infrastructure facilitates farmers' utilization of technical and administrative assistance and the marketing of crops and livestock. Household adoption of standardized practices of production and marketing overseen by government agencies affords administrators and planners a vehicle for achieving wide-ranging development goals. Foremost is the boosting of rural employment and income levels through expanded agricultural activity combined, ideally, with safeguards for conserving natural resources. Group-farming systems, again in principle, achieve these targets by giving farmers and landless laborers affordable access to credit, extension, marketing, and health and educational services. They also make land reforms easier to enforce. Such systems provide structures for centralized procurement and redistribution of grain surpluses, and coordinate public works construction during slack seasons of the agricultural/herding cycle. Potentially, they furnish tools for political indoctrination in the name of nation-building, and for law enforcement. Schemes which eliminate middlemen from the production process and demand strict limitations on land sales can arrest social-class formation and encourage security of land tenure.

Group-farming systems seldom work as planned, however and failures in their major objectives are extensively documented (Widstrand, 1970; Worsley, 1971; Dorner, 1975; Bhaduri & Rahman, 1982). Low levels and rates of farmer participation pose an overriding problem. Where state bureaucracy takes charge of most or all major production and marketing decisions and/or where crop production and prices are not adequately subsidized, farmer incentive and self-reliance may be low. Programs of land redistribution which do not thoroughly take into account variations in land quality and access to permanent water points can create economic inequalities among households. 'We–they' opposition dividing management and workers obstructs productive efficiency. Problems of debt repayment on loans from government agencies and marketing outside of official channels erode project earnings and penalize farmers who comply with project bylaws. Persisting tribal inequalities and sectional rivalries interfere with collective activity and goals fostered by scheme developers.

What bearing do group farming experiments have on drought management and food rationing? Overall, it is correct to maintain that, *even if planners do not feature the group projects they design in semiarid regions as instruments for minimizing adverse drought impacts, this is precisely a primary function these projects assume.* In the special context of dry zone land use, assuring food sufficiency through technical and organizational innovation means making a development scheme and the area it impacts resistant to drought. Plainly, a scheme can fail completely without laying down provisions for drought protection. Drought resiliency rarely is addressed in the research literature on group-farming schemes in Africa (for exceptions see Swift, 1977; Merryman, 1983). Weaknesses and strengths likely to appear during drought emergencies can nevertheless be inferred from case studies of less extreme situations. We will consider drought resilience of group farms in several respects and then examine separately possible emergency food-rationing functions that such schemes may subsume.

We need several criteria for evaluating how well group schemes access and ration food under emergency conditions. One set of factors discriminates internal and external costs and benefits. Can what is good for the scheme be harmful to neighboring communities? Another set differentiates the scheme's effectiveness at helping itself during an emergency and its ability to obtain help from government agencies.

A serious potential problem involves regional economic stability affected by interactions between scheme households and bordering communities. Often governments favor scheme over non-scheme farmers in the allocation of economic resources (e.g. Reed, 1975, 365). For obvious reasons, favoritism will probably redouble during an extended drought. First, we can appreciate the advantages administrators calculate of disbursing food and money, gathering information, and implementing public works projects among aggregated populations supporting a managerial infrastructure already assimilated with government bureaucracy at the local level. This situation contrasts with the difficulties of reaching dispersed and autonomous hamlets and villages. Second, insofar as scheme experiences politicize group farmers and persuade them of a preordained entitlement to public resources, participants in the scheme may rally and exert considerable pressure on government officials. By ably lobbying for aid, scheme members may earn for themselves a distinct advantage in regional competition for official emergency assistance. Third, governments funnel more sizable investments into group farming settlements than into 'underdeveloped' neighboring agricultural communities. So, administrators have a stake in protecting these expensive investments from losses.

Other inequalities between scheme and non-scheme residents can be inferred from available studies. Irrigated projects may assure crop production through the length of a drought. Generally, authorities cannot completely prevent surplus crops produced by these schemes from passing illegally into local markets and selling at steeply inflated prices which contribute to a reduction in food availability to poor households living outside the scheme. Further, scheme farmers with steady incomes through the drought may seek distress purchases of land and livestock owned by neighboring farmers whose livelihoods are put in jeopardy because of such transactions.[13]

Another little-explored dimension of regional imbalances concerns the capacity of a group scheme to incorporate drought refugees from neighboring villages and adjacent districts and regions. Where land-tenure laws limit ownership or leasehold rights to a very restricted range of persons, the group-farm system may not be as absorptive of refugees as non-scheme villages, except perhaps if the scheme fails to meet its enrollment quotas. Yet, scheme households may eagerly exploit the labor of the refugee population (Yeld, 1968; Johnson, 1979).

Some drought-related problems within the scheme can be postulated. If India's experiences with rural credit cooperatives is any indication, many households will almost certainly delay or default on the repayment of credit from institutional lenders because of crop and livestock losses. Another problem mentioned frequently is where farmers defer important decisions to managers and other government agents. Households which do not participate fully in production decisions within the scheme and whose members perceive their input as unimportant may not take adequate precautions against famine, but rather count on the government to assume all of the risk and responsibility, which authorities may not be equipped to do during a food emergency.

Self-provisioning of schemes through a drought crisis has not been the subject of published case studies. The probability of projects sustained by irrigation technology surviving a drought has been indicated already. In areas of rainfed agriculture and in pastoral settings, schemes that build up contingency reserves of food and income with the benefit of development planning may be able to get by on less emergency assistance than non-scheme communities in drought years. Indeed, food storage and monetary reserves should constitute a cornerstone of group scheme designs for climatically harsh environments. Many such risk management strategies are possible. Swift (1977), for instance, envisages for pastoral cooperatives that will serve Somali herders a program in which the cooperative buys animals from nomads at subsidized prices in years of drought so that members can earn cash with which to buy grain. Livestock released for sale could presumably be kept in good condition and sold back to nomadic buyers at fixed prices after the drought. The cooperative might operate a fund issuing emergency credit and grow and store food crops and fodder for lean periods. Swift (1977, 304) speculates that

> the cost of such [operations] would have to be carried by cooperative marketing levies and other funds, with central government help in bad years. The cost of this would almost certainly be less, even in cash terms, than the high cost of relief operations as that in the 1974–75 drought. In addition, the social benefits would be incalculable.

A combination of self-help and government sanctions can enhance viability of group schemes during episodes of drought. That brings us to the issue of rationing. For all their documented shortcomings, group schemes in many cases have in place a distributive infrastructure adaptable for community-wide involuntary rationing of emergency supplies. Compared with the power that it yields over non-scheme villagers, state bureaucracy generally would impinge substantially on policies made within the jurisdiction of leaders (including managers) of group-farming settlements. It is likely, therefore, that authorities can apply comparatively more pressure on scheme leaders to comply with officially sanctioned, need-based norms of emergency rationing. That is, administrators more closely approach their avowed goals of ensuring 'fair' distribution grounded in need, when they operate relief programs in scheme communities. The state will also get further rationing cash and/or food through relief works projects where group schemes are involved, given that members and managers probably have acquired experience working on bureaucratized projects for building roads, bunds, dams, irrigation canals and the like, associated with the construction of their settlements.

The food-procurement capabilities of these schemes may also enable voluntary rationing within small household networks. Arguably, cooperative task sharing does not itself foster altruism and generosity. Still, we can reasonably expect that a household's propensity to ration out its wealth will correspond with the security of access to food that its members perceive themselves as having over the anticipated duration of a drought. The more confidently household heads foresee the prospect of replenishing depleted stocks, whenever the larder approaches some critical level, the less anxious they will be about sharing their stocks with other households. This sense of security hinges in part on the availability and effectiveness of scheme programs which accumulate and store reserves of grain and other consumable wealth. Such systems may entail a surcharge assessed periodically on family earnings, harvests and livestock production, and enlist methods which protect these reserve assets against losses from animal pests, thieves and other agencies.

The rationing incentive will depend also upon the strength of the household head's belief in the premise of economic mutuality binding scheme participants. All other considerations aside, a farmer will be more prone to ration supplies with other scheme residents, if he views his contribution as an investment in his own long-term welfare. This reasoning will be most evident, perhaps, in schemes in which profits and finances are highly collectivized.

Certainly, voluntary rationing may not be a dominant factor of distribution, and the complacency of scheme members about their own planning for the worst may be a real problem. The scheme may run dangerously low on food reserves and finances. All of these possiblities considered, group farms may still come out ahead of traditional non-scheme communities with the selective advantages they have in procuring outside assistance rationed on a need-determined basis.

This section has concentrated on exploring issues that establish group farms as a factor affecting community and regional adjustments to food emergencies set off by droughts. There has been no attempt to make the investigation of connections among group-farming systems, droughts, and food-distribution methods either systematic or complete, and many difficult questions remain. The connections suggested do not take into account significant variations from one scheme to another in the type of agricultural operations collectivized, the degree of collectivization attained, and the socio-economic characteristics identifying project participants. Differences from case to case in the severity of drought and in the human and fiscal resources of central governments will also qualify the generalizations put forward. Admittedly, a great deal of research lies ahead, but at this stage it is probably sufficient to illustrate briefly what types of issues are involved.

Conclusions

This chapter examines how proficiently societies get supplies of food to where they are critically needed during periods of acute drought. We investigated this issue by comparing the food rationing capabilities of stateless societies and modern nation-states. Emergency food rationing (EFR) programs, when operated properly, excel primitive rationing systems, as far as sustaining broad sections of a community or larger population unit over prolonged periods.

It is doubtful that any sub-Saharan countries have achieved high EFR proficiency. This requires bureaucratic structures coordinated efficiently in short order and on a large scale. Post-war India, of course, stands out to illustrate the EFR performance levels that governments can attain. In comparing primitive rationing and EFR systems in African contexts, however, our attention should not linger on known shortcomings of the latter, although for other purposes this matter becomes critically important. The point is that EFR systems harbor a potential which exceeds adaptive limits restricting peasant and tribal institutions in environmentally extreme situations.

How might EFR systems fulfill their potential in drought-vulnerable areas of Africa and in other arid regions in the developing world? This question inspired our exploration of links between EFR mechanisms and some newly evolving forms of community organization absorbing farmers, herders and the dispossessed. We considered mechanisms by which group farming schemes can enhance EFR effectiveness. This brief analysis indicated that (despite their many weaknesses) these schemes, inasmuch as they couple with EFR operations, may prove to be an evolutionary advance in community management of food emergencies. This sweeping hypothesis will certainly demand extensive testing and refinement. For now, however, it may point out at least one potentially fruitful direction for researchers just beginning to take notice of the little-explored interface between famines and economic development.

Thinking through the merits of group-farming schemes has brought into focus some side effects conceivably produced by such adaptations on families not participating in these schemes but likewise beset by drought. A scheme might monopolize regional allotments of government aid and use its more bountiful resources for impoverishing famine-stricken neighbors. Following this reasoning, we might add that it would not be surprising to find instances of politically powerful scheme farms domineering in much the same way the smaller or less prosperous collective settlements co-existing with them in a multi-scheme region. If group farms can actually exercise such leverage, then administrators, planners and researchers will have to think carefully about the complex problem of putting the concentrated resources of a scheme to constructive use during emergencies, so that they can serve the needs of villages less abundantly endowed with development and emergency assistance.

Notes

1 I owe this expression to Douglas (1982).
2 Throughout, the arguments that apply to tribal and band (i.e. stateless) societies generally extend to the peasantry of state-level societies as well.
3 'Social storage' is a term adopted from Halstead (1981) and Halstead & O'Shea (1982). By their usage, items stored are material tokens of food that 'can later be re-exchanged for food' (1982, 93). My application of the concept is more in the line of Wiessner's less specific definition, denoting 'a social method of pooling risk through storage of social obligations' (Wiessner, 1982, 85).
4 This section is adopted from Torry (1986).
5 Imagine that a drought takes hold of areas A, B and C, but spares an adjacent zone D, occupied by family P. Many of P's *hxaro* partners live in the affected areas. P anticipates from past experience that some of these partners will come to it and request permission for extended visiting privileges, but not every would-be visitor will come at once. Assume further that P does not know in advance who these partners will be. It finds that out, once they arrive. In view of the incompleteness of P's information about this potential queue, it will likely accept petitions for *n!ori* visiting space on a first-come-first-served basis instead of by order of preference. Of course, it would be interesting to ascertain if, as its *n!ori* space fills, P begins to turn certain partners away, saving space for preferred associates whose *n!ori* it knows is in trouble.
6 Popkin (1979) extends much the same argument to peasant societies in general, and evidence cannot be found where it would not apply to tribal and band societies as well, aside from partnerships linking close consanguineal kin.
7 Again, very much of what follows also fits the pattern of peasant existence in archaic and modern states.
8 Examples come from Sterling (1950), Bohr (1972) and Miers & Kopytoff (1977).
9 For an idea of the variety of procedures that food rationing systems exhibit in different countries, compare China (Oi, 1983), India (Chopra, 1981), and Egypt (Alderman, von Braun & Sakr, 1982).
10 An exception to this point may be the edge that big farmers have in access to production credit (Torry, 1986).
11 To be more accurate, the emergency thrust of a rationing system comes to a halt. In countries or regions where ongoing ration programs are absent and EFR alone does this job, the ration system is in fact immobilized when a general scarcity expires. The situation is somewhat different where rationing components are ever-present. In these cases, the number of fair price shops and public works projects, for example, expanded during the emergency, diminishes by government decree and quotas may be lowered (Torry, 1986). Other components, including feeding programs and emergency credit, shut down.
12 For convenience the term 'group farms' will also designate group ranches.
13 This generalization has to be qualified for herders whose problems may be not finding buyers, owing to the glut of livestock available for sale and the poor condition in which they tend to be during a drought.

References

Alderman, H., von Braun, J. & Sakr, S. A. (1982). *Egypt's Food Subsidy and Rationing System: A Description.* Research Report 34, Washington, DC: International Food Policy Research Institute.

Baxter, P. T. W. & Almagor, U. (eds) (1979). *Age, Generation, and Time: Some Features of East African Age-Organizations.* New York: St Martin's Press.

Bhaduri, A. & Rahman, M. D. A. (1982). *Studies in Rural Participation.* New Delhi: Oxford and IBH Publishing Co.

Bohr, P. R. (1972). *Famine in China and the Missionary: Timothy Richards and Relief Administrator and Advocate of National Reform, 1876–1884.* Harvard East Asian Monographs 48. Cambridge: Harvard University Press.

Breman, J. (1974). *Patronage and Exploitation: Changing Agrarian Relations in South Gujarat, India.* Los Angeles: University of California Press.

Chopra, R. N. (1981). *Evolution of Food Policy in India.* New Delhi: MacMillan India Ltd.

Dahl, G. (1979). *Suffering Grass: Subsistence and Society of Waso Borana.* Stockholm Studies in Social Anthropology. Sweden: University of Stockholm.

Dalton, G. (1981). Symposium: Economic anthropology and history: the work of Karl Polanyi (a comment). *Research in Economic Anthropology,* **4,** 69–93.

de Scitovsky, T. (1942). The political economy of consumer's rationing. *Review of Economic Statistics,* **24,** 114–24.

Dickemann, M. (1979). The ecology of mating systems of hypergynous dowry systems. *Social Science Information,* **18**(2), 163–95.

Dorner, P. (ed.) (1975). *Cooperative and Commune: Group Farming in the Economic Development of Agriculture.* Madison: University of Wisconsin Press.

Douglas, M. (1963). *The Lele of the Kasai.* London: Oxford University Press.

Douglas, M. (1982). Primitive rationing. In *The Active Voice,* ed. M. Douglas, pp. 57–81. London: Routledge & Kegan Paul.

Dumont, L. (1980). *Homo Hierarchicus: The Cast System and its Implications.* 2nd ed. Chicago: University of Chicago Press.

Epstein, S. (1967). Productive efficiency and customary systems of rewards in rural South India. In *Themes in Economic Anthropology,* ed. R. Firth, pp. 229–52. London: Tavistock.

Firth, R. (1959). *Social Change in Tikopia.* New York: MacMillan.

Government of India (1901). *Report of the Indian Famine Commission 1901.* Calcutta: Office of the Superintendent of Government Printing.

Gray, R. F. (1963). *The Sonjo of Tanganyika.* Oxford: Oxford University Press.

Halstead, P. (1981). From determinism to uncertainty: social storage and the rise of the Minoan Palace. In *Economic Archaeology: Towards an Integration of the Ecological and Social Approaches,* ed. A. Sheridan & G. Bailey, pp. 187–214. Oxford: BAR.

Halstead, P. & O'Shea, J. (1982). A friend in need is a friend indeed: social storage and the origins of social ranking. In *Ranking, Resources and Exchange,* ed. C. Renfrew & S. Shennan, pp. 92–9. Cambridge University Press.

Harper, E. B. (1968). Social consequences of an 'unsuccessful' low caste movement. *Comparative Studies in Society and History,* Supplement 3, pp. 36–65.

Helland, J. (1982). Social organization and water control among the Borana of southern Ethiopia. In *East African Pastoralism: Anthropological Perspectives and Development Needs,* pp. 117–38. Addis Ababa: International Livestock Center for Africa.

Hjejle, B. (1967). Slavery and agricultural bondage in South India in the nineteenth century. *Scandinavian Economic History Review,* **15**(1&2), 71–126.

Johnson, T. (1979). Eritrean refugees in Sudan. *Disasters,* **3**(4), 417–21.

Levi-Strauss, C. (1949). *Les Structures Elementaires de la Parente*. Paris: Presses Universitaires de France.

Lewis, O. (1958). *Village Life in Northern India: Studies in a Delhi Village*. New York: Vintage Books.

Mencher, J. (1982). Agricultural laborers and poverty. *Economic and Political Weekly*, 17(1&2), 38–44.

Merryman, J. (1983). *Ecological Stress and Adaptive Response: The Kenya Somali in the Twentieth Century*. PhD thesis, Department of Anthropology, Northwestern University.

Miers, S. & Kopytoff, I. (eds) (1977). *Slavery in Africa: Historical and Anthropological Perspectives*. Madison: University of Wisconsin Press.

Oi, J. C. (1983). *State and Peasant in Contemporary China: The Politics of Grain Procurement*. PhD thesis, Department of Political Science, University of Michigan.

Ottenberg, S. (1971). *Leadership and Authority in an African Society: The Afikpo Village-Group*. Seattle: University of Washington Press.

Polanyi,. K. (1944). *The Great Transformation*. New York: Rinehart.

Popkin, S. L. (1979). *The Rational Peasant: The Political Economy of Rural Society in Vietnam*. Berkeley: University of California Press.

Rao, N. V. (1974). Impact of drought on the social system of a Telagana village. *The Eastern Anthropologist*, 27(4), 299–315.

Reed, E. P. (1975). Introducing group farming in less developed countries: some issues. In *Cooperative and Commune: Group Farming in the Economic Development of Agriculture*, ed. P. Dorner, pp. 359–79. Madison: University of Wisconsin Press.

Sahlins, M. D. (1972). *Stone Age Economics*. Chicago: University of Chicago Press.

Scott, J. (1976). *The Moral Economy of the Peasant*. New Haven: Yale University Press.

Sterling, E. O. (1950). Anti-Jewish riots in Germany in 1819. A displacement of social protest. *Historica Judaica*, 12, 105–42.

Stewart, F. H. (1977). *Fundamentals of Age-Group Systems*. New York: Academic Press.

Swift, J. (1977). Pastoral development in Somalia: herding cooperatives as a strategy against desertification. In *Desertification: Environmental Degradation in and around Arid Lands*, ed. M. H. Glantz, pp. 275–305. Boulder: Westview Press.

Tobin, J. (1952). A survey of the theory of rationing. *Econometrica*, 20(4), 521–51.

Torry, W. I. (1986). Drought and the government-village food emergency system in India. *Human Organization*, 45(1), 11–23.

Turnball, C. (1978). Rethinking the Ik: A functional non-social system. In *Extinction and Survival in Human Populations*, ed. C. D. Laughlin & I. A. Brady, pp. 49–75. New York: Columbia University Press.

Turton, D. (1977). Response to drought: the Mursi of southwestern Ethiopia. In *Human Ecology in the Tropics*, ed. J. P. Garlick & R. W. J. Keay. Symposia of the Society for Human Ecology, 16. London: Society for Human Ecology.

Widstrand, C. G. (ed.) (1970). *Cooperatives and Rural Development In East Africa*. New York: Africana Publishing Corporation.

Wiessner, P. (1977). *Hxaro: A Regional System of Reciprocity for Reducing Risk among the !Kung San*. PhD thesis, Anthropology Department, University of Michigan.

Wiessner, P. (1982). Risk, reciprocity and social influences on !Kung San economics. In *Politics and History in Band Societies*, ed. E. Leacock & R. B. Lee, pp. 61–84. Cambridge University Press.

Worsley, P. (ed.) (1971). *Two Blades of Grass: Rural Cooperatives in Agricultural Modernization*. Manchester: Manchester University Press.

Yeld, R. (1968). The resettlement of refugees. In *Land Settlement and Rural Development in Eastern Africa*, ed. R. Apthorpe, pp. 33–7. Kampala: NKANGA Editions No. 3.

15

The role of non-government organizations in famine relief and prevention*

MICHAEL F. SCOTT
Overseas Programs, Oxfam America, Boston

Introduction

Two acronyms have come into common usage in relief and development circles: 'ngo' and 'pvo', meaning, respectively, non-government organization (as they are commonly known in the United Nations and Europe) and private voluntary organization (as they are commonly known in the United States). The two are synonymous, referring, as their collective weight suggests, to non-state societies created and sustained by the good will of their members, constituencies or supporters. Ngos, as I will call them, are to be seen in contrast to state or official agencies that are created and sustained by government. On the same score, ngos also differ from agencies created and sustained by multistate organizations, such as the United Nations. While ngo, bilateral and multilateral agencies are the most commonly mentioned types of organization, a fourth variety, the solidarity organization, also deserves mention.

Usually, when ngos are mentioned, one hears the names Catholic Relief Service, Christian Aid, CARE, Save the Children, Oxfam and dozens of other Western agencies that are engaged in international relief and development activities. Less well known but no less important are African ngos, such as the Organization for Rural Progress (ORAP) in Zimbabwe, the Community Development Trust Fund (CDTF) in Tanzania, the Christian Relief and Development Association (CRDA) in Ethiopia, the Somali Unit for Research and Education for Rural Development (SURERD), and many more.

Despite their many differences, Western and African ngos are formal societies operating as legal entities with boards of directors, paid staff, budgets and plans, offices and telephones, and so forth. Although they

---

* The views expressed in this paper are those of the author, and do not necessarily reflect those of Oxfam America or any of the other Oxfam organizations.

often depend upon volunteers and are themselves voluntarily created, they are a recognizable type of organization. As such, they are different – though perhaps only further evolved – from the wellspring of their continuing inspiration, which is the 'self-help' society of peasants or pastoralists or urban dwellers or workers or others who, reaching beyond their individual households, have decided to solve common problems through collective effort. Many examples come to mind, such as 'shramadana' in Sri Lanka and 'minka' in the Quechua-speaking areas of the Andean highlands. In West Africa, the pastoralist Fulani custom of exchange 'habbanaae', whereby female livestock are 'lent' until three calves are produced for the borrower, is another example of indigenous self-help (Scott & Gormley, 1980). Village-based saving (and lending) societies, with their 'cash box', offer a further example of local self-help societies from Africa.

This inspiration for formal ngos is part truth and part myth. It is true in the sense that some ngos identify, encourage, and even help kindle (or re-kindle) such self-help activities; some even building their approach on the basis of these less formal societies. It is myth, however, in the sense that these self-help exchanges and societies may bear only nominal relation with the dominant or common mode of ngo operation.

There is quite a range of approach and organization among ngos, regardless whether they are African or First World. Ngos include both secular and religiously based organizations – Islamic, Protestant, Catholic. There are ngos which function primarily as funders and others which are primarily operational. Some ngos are specialized in providing food aid, while others provide medicine. Some ngos focus largely on relief, while others set a priority on development activities. Some work closely with the state, some receive substantial revenues from the state, and some others work independently of the state. There are further differences in the domestic programs of international ngos. Some international ngos undertake virtually no domestic education or lobbying activities, while others are more visible at home than they are in Africa. In short, there is such a substantial range of ngos that generalizations are hard to come by but, daring contradiction, I will sketch out some ngo activities that can help to deny famine a future.

This chapter is divided into five sections. The first reviews the range of activities that ngos undertake to deny famine a future. The second section addresses the question of drought and famine, differentiating areas of chronic and periodic famine and supporting the analytic framework of food systems as a means of achieving better understanding. The third section discusses the comparative advantages of ngos. The fourth discusses the obstacles and difficulties encountered by ngos, focusing particularly on the relationship between ngo and 'home' and 'host' governments. The last section discusses food aid and the need for greater autonomy in delivery of this assistance as well as the need for more development assistance. Overall, the chapter argues that ngos have a major role in defending the battered concept of humanitarian assistance – defined as relieving human suffering regardless of political considerations – an essential element to denying famine a future.

Denying famine a future

Ngos can work on two broad fronts to deny famine a future. The first is the provision of resources to those people who are vulnerable to or are already suffering from famine. Commonly, ngos operate relief programs by providing food, medicine, medical supplies, shelter, water and transport, along with the technical knowledge to use these resources. By far the largest share of ngo resources is allocated for relief programs and is consistent with the mainstream public view of ngos as relief organizations. The feeding centers that have figured so prominently in the news from Africa are the most visible elements of ngo relief activities, despite the estimate that for every hungry person who manages to arrive at a feeding center, there are probably two persons who do not.

Relief is very costly in lives, as it often occurs too late for many in need. It is also costly to deliver, if one considers that in the airlifting of food supplies, while essential to saving lives in many situations, more is spent on transportation than on the goods being transported. As often as not, dramatic 'mercy' flights are a measure of the tardiness of adequate response to a human catastrophe long in the making, as well as a gross measure of the failure of development programs and assistance to address the causes of famine in the first place. Invariably and correctly, the observation is made again and again that an ounce of prevention (i.e. support for development programs) is worth a pound of treatment (i.e. famine relief). The only problem with this morally and substantively correct homily is that the cost ratio of relief versus development is probably far higher than the pound to an ounce, which only adds further substance to the observation.

Less commonly but no less importantly, ngos also assist in reconstruction or rehabilitation activities, such as re-capitalizing peasant or pastoral households through the provision of loans for draft animals or loans for reconstituting pastoral herds. For example, ngos provide seeds, tools and other essential inputs in the form of subsistence agricultural packages to drought-stricken peasants, either before they have been forced from their homes or when they seek to return to them from food-distribution centers. Similarly, ngos also provide badly needed cash for food purchases to vulnerable groups, such as pastoralists in a severe drought situation, by buying their products (for example, buying dried meat from slaughtered livestock), thus helping to avert an impending crisis and to keep people from the food-aid dole.

Though usually small in scale (and perhaps this is both trademark and saving grace), ngo relief, reconstruction and development activities can experiment with useful strategies and approaches that can be replicated on a larger scale by government, bilateral and multilateral agencies. Thus, ngos which are often less fettered by constraints can serve as a testing ground for larger agencies and, perhaps, can have an impact on national policy and programs. For example, ngo support for the dissemination of single-ox plow technology in the highlands of Ethiopia, may offer significant savings both in capital and running costs for peasant households. This program could be replicated on a broad scale, beyond the several hundred households where it is currently being tried.

While ngos are often correctly criticized for having 'short institutional memories', other institutions seem not much more advanced in this respect. In either case, there is evident need to develop a better understanding of the processes that result in famine and of the strategies to overcome it. That is to say, beyond the wide sea of ideological debate and polemic, there are precious few empirical data or analyses on African peasants and pastoralists which could guide approaches to the resolution of their problems. Ngos can have a modest role in chipping away at this mountain of ignorance, as, for example, in the case of the application-oriented research of Jim McCann, whose chapter on a draft animal loan program in Ethiopia is part of this volume.

There is also an important range of activities that ngos can undertake in their home countries that, when combined with effective overseas work, constitute a broad approach to famine and poverty issues. Undoubtedly, the most important home-front activity is mobilizing support to finance the work overseas. To do this requires informing the public, or at least one's constituency, of the problems and proposed solutions. To one extent or another, all ngos receive their revenues from individuals, churches, interest groups and other associations. While many ngos receive a significant portion of their income from governments, there is inevitably a call for giving from other donors.

Media coverage of a specific problem is a major determinant of donor response. The response to an appeal for aid to sub-Saharan Africa in the summer of 1984 was modest compared to the response following the media's coverage of Africa in October 1984. Yet, in fact, the 'ground truth' in Ethiopia or Mali had shifted only slightly between the summer and fall, moving in an inexorable direction taken years before. The media made the difference. Often, ngos are particularly well informed and are consulted by the press for an unofficial and presumably more substantive description of the problems, or at least an insight into some of the difficulties and nuances. In this respect, ngos can play a key role in improving the level of understanding of the press and, hence, of the public. Some ngos regularly disseminate press bulletins, meet with editorial boards. Some even help to arrange press visits to their places of work in Africa.

Independent of the press and on the margin of fund-raising appeals, ngos also produce useful descriptions and analyses of development issues. Using newsletters and special publications, as well as audiovisual materials, many ngos offer an important, though less dramatic, presentation of national, regional and even village-level issues. Usually based upon first-hand data, these 'educational' materials frequently offer useful insights into problems that at best make the headlines only infrequently.

Also at home, ngos can mobilize support for, or criticism of, policies of their own governments. For example, many ngos recently lobbied the United States government to provide additional food assistance to Ethiopia, as well as to other African countries, whose immediate needs surpassed their ability to fulfill them. It is not unlikely that a similar call will be required in the near future, when the question of continued food aid to Africa will once again be raised. In this regard, ngos can present an independent and, in most cases, knowledgeable view to decisionmakers in government, complementing or contrasting with the views provided by various government agencies.

Famines and droughts

Relief is only a palliative, when it is successful. Beyond relief, the role of ngos in overcoming famine depends in part on one's analysis of the causes of famine. Usually, there is no single cause of famine, although there is no lack of differing schools of single causation – the 'naturalists' seem to favor drought, ecological complexities or plant diseases; the 'ideologues' seem to favor political systems as culprits; the 'partisans' point to those on the other side of the conflict; while the 'internationalists' say it's too much of the 'wrong' and not enough of the 'right' kind of international assistance; the 'colonialists' say it is the past; the 'neocolonialists' say it is the past and the present, and so forth. Rather, there are usually multiple causes of famine, with key natural, political, economic and agronomic factors operating in a common negative pattern.

It is also true that there are different kinds of famines, reflecting differences in the long- and short-term capabilities of various food systems. There are areas of Africa with chronic food-availability problems, such as parts of northern Ethiopia and the northern tier of the West African Sahel. In these areas, the frequency of 'bad' harvests is usually high, the level of poverty is substantial, the ecological situation is often described as precarious, rainfall patterns are quite variable and seemingly deteriorating and ethnically the local population is often poorly (if at all) represented in the central government; indeed, governance itself is a question, and in some cases there are military conflicts, to mention a few of the factors that appear to cluster in areas of chronic famine or famine potential.

These chronic famine areas in Africa are often represented on a map as a thin elongated crescent, often color-coded in red, stretching across much of the northern part of the continent, with occasional discontiguous blotches further south. These areas, however, are very different from, for example, parts of Kenya or Zimbabwe where at some point last year the spectre of famine was also raised. One difference in these latter cases is the absence of chronic famine; rather, famine in parts of these countries is recurrent but not chronic. Such distinctions are critical, as the responses must vary to meet the causes of the situation.

Natural causes that trigger famine in an already vulnerable or poor area or country, will not have the same effect or sequence of effects as in a less vulnerable or richer area or country. Between such natural factors as unseasonal or inadequate (or excessive) rainfall and the associated inability of certain African cultivators and pastoralists to feed themselves, there are usually a cluster of social, economic and political dimensions of poverty and underdevelopment that combine to create famine. Yet, famine is not an abstract phenomenon. It is the result of concrete factors operating in a particular situation. At the very least, one must distinguish areas of chronic and recurrent famine (see, for example, the United Nations Research Institute for Social Development [UNRISD] research and publication series on food security, e.g. Pearse, 1980), being careful to observe that such areas are often not coterminous with nation-states. One must also bear in mind that few famines affect everyone equally: social position, gender, age and ethnicity often determine who lives and who dies in a famine. A more finely drawn famine map of Africa must go beyond nation-states and rainfall isohyets to distinguish the vulnerable social groups and their members.

Seeking solutions to famine implies value judgments about the nature of a food-secure society about which there is considerable debate. One view of food security may envision a society which is largely dependent upon imported food from Western Europe and North America. Other views may emphasize strong state or, alternatively, strong private sector participation to improve local food production and distribution. There are also contrasting views on the relative importance of agricultural production for export or for domestic consumption. My own view is that there is no universally applicable blueprint for achieving food security, and that there are many ideologically inspired false starts and culs-de-sac in the African experience and elsewhere.

Nonetheless, there are several essential features for building food security, including vigorous peasant- and pastoral-based policies and programs. Food security for the majority implies broad popular participation by the majority in defining policy and carrying out programs of social and economic development. In this sense, food security cannot be measured solely in terms of food production increases or in per capita income gains, as these do not necessarily mean more food for the hungry; rather, food security means 'minimum' access to staple foods for all persons and groups. This implies not only an increase in the production of basic food grains and an increase in income but also a relatively egalitarian distribution of income. The tensions between providing the material incentives for greater food production and a sufficiently egalitarian structure that provides (economic) access to this food are basic issues that need to be addressed in order to build food security.

Strategies to attain food security in areas of chronic famine, as the discussion above suggests, will probably differ from strategies in areas of greater production potential. They would probably include special relief provisions, such as *in situ* grain stocks, programs of soil recovery and pasture conservation and regeneration, as well as radical measures such as temporary or permanent population resettlement. Strategies, however, must be geared to specific areas and problems, as these differ considerably. For example, in an area of chronic food insecurity where war is among the causes, there is little scope for addressing agronomic problems while blood is being spilled.

The concept of food system is a useful descriptive and analytic tool for understanding food security. It can be used at local, national and international levels as well as help explore the interrelation between these levels. For example, a change in the international price of coffee has both national and local impacts upon food security. Building food security implies changes in local, national and international systems which, in turn, suggests that ngos addressing famine issues must operate at each of these levels, both practically and in terms of policy and program advocacy. For example, an increase in the price paid to food producers will not serve as a production incentive, unless there are consumer goods the producers can purchase with their added income. In the absence of national production of consumer goods and without adequate terms of trade to import them, improvements in food production pricing policies will have little impact.

Achieving food security in much of Africa is a long-term proposition that will not be accomplished in years but rather in decades, if there is sufficient political will to address the problems, which are as much political as technical. Achieving food security is also a complex task involving changes, in some cases dramatic changes, in domestic, regional and international spheres. As a touchstone of optimism in an otherwise pessimistic prognosis, it may help to remember that as recently as the 1930s, and for decades before, China was known as 'the land of famine'. Relief efforts there were considered relatively effective when 'only a half million Chinese died', as Lillian Li's contribution to this volume indicates. This unrelentingly dismal picture has largely changed (and of course is undergoing further change) so that now most Chinese, most of the time, enjoy what they describe as an unbreakable or 'iron' rice bowl, as opposed to a breakable or 'porcelain' one (Croll, 1982).

Ngos

In the context of building food security in Africa, what is the role of ngos? What should these organizations do more or less of in defining the problems and seeking solutions? What dilemmas and practical problems do they face and what prospects do they offer?

While ngos are financial 'lightweights' in the foreign aid picture for Africa, accounting for a mere fraction of total dollar aid compared with the bilateral and multilateral donors, this broad comparison is deceiving in several respects. At times when world attention is focused on famine, as has recently been the case in Africa, the scale of ngo operations increases dramatically. For example, consider that by April 1985, American ngos raised over $110 million for Africa, while in the same month the US Congress in its final action on the African Famine Relief Supplemental Bill provided only $175 million in non-food assistance (and $625 million in food aid) for sub-Saharan Africa. What is remarkable about this is the closeness of the amount of ngo and government non-food assistance, as well as the dominance of food aid in the government aid picture. Thus, the apparently marginal amount of ngo aid is, in fact, fairly substantial when compared with US official assistance. Normally, however, ngo assistance is modest in scale compared with other donors.

In other respects, ngo assistance may have certain advantages over official aid. Unlike much official aid, ngo financial assistance is not tied to purchases in the donor country (with the major exception of food aid, of course), thus permitting use in, and hence bolstering the economy of, the country or locale where it is spent. For example, grains, seeds, oxen and other goods needed in an afflicted area of one country can often be purchased in more prosperous areas of the same country. Recently, Oxfam America has made such purchases in Ethiopia, Sudan, Zimbabwe and Mali, as well as elsewhere in Africa. This has several advantages, including bolstering the local economy, diminishing transport costs, lowering the risk of providing inappropriate technology and diminishing the risks of importing plant and animal diseases. 'Untied' assistance means greater flexibility in meeting needs and often means better value for the money.

At its best, ngo assistance is distinguished as rapid and expeditious. Often it is the first available foreign assistance and sometimes the only assistance in a given locale. For example, in late 1984 several European ngos and one American ngo organized an emergency food shipment to the Horn of Africa, which took about 6 weeks total time from the start until its cargo was offloaded at its destination. By comparison, it is said, the turnaround time for official food aid can be 6 months, and the startup times for official aid projects average more than 2 years. While ngos will never be able to substitute for the volume of official emergency food aid or development assistance, they are certainly able to help fill some alarming gaps and time lapses in delivery.

Alacrity in taking action may have another dimension of considerable importance, which might be described as 'pump priming' or the 'embarrassment factor'. During the summer of 1984, in another English-speaking country, ngos raised more funds from the public than the home government was willing to commit to famine relief in Africa. It became a public embarrassment to the government that its people were more generous with their voluntary contributions (after taxes) than government was with its tax revenues and, with prodding from the ngos, this served to prime the pump of needed official aid which reluctantly followed. Unfortunately, this sort of leverage is all too rare, owing in large measure to the level of public attention provided in the press, as there are probably many more instances of failure than success in 'pump priming'.

The concept of humanitarian aid is under attack from several quarters. Recently we have seen foreign states providing 'humanitarian' assistance to military organizations at war with the prevailing regime. By providing what is termed 'non-lethal' forms of assistance, there is a thinly veiled attempt to cloak what is simply partisan aid to political-military fronts, which might languish and disappear if it were not for foreign aid. Given the political-military intent and nature of the receiving organizations, this is a perversion of the concept of humanitarian assistance. There are also more subtle perversions of humanitarian aid, such as when solidarity organizations with lopsided political agendas appeal for funds without revealing their affiliations.

This chapter argues that ngos have a three-part role in building food security in Africa. First, ngos have the moral obligation to make and defend the case for humanitarian assistance, broadly defined as relieving human suffering, regardless of political considerations. This is a task involving local, national and international spheres; particularly in the African context, this raises issues of a state's relations with non-state organizations. It is an uphill battle for many reasons, not the least of which is the fact that food availability for nations or groups within nations is not governed so much by that nation's or group's needs for sustenance as by its ability to offer goods and services, or cash, in exchange for food. Second, ngos must actually provide relief, reconstruction and development assistance in ways that are consistent with building long-term food security. This task raises many programmatic questions, including issues of food aid, self-reliance and access. Third, ngos should defend the dignity and the cultural integrity of those people with whom they work, paying particular attention to low-status groups defined ethnically, politically, by gender, or by other criteria.

Ngos and the state

Among national, multilateral, and solidarity organizations, ngos have perhaps the greatest potential for providing humanitarian assistance, because they are usually more independent of the state. The stark reality is that political and military security is always the top priority of governments, or would-be governments, and that relieving or overcoming famine is, at best, a secondary priority. While political considerations are always paramount to the state – and this is no less true of African states than it is of others – they must always be secondary for ngos operating on humanitarian principles. The West, for example, will only reluctantly and under pressure provide limited famine assistance to Marxist governments in Africa, because these states are perceived either as a threat to Western interests or as a 'problem' of the other bloc. Likewise, the Eastern bloc's contribution to African famine relief and prevention – if they can be learned at all, as they are often shrouded in secrecy – follow its African political agenda, which relates only tangentially to an assessment of need. The apparent exceptions to this bipolar division of the world, on reflection, seem only to reinforce the basic political axiom of famine relief and prevention. Thus, according to some observers, Mozambique's recent dramatic increase in the receipt of Western food aid – the proverbial carrot – was more a measure of Western interests in securing the Nkomati Accord between Mozambique and South Africa, than a reflection of dramatically increasing food needs or a change in what has otherwise been considered an Eastern bloc problem.

Perhaps the most problematic ngo relationship is with the state, particularly when the state wants to control and define humanitarian assistance, as increasingly appears to be the case both in the United States and abroad, including many parts of Africa. Sovereign states have many legitimate rights, particularly in dealing with international organizations that they host. In at least some instances, however, there appears to be a collision course between ngos seeking to provide famine relief on the one hand, and security interests of the state on the other. This contradiction is expressed in very practical terms, such as gaining physical access to famine areas, as well as programmatic issues such as the use of transportation facilities. In a famine situation, food or access to food is power, which contending parties in a conflict will seek to use for their own purposes, with each accusing the other of misusing assistance. Increasingly, it appears that technical and administrative issues of famine relief and prevention are lodged at the highest levels of state political authority. This is not necessarily a promising development for fulfilling needs. It is evident in several African countries that relief to certain famine areas is subject to the political-military agenda. Ngos, however, may have more latitude in this situation than they suppose, as suggested in Shawcross' (1984) thorough-going review and critique of the humanitarian relief programs in the politically and militarily conflicted Kampuchean (Cambodian) famine 5 years ago.

In the case of international ngos, there are two potential sets of dilemmas, as there are both 'home' and 'host' governments. Some ngos are faithful to an implicit or explicit humanitarian charter, making their assistance as politically neutral as possible. They strive to deliver aid on the basis of need and the opportunity of meeting needs, using whenever possible non-partisan channels. Others, particularly solidarity organizations, have consciously or otherwise taken a partisan role, in some extreme cases acting in simple partnership with and virtually representing a government or would-be government to the public at home. Thus, in the African context, we continue to hear the predictable litany of the virtues of government X or of liberation front Y and the equally predictable one-sided corollary of how the 'other' (read wrong) side is misusing food and other development assistance for political purposes, while the 'good guys' (there's no doubt in the agenda at this point) continue their selfless struggle, always providing food for the poor first and guns for themselves second. While laughably simplistic and romantic, this partisanship is also perverse, as it cloaks its political agenda in humanitarian trappings, thereby discrediting the concept of legitimate humanitarian assistance.

On the home front, humanitarian aid may run counter to the home government's foreign policy which, consequently, may legally constrain or simply block ngo operations. More commonly, the home government's foreign policy interests are expressed in the amount of resources (food or funds) it makes available for ngos to use in a given country. As many ngos are partially or largely dependent upon their home governments for resources (which itself raises a serious question about their autonomy, a necessary factor in providing humanitarian aid), an absence of, or even modest, home government funding may mean no, or modest, ngo activity; it means aid responds to needs only insofar as those are reflected in foreign policy considerations.

Similar issues are raised between ngos and 'host' governments, especially when national political authority may be at variance with solutions to famine. This dilemma is perhaps most acute in the context of civil strife or outright war, when the state is one of the parties to the conflict. It is no accident that war and famine go hand-in-hand in several famines in Africa today. One needs only to mention that such is the case in Angola, Chad, Ethiopia, Uganda and Mozambique. It is the nature of states and would-be states to put military security above food security, thus making the needs of non-combatants all the more pressing in addition to complicating the delivery of assistance.

While 'hot' wars are one problem, the 'cold' war between East and West, often feeding on local conflicts, is a further serious obstacle to food security. Financial resources are drained for arms purchases, and hollow slogans of 'state' versus 'private' modes of development are bandied about to justify either the provision or the curtailing of foreign aid. The cost of national security in Africa today, as measured in arms purchases alone, is staggering: $3.5 million for a battle tank, $35 million for an advanced fighter aircraft. Perhaps no other continent in a similar position of poverty and food insecurity pays such a high price. Given the direct and indirect costs of war and its contribution to the lack of food security, ngos must emphasize peace and reconciliation.

Ngos need to emphasize their non-government and non-partisan purpose to live up to humanitarian standards, particularly where civil conflict is a cause of the lack of food security. According to the Geneva Conventions, organizations distributing aid should be independent from the parties to the conflict. Aid should be distributed directly on the basis of need and only to non-combatants. Aid should be offered impartially to all affected civilians on both sides of the conflict. In providing famine relief and famine prevention assistance, however, ngos may face the moral and political dilemma of either providing assistance through one or both parties to armed conflict, as there are no other means of reaching those in need, or alternatively, not responding at all. This dilemma is strongly evident in the current African situation, though it is neither new nor unique to Africa.

One important role for ngos is to criticize constructively the political order of priorities and to appeal to the other interests of states to respond to human suffering. For example, the United States government has restricted ngo assistance to the provision of food, medicine and clothing, and excluded from permission the provision of agricultural inputs, training and other forms of assistance in certain countries. (In particular, the US State Department's interpretation of the Trading With the Enemy Act makes this distinction for Vietnam, Kampuchea (Cambodia), Cuba and North Korea, as described by Charny & Spragens, 1984).

The argument given by government is that assistance such as training and agricultural inputs could have a longer-term impact on organizational capabilities and, hence, is not humanitarian aid. In other words, the very objectives of overcoming basic food production and distribution problems – famine prevention as opposed to famine relief – are defined as inappropriate. Famine relief qualifies as humanitarian assistance, but famine prevention does not, if the latter is intended for countries that are not considered 'friendly'. Certainly this is a short-sighted and politically motivated interpretation of humanitarian aid. It is also costly in terms of lives and resources, considering that the long time spent in delivery and the large sums spent on transportation of emergency food supplies would be given better value in preventive measures of longer-term food production assistance. I hasten to add that the United States is cited here only as an example, there being countless examples of similar predicaments both East and West and, of course, in Africa itself.

Similarly, ngos must appeal to the industrialized countries, both East and West, to increase rather than decrease (as is currently contemplated in many countries) financial support for multilateral assistance programs dealing directly with long-term food security in Africa. This means support for such organizations as the International Fund for Agricultural Development (IFAD), the United Nations Food and Agriculture Organization (FAO), and the World Bank's African Development Fund. These multilateral programs tend to be insulated from the vagaries of national political agendas.

In general, there are few African ngos in comparsion with Latin America, the Caribbean and Asia. While there are significant exceptions, for example, Zimbabwe, usually international ngos operating in Africa find few ngo counterpart organizations. Accordingly, ngos take a more operational role and usually find themselves in the dilemma of working closely with government. This leads to the need to emphasize the building of local counterpart organizations, not as mirror images of the international ngos but as African organizations that will define food security in their own terms and be able to implement programs that, in turn, strengthen the participation of vulnerable groups.

Food aid, recovery and development

There are substantial programmatic concerns for ngos in implementing humanitarian assistance in ways that are consistent with building long-lasting food security. Many ngos depend to a greater or lesser extent on the channeling of food aid from their home government. This raises many concerns. For example, when does food aid represent needed short-term assistance and when does it mean increased dependency, pauperization and disruption of local food systems? When should it be given away or sold or exchanged for labor? Should food aid be replaced with cash assistance? How much short- versus long-term aid should be given? Where or how should assistance be targeted? Will local organizations be strengthened or weakened by the provision of assistance?

Food aid is a disproportionately large component of Western ngo aid to Africa. The official United States emergency contribution to sub-Saharan Africa was 1.8 million metric tons of food aid, valued at $780 million and $90 million for internal transportation. When added to the rest of food aid contribution during fiscal year 1985, the total was 3 million metric tons, valued at $1 billion, making the United States the most generous of all of the countries donating food aid. By comparison, US official development assistance for Africa runs at $500 million a year, making about a two-to-one ratio between food aid and development assistance allocations over the last year. But already in 1985, on the first anniversary of the October 1984 revelation of massive famine in Africa, even official food aid is being cut, despite ngo and United Nations assessments indicating that, for Ethiopia at least, an equal amount of food aid – about 1.2 million metric tons – is required over 1986.

Official food aid is commonly delivered through one or more of three channels: (*a*) a specialized United Nations agency, the World Food Program; (*b*) the home government's own bilateral food agency; and (*c*) through ngos that participate in the home country's food aid program. In all cases, the donor state makes the decision about the appropriate channel for food aid delivery. Ngos are often used as intermediaries by government to channel food aid, when the donor state is reluctant to deal directly with the recipient government.

With all the food aid that poured into Africa in 1985 alone, it is worth considering whether food aid is addressing systemic problems of food insecurity, or whether it has only been a stopgap relief measure or, at worst, an obstacle in the struggle to end hunger. In light of this, the question of the appropriateness of food aid as a development tool – as a means of both increasing food production and strengthening the peasant producer sector – becomes most pressing. Africa, the only continent in which per capita food production has been declining, is fast becoming a test case for the effectiveness of food assistance.

There are two motivations behind US food aid (Public Law 480) that adversely affect its role in development. The first is that expansion of US international trade and the opening of new markets for US farm product exports is still a major priority and rationale. From a food-security perspective, the need to increase local food production and to promote greater self-reliance often runs counter to this. A large proportion of food sold under Title I (concessional sales) goes to the urban middle class who can afford to buy it. This cultivates a taste for US food exports, decreasing urban demand for locally grown food, while increasing demand for imports. Food prices are kept low, which again benefits the urban population but at the same time acts as a disincentive for producers and does little to help the rural poor. Food aid promotes a false sense of food security, allowing governments to postpone funding programs aimed at increasing national food production (Jackson & Eade, 1982).

The second rationale is the promotion of US foreign policy. A review of the list of recipient governments shows that P.L. 480 has gone predominantly to nations either strategically important to US interests or friendly to US government policy. Thus, policy interests, not food needs, become the overriding concern in providing food aid. For example, Egypt, a country of major importance to US interests, receives a major proportion of US food aid. At the same time, Mozambique and Ethiopia, until recently, received no Title I and very little Title II assistance donations, despite serious and well-documented needs since at least the fall of 1982.

The food shortage in Africa will not be solved solely by providing food. In an emergency relief situation, food aid can be effective, contingent upon its timely delivery once the need is recognized, not years afterwards as in the case of Ethiopia. Relief food must be accompanied by logistics support, often at the relatively high value and foreign exchange costs of trucks, maintenance and repair systems, and petroleum. For example, when mercifully there was a response to the Ethiopian famine of 1984–85, food aid at times piled up on the docks, or in ships waiting to be unloaded for lack of sufficient trucks, fuel and spare parts. This bottleneck was anticipated, and requests for logistics support were made well in advance, but generous grain donors were stingy when it came to helping move the food. It is equally simple but no less true that relief must be accompanied by recovery assistance, such as seeds, tools and draft animals, as part of an overall development plan. A reasonable but flexible cutoff date for the provision of emergency food aid is also essential to building local food security. To ensure that its humanitarian use responds as much as possible to need rather than to political agendas, emergency food aid should be lodged in agencies as independent of foreign policy considerations as possible. There should be greater use of the United Nations food aid agency and greater autonomy given to the ngos in allocating official food aid.

Ngos should focus more on longer-term food security problems and should move as rapidly as possible to reconstruction and development activities. Relief assistance is most effective when it is part of a recovery plan. Relief assistance unaccompanied by seeds, tools, draft animals or livestock can act as a disincentive for displaced persons to recover their preferred livelihoods and return to their homes as peasants or pastoralists.

In general, there are more economic resources available for relief than there are for reconstruction and development activities, but ngos can help refocus relief efforts by casting them in a reconstruction mode.

There is much to be learned and much to be done in supporting peasants' and pastoralists' strategies for improving their productivity. By working at the peasant association, service cooperative, or village level (or its equivalent among pastoralists), ngos can directly assist peasant producers and help build a body of knowledge that will assist in the design and implementation of larger-scale programs. It is at this level where the technologies and procedures for improving short- and long-term food security need to be addressed and implemented. By working in this manner, ngos can help avoid the high and often inappropriate technological 'solutions' – the almost predictable tractor request – that seem to abound in national development plans.

As an example of longer-term food security problems that ngos are in a particularly good position to address, consider the often ignored role of women in the food system. In much of Africa, women are the primary food producers as well as the primary actors in local food distribution and, at the household level, they are responsible for food storage, processing and preparation, and usually the provision of water for domestic consumption. This latter task alone can require a third of a working day. All too often, assistance programs are developed and implemented with a gender bias that ignores the key roles of women. Accordingly, men may be consulted about appropriate varieties of hoe heads – the major agricultural implement in many areas – but it is women who use them to produce food. For example, Oxfam America was one of several ngos that recently helped provide several hundred draft animals as well as seeds and agricultural implements to poor peasants in a particularly hard-hit area of northern Shoa province in Ethiopia. Important and unusual as this assistance was, we learned to our chagrin that no female-headed households came forward to request draft animals, despite the fact that they formally qualified for the loan program. We also learned that cows are used as traction animals in parts of the project area. These experiences suggest that, in future modifications of the project, cows can be used instead of only bulls, as they will provide an immediate source of food as well as needed draft power. The experience also suggests that to provide basic capital (draft animals) to female-headed households will require modification of program design and implementation.

Conclusion

On balance, ngos have several clear advantages, among their limitations, but there are also key choices as to the humanitarian or partisan character of the role that they play. Ngos need to distinguish themselves from acting as surrogates for states and from solidarity groups, when the partisan agendas of these are at variance with meeting needs. Independence of action is probably the most important characteristic for ngos to maintain, if not bolster. Ngos tend to make the 'aid game' more honest when they express themselves with a critical and humanitarian voice. They can contribute to making a better informed public and more responsible home and host governments, by taking an active and legitimately humanitarian role in addressing famine and food insecurity through relief, recovery and development programs.

References

Charny, J & Spragens, J. Jr (1984). *Obstacles to Recovery in Vietnam and Kampuchea: US Embargo of Humanitarian Aid*. Boston: Oxfam America.

Croll, E. (1982). *The Family Rice Bowl: Food and the Domestic Economy in China*. Geneva: United Nations Research Institute for Social Development, Food Systems and Society Series.

Jackson, T. & Eade, D. (1982). *Against the Grain: The Dilemma of Project Food Aid*. Oxford: Oxfam.

Pearse, A. (1980). *Seeds of Plenty, Seeds of Want: Social and Economic Implications of the Green Revolution*. Geneva: UNRISD and Oxford: Clarendon Press.

Scott, M. & Gormley, B. (1980). The animal of friendship (Habbanaae): an indigenous model of Sahelian pastoral development in Niger. In *Indigenous Knowledge Systems and Development*, ed. D. Brokensha, D. M. Warren & O. Werner, pp. 92–110. Washington: University Press of America.

Shawcross, W. (1984). *The Quality of Mercy: Cambodia, Holocaust and Modern Conscience*. New York: Simon and Schuster.

PART IV

Lessons for the future

16

Food self-sufficiency in Malawi: are successes transferable?

J. GUS LIEBENOW

Indiana University

Hunger and malnutrition are among the most persistent indices of poverty in developing countries. Malawi (Fig. 16.1) one of the poorer countries in Africa, constitutes a significant exception to this generalization. While the media has been filled in 1984–85 with stories of mass starvation in Ethiopia, Mozambique, and various parts of West Africa, Malawi has stood out as an underdeveloped country which has not only been able to feed itself but to export substantial quantities of maize, fish, sugar, and other commodities to its neighbors who were short of food. In 1984, when even South Africa was importing food, Malawi exported over 100 000 tons of white maize. During the past decades, it has been one of about 7 African nations – out of 50 states in the sub-Saharan region – that have been relatively consistent in meeting their goals of self-sufficiency in food production (Liebenow, 1982b). In a continent where agricultural programs are failing miserably, moreover, Malawi's agricultural achievements account for the major share of a remarkable growth rate of approximately 6% per year during the decade from 1973 to 1983 and 7.6% rate in 1984 (Lele, 1981). Its exports of food as well as such non-food crops as tobacco and tea provided it in 1984 with one of Africa's few favorable balances of trade.

What is even more remarkable about the preceding statements is that Malawi stands tenth from the bottom of the list of Fourth World African states in terms of the World Bank's basic economic indicators that are used to differentiate developed from less-developed states (International Bank for Reconstruction and Development, 1984). The per capita GNP in 1982 was US$210, and the life expectancy was 44 years. There was one doctor for every 40 000 citizens in 1980. Malawi, moreover, has none of the valuable natural resources which have given southern Africa the label of the 'Persian Gulf of the mineral world', and it lacks within its boundaries the coal, iron, oil and other ingredients of industrialization. Roughly 83% of its more than

6 million inhabitants are engaged in agriculture, fishing, forestry and hunting, and more cultivators spend their working hours in subsistence rather than cash economic activities. Although considerably reduced from the pre-independence figures, approximately 25 000 young adult males are absent from the country in any given year, seeking employment in the mines and other enterprises in South Africa and other nearby countries. Being landlocked, moreover, Malawi must use some of its hard currency to pay Mozambique the rail, port and road charges required to export its agricultural commodities through Beira and the newer port of Nacala, as well as to import the oil and other commodities needed for development.

The relative success of Malawi's agricultural program in the face of the foregoing adversities cannot be simply explained in terms of advantages in rainfall, altitude or other factors of the physical environment, even though rain did appear to make a difference in 1984. Rainfall patterns for the arable

Fig. 16.1. Malawi.

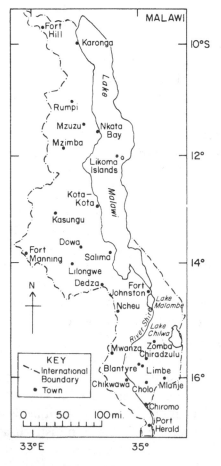

areas of Malawi over the past decade are not radically different from conditions in the arable areas of neighboring Zambia, Tanzania and Mozambique, all of which have experienced severe and recurrent food shortages during that same decade. With reference to Mozambique, which shares a common border with Malawi over half of the latter's perimeter, Malawians are fond of stating that 'the rain does not fall on one side of the road and not the other'. Indeed, national average rainfall figures mask the fact that much of Malawi's domain which received copious quantities of rain consists of mountains, swamps and other non-arable land (Pike, 1968, 184.). One-sixth of its political jurisdiction is covered by Lake Malawi. Taking into account all these factors reveals that the pressure on cultivatable land with adequate rainfall is far more desperate than is suggested even by the gross data. The factor places Malawi's 59 persons per km^2 as the third highest population density in Africa.

Traditional cultural factors also provide only limited explanations for Malawi's advances in agriculture. It is true that both the historic record of early travelers and the commentaries of World Bank and other contemporary observers make continued reference to the industriousness and cooperativeness of Malawian people. It is as difficult, however, to handle in a scientific manner positive stereotyping as it is negative labeling of national traits. In addition, there are many traditional values and practices among Malawian farmers that actually inhibit modern agricultural development (Pike & Rimmington, 1965, 173–8). These include the emphasis on subsistence agriculture as opposed to trade and crafts; the persistence of the wasteful practice of shifting cultivation; and the low level of traditional technology, epitomized by the hand hoe and even the digging stick. The land tenure system, while furthering the interest of social justice, has left most contemporary Malawian farmers in control of no more than 1 – 2 ha per family, with most of this communally owned land consisting of widely scattered fragments which vary in quality and in distance from each other, from water sources and from villages. Finally, the traditional attitudes toward cattle among the northern pastoralists result in cattle remaining largely an element in the social and political prestige system rather than constituting either a significant factor in the exchange economy or a source of much-needed protein for the Malawian people. At the time of independence, for example, there were 401 187 cattle recorded on land under customary tenure, with 39 613 beasts being slaughtered. A decade later the number of cattle on customary land increased to 700 475, with 65 561 being slaughtered for food purposes (Malawi, National Statistical Office, 1977, 16). Thus, although the off-take remained constant at about

10%, there were many more – and healthier – head of cattle occupying scarce arable farmland and contributing to the erosion of mountain slopes, without this making a substantial difference in the protein intake of Malawians. Since grazing lands are not privately owned, soil conservation is not a high priority among pastoralists.

On balance, furthermore, Malawi's colonial experience was not more favorable to agricultural development than the situation faced by its neighbors, even though the far smaller percentage of European settlers (less than 1% of total population) resulted in only roughly 3% of Malawi's land being effectively reserved for Europeans (although three times that figure was technically reserved for non-Malawians). This advantage was more than offset by the fact that Nyasaland (as Malawi was called) came to be regarded as a pool of cheap labor for the tea, tobacco and other plantations in Malawi as well as for the European-owned mines in southern Africa (Chanock, 1977, 396–408). Those young adult males who remained at home rather than engaging in long-distance migratory labor had to meet their tax obligations by working as tenants on European estates or engaging in cash crop cultivation, which diverted their attention from domestic food production (Kandawire, 1979).

Like other territories in Africa, innovation in African agriculture came as a consequence of coercion, with threats of fines or even imprisonment being more persistent instruments of change than were education, demonstration or significant monetary rewards. Africans were prevented by law or by pressure from European settlers from independently cultivating crops such as coffee or flue-cured tobacco, which were in competition with those cultivated by the Europeans (Vail, 1982). Even where coercion could be rationalized by the Europeans to be in the interests of Africans, such as the mandatory planting of cassava and other 'starvation' crops, coercion created an attitude of hostility towards agriculture in general.

The political repercussions were such that one of the first actions of the Banda government in 1964 was the elimination of all compulsory planting rules. Coercion in achieving agricultural change under colonial rule was matched by caprice and an almost whimsical experimentation by colonial officials. This is demonstrated by half a century of experience with maize production in which Malawians were, in an almost continuing cycle, vigorously encouraged and then severely discouraged from planting maize as a cash crop. The alternating pressures came not in response to situations which were evident to Malawian farmers but rather in response to fluctuating global or British imperial interests (Nyasaland Protectorate, 1911–1963). Similar caprice was evident in the often contradictory planting rules with regard to African agriculture. In this respect the Malawian experience differed little from that of its neighbors.

It is not the intention of this chapter to present Malawi as a 'miracle case' with respect to economic development generally. As the preceding commentary suggests, it is a poor country and has a long way to go in overcoming not only poverty but some of the specific problems relating to food production and nutrition. Many Malawians lack the proteins, carbohydrates and fats needed for healthy development. Even in areas of relative abundance, people suffer from dietary imbalances. Nevertheless, compared to many of Malawi's more prosperous or potentially more prosperous neighbors, Malawi since independence has made remarkable strides in coping with the problems of self-sufficiency in food production. Hence, many of the policies and programs are worthy of examination and emulation in other parts of a continent which has generally failed where Malawi has succeeded.

Banda's political strategy for agricultural development

In seeking to account for Malawi's relative success in achieving food self-sufficiency and examining its relevance to agriculture elsewhere in Africa, more satisfactory explanations than weather alone can be found in an analysis of the agricultural policies and programs of the Banda government. These policies and programs are established within a broad set of political parameters which tend to set Malawi apart from many of its African neighbors. A major element in that political framework is the relative political calm and stability that Malawi has experienced in contrast to the interracial, interethnic and civil–military conflict experienced in a majority of African states. Whatever one's view of Malawian domestic politics, stability must certainly be reckoned as an advantage in terms of growth and development.

A further contrasting feature is Banda's parting company with most African national leaders on the issue of ideology, and socialism in particular. Finding little evidence of success on the continent in the high risk and often inappropriate socialist strategy of development (Eicher, 1982, 161), Malawi's leaders have opted for a course which defies neat ideological labeling. There is no single document or series of state papers which provide a blueprint for a well-integrated, internally consistent ideology.

As will be evident later in this chapter, the Malawian strategy of development is a mix of commitment to the preservation of many traditional values and institutions (family, chieftainship and land tenure); the furtherance of the institutions of private ownership of property and free enterprise initially manifest among the European settlers and Indian merchants; the continuation of the form of state capitalism which was a legacy from colonial rule; the acceptance of modern forms of cooperativism; and even the adoption of some aspects of socialism. Although Banda avoids labeling his development approach, it could be referred to as 'incremental eclectic pragmatism'. That is, his program involves a constant, step-by-step search for workable answers to problems irrespective of ideological origins. If an innovation works within the Malawian context, it is implemented as policy; if it does not, it is discarded. There is no *a priori* inclusion or exclusion of ideas on the basis of ideological predilections. The test is workability. Thus, the ideology is derivative rather than prescriptive.

With respect to a further element in Banda's political strategy, he differs not so much in rhetoric as in action from many of his African counterparts. This refers to his emphasis on the central role of agriculture in Malawian development. Admittedly his rejection of the lure of instant industrialization as the key to modernization may be based upon the paucity of mineral resources needed for industrialization, but the latter situation in itself has not deterred other African leaders. Whatever the rationale, Banda goes beyond rhetoric and symbolism, as evidenced by his retention of the portfolio as Minister of Agriculture during his more than two decades of governance. He is an inveterate personal inspector of farms, appearing unannounced and constantly exhorting farmers to diversify their crops, to abandon the practice of shifting cultivation, and to stop the wanton cutting of trees. His commitment to education rather than coercion in agriculture is evidenced by the prestige which Bunda College of Agriculture of the University of Malawi enjoys not only nationally but throughout the continent.

Although critics may insist that the smallholders are still not adequately compensated for their produce, Malawi is nevertheless one of the few African states that has permitted the adjustment of food prices, particularly maize, to favor the interests of the rural producer rather than allowing prices to subsidize the food purchases of the more politically volatile urban minority (Bates, 1981, 11–45). The Malawi government, moreover, is one of the few in Africa that has succeeded in slowing the rate of rural-to-urban drift by insisting not only on the more orderly growth of urban centers but more importantly by making the rural areas attractive places to live. Providing clean water to the rural villages, more equitable distribution of educational and health facilities throughout the countryside, and more effective involvement of rural areas in the national economy have been significant in this respect.

The third aspect of Banda's political strategy constitutes a compromise between modern and traditional forms of political organization. Paralleling developments elsewhere, the Malawi Congress Party (MCP) became not only the modern vehicle for achieving independence in 1964 but it has also assumed the primary task of overcoming poverty. Unlike most of its counterparts (with the obvious exception of Tanzania's dominant party), the MCP has not only achieved a solid base in the countryside, but it has been involved in actively encouraging rural cultivators to innovate with new crops, to participate in self-help projects in natural resource development, to maintain feeder roads needed to get produce to central markets, and to engage in other rural development schemes. On the other hand, far from assaulting traditional chiefs and elders (many of whom had earlier been accused of collaborating with the colonial authorities), Banda has relied heavily on the bonds which link those authorities and their people in advancing the goals of agricultural improvement. Chiefs play a continuing role in the allocation of land under customary tenure as well as being the organizers of village self-help projects. Moreover, instead of having to dismantle cooperatives and other modern structures (as has occurred in Tanzania, for example), the MCP encourages the formation of credit groups and other modern collectives.

In still another facet of Banda's strategy he parallels the approach of his political nemesis, Julius Nyerere of Tanzania. I refer to Banda's insistence on balanced national development which attempts to ensure equity for districts and regions which were economically neglected during the colonial era. Colonial Nyasaland had been settled by the British from the south, with the consequence that not only were the administrative and commercial capitals established at Zomba and Blantyre-Limbe but most of the tea, tobacco and other agricultural enterprises were also concentrated in the southern region (Fetter, 1982, 82). Although the northern third of the country was the target of some mission activities as well as a source for mine labor, its relative neglect was evidenced by the popular reference to the 'dead north'. To correct this 'pear-shaped' concentration of agriculture, commerce, education, health facilities and other enterprises in the south, Banda did several things. The most dramatic (and the most underestimated) was the creation of a new capital at Lilongwe in the Central Province. Unlike the less-successful efforts at capital relocation in Tanzania and Nigeria, the Malawian experiment has worked. The location of the capital closer to the geographic center has shifted the focus of agricultural production and marketing significantly to the center and the far north. This is confirmed by purchasing figures of the Agricultural Development Marketing Corporation (ADMARC).

Also of significance in achieving balanced national development have been the extaordinary outlays – which in some years amounted to 25% of the national budget – for development of major highways and the construction of all-weather feeder roads. Prior to independence, there were only 200 miles of paved road in the country – mostly in Blantyre-Limbe and Zomba. Most areas of the country now are in contact by rail, road or lake steamer with Lilongwe and are involved in a national political, economic, educational and cultural system. Finally, a critical plank in the Industrial Development Act of 1966 was the requirement that, where feasible, new economic activities should be dispersed around the country, rather than being concentrated in the Blantyre-Limbe area. Thus, the major milling of rice, the production of fertilizer, the creation of new lumber industries and pulp mills, the processing of fish and other new economic ventures have been spread fairly widely around the country.

The last element in the political strategy of development has to do with sources of foreign aid. In keeping with Banda's rejection of socialism, Malawi has consciously avoided seeking development aid from any Eastern bloc country. The most significant providers of aid have continued to be the Western bloc nations, that is, Great Britain, the United States, the European Community as a whole, and the Scandinavian countries (Morton, 1975, 169ff). The second bloc of significant donors are the states which have sometimes been referred to as the 'pariah' states because they are often the target of UN censure or are at odds with one or more major blocs within that organization. Increasingly, for example, Israel, Taiwan and South Korea have been involved in important economic relations with Malawi. The most significant of the 'pariah' group, however, has been South Africa. Banda is the only leader in Africa to have maintained diplomatic relations with South Africa, a situation which has earned him the title of 'Africa's Odd Man Out'.

Curiously, the intensity of economic and other relations that have been maintained between South Africa and Zambia, Mozambique and, indeed, most of Banda's southern African critics, makes a mockery of their criticism of Malawi. Banda, after all, only approached the South Africans for assistance in building the new capital at Lilongwe and in financing the shorter rail link to the port of Nacala after Western sources and the World Bank had rejected his appeal. This parallels Nyerere and Kaunda turning to China for aid in constructing the Tazara railway, following similar rejections from the West. The South Africans, on the other hand, were more than willing to oblige the Malawians since they sought to end their growing diplomatic isolation on the continent. Ironically, the construction of both the capital and the railway have actually made Malawi less dependent on South Africa than many of its southern African critics. It was this mutual realization by Malawi and its Black African neighbors that led to Malawi's inclusion in SADCC, the Southern African Development Coordinating Conference (Liebenow, 1982*a*). The latter regional organization seeks to disengage nine southern African states from their economic dependence on South Africa.

The two-pronged approach to agricultural development

At the time of independence, Malawi had two distinct agricultural sectors. The more modern of the two was the estate, or plantation, sector which was largely in the hands of the less than 1% of the population that was European in origin. The primary thrust of this sector was the production of tea, tobacco, cotton and other non-food crops intended for export. It accounted for roughly two-thirds of the country's export earnings and continues to do so today. This sector received the bulk of the colonial government's attention. Malawians were involved in this first sector primarily as grossly underpaid wage laborers, most of whom drifted back and forth between the estates and the other sector. The second sector, which involved the energies of over 80% of the Malawian population, consisted of smallholders who cultivated small tracts averaging about 1–2 ha per family unit. The smallholders were engaged largely in subsistence cultivation of traditional food crops such as sorghums, millet and cassava or the introduced staples of maize and rice. Increasingly, subsistence agriculture was being supplemented by both food and non-food crops being grown for local sale or for export to Zambia and other food-short countries.

Following a course pursued earlier by Kenya and more recently by the leaders of independent Zimbabwe, Banda in 1964 opted for a continuation of the two sectors, but with several significant modifications (Morton, 1975, 169ff). Increasingly, for example, the expatriate owners of the private estates have either been displaced or supplemented by Malawi estate owners. The smallholder sector has also changed, with more and more farmers engaged in cash crop production, although recent estimates suggest that in many areas subsistence cultivation still dominates and as much as 30% of the Gross Domestic Product (GDP) is derived from that source. The distinctions between the estate and the smallholder sectors, moreover, are no longer as sharp with respect to the type of crops produced. Smallholders are growing edible export crops such as groundnuts, macadamia and maize as well as tobacco, coffee and other non-food crops. The estates, on the other hand, have dominated with respect to one new food crop, namely sugar.

The most dramatic change since independence in the two-pronged approach has been the relationship between the two sectors. Clearly, in the colonial era the estate sector was the privileged one in terms of tax concessions; allocations of choice land; the outlay of government expenditures for roads, railroads and other infrastructural facilities; and the expenditures of funds on agricultural research. The smallholders, on the other hand, were required to pay hut and poll taxes and provide lowly compensated labor on the tea and tobacco plantations. They were compelled to plant cash crops which were sold to the government or the estates at non-competitive prices but received comparatively little assistance from the colonial administration in terms of agricultural advancement.

There are those observers who argue that the smallholder sector continues to subsidize the expansion of the estate sector (Kydd & Christiansen, 1982). I think the evidence, however, suggests a positive interdependency. First of all, there is a continuation of the interdependence of the two sectors, particularly with respect to the reliance of the estates on food supplies from the smallholders. Second, the government programs designed to bring about more effective utilization of land, diversification of crops and other innovations in agriculture are directed to improving both sectors. Third, it has been the growth of the estate sector which currently accounts for roughly two-thirds of the large agricultural export earnings that have, in large measure, made possible the expansion of the road and rail networks, the creation of processing and import substitution industries which utilize locally grown crops, the purchase of fertilizers and new seed varieties, the improvement in education and health care and other innovations in the Malawian economy.

Both sectors have been the beneficiaries of these modernization efforts. Earnings from the smallholder sector, which is still struggling to improve efficiency, could not in themselves account for these major infrastructural innovations, and both sectors should be expected to contribute to overall national development. Finally, the dramatic food production increases from the smallholder sector in response to upward price adjustments strongly suggests that the government is perceived by the smallholders to be acting in their interests.

Improved use of land and other surface resources

The two-pronged strategy is certainly in evidence when it comes to government-sponsored efforts to improve the use of land and other surface resources. Both sectors have benefited from these efforts. There are several aspects to improved land utilization, beginning with the more efficient use of land already under cultivation. The use of fertilizers, insecticides and new hybrids in great measure account for the dramatic increases in yields on both smallholder plots and the large estates. The smallholder sector, for example, 4 years prior to independence utilized only 183 short tons of fertilizer. By 1967 – 3 years after independence – use had expanded to 10 834 short tons, and by 1980 the smallholders were using 43 939 short tons (Malawi, National Statistical Office, 1977). Increasingly, smallholders have been encouraged through extension agents to engage in green manuring and to utilize the wastes from oxen, cattle and other domestic animals. More recently, the use of imported fertilizers has been supplemented by the development of a local fertilizer plant at Nsuka Falls. Fortunately, one of the few exploitable minerals found in Malawi is dolomite, a significant ingredient in chemical fertilizers. With respect to new seed varieties which have expanded the yield per hectare, strides are being made in the major staple of the Malawian diet, maize, as well as rice, which is the staple of the roughly 10% of the population living in the north. Unfortunately, much of the research associated with the Green Revolution elsewhere in the Third World has been devoted to wheat and other grains which, with the exception of rice, are thus far only marginal to the Malawian diet (Eicher, 1982, 162–3). The consumption of wheat by expatriates and urbanized Malawians has led to increased experimentation and production of that crop, but most needs are filled by a modest 30 000 tons per year of imported wheat.

Despite many difficulties, Malawi is one of the few African states that has experienced steady increases in production with respect to a broad range of food and other crops. Admittedly, assessment of production constitutes an imperfect art, due to the high percentage of food that is retained within the subsistence sector or exchanged in local markets. Informed estimates, nevertheless, can be made on the basis of recorded crop purchases by a state marketing board, particularly where that board maintains a near monopoly on crop purchases. In Malawi, the current ADMARC has enjoyed a virtual monopoly over crop purchases outside of localized transactions. Its purchasing figures since independence reflect the government stimulation of smallholder innovation and production. The following pre-1977 production figures are instructive in this regard (Malawi, National Statistical Office, 1977).

Maize

In the 17 years preceding independence there were only 5 years in which maize purchases by ADMARC exceeded 30 815 tons. In the 17 years after independence only twice did the purchases fall below that figure, and during 12 of the years purchases were double or triple that amount. In the 2 years prior to the 1980 drought, purchases reached 127 900 and 90 833 short tons, respectively.

Rice

In 15 years prior to independence ADMARC rice purchases averaged 4450 tons per year. From 1971 to 1981, average yearly purchases exceeded 22 600 tons.

Groundnuts

Groundnuts were only slightly encouraged as a food crop during the colonial era, sales to ADMARC reached a high mark of 17 400 short tons in 1964. In most years since independence, ADMARC purchases have been double to quadruple that figure.

Sugar

From a net importer prior to independence, by 1980–81 ADMARC sugar purchases of 145 000 tons had made Malawi a significant exporter.

Fish

Four years prior to independence, only 6400 short tons of fish were commercially caught in Malawian waters. During the years 1971–75 the annual average was 83 000 short tons.

Non-food crops

For non-food crops such as tobacco, tea, coffee, and cotton, similar dramatic rises in tonnage purchased by ADMARC were evident in the two decades since independence compared to sluggish performance under the colonial regime.

Some of the increased yields are attributable to the second aspect of improved land utilization, namely expanding areas brought under cultivation. The FAO at the end of the colonial era estimated that only 10 000 square miles (20 560 km^2) in Malawi out of a total land domain of 36 000 square miles (92 160 km^2) was suitable for arable agriculture (Pike, 1968, 173). Most of the land mass consisted of forest, swamps, and floodplains, steep escarpment, hillsides and rough grasslands. A good portion of the land was infested with tsetse fly, the host carrier of sleeping sickness. With respect to swampland and flood plains, significant changes have taken place within the past decade concomitant with an expansion in both sugar and rice production. The number of hectares under sugar production rose from 919 in 1968 to 3717 a decade later. Malawi was transformed from being a net importer to a major African exporter of sugar. The expansion in rice production is even more dramatic, going from 48 ha under cultivation in 1968 to 1400 ha in 1978. By 1982, with the assistance of the Taiwanese government over a two-decade period, some 6465 ha, involving over 6000 smallholders, have radically altered the production as well as consumption of rice throughout Malawi and neighboring countries. A yield of 62 bags (4000–to 5000 kg) per hectare on the Taiwanese plots brings results close to those being achieved in many parts of Asia under the Green Revolution. The one dark side of this otherwise rosy picture is that expanded rice cultivation does increase the risk of schistosomiasis, or bilharzia, for the farmers directly involved. This is one of the problems of economic versus social tradeoffs.

Land, however, is not the only element in Malawi's surface resources. Roughly one-sixth of the area under Malawian political jurisdiction is Lake Malawi, Africa's third largest lake. There are smaller lakes as well as rivers and streams that relate to the goal of food self-sufficiency. Considering the limitation that the tsetse fly and cultural attitudes imposed upon the intake of animal protein, it is surprising how little the fish resources of the country were developed and utilized under the colonial administration. Once again, in pursuit of a two-pronged strategy, the government has not only established Maldeco, a parastatal which carries out large-scale trawling on the lake, but has encouraged small fishermen. As a consequence, the production and consumption of fish in Malawi has altered significantly. At independence, fish provided 2% of the protein in the Malawian diet. By the early 1980s fish constituted 17% of the amino acid needs. Goats, sheep, chickens, pigs and cattle account for the other sources of animal protein. Through the use, moreover, of a wide range of preservation methods, including sun-drying, canning and refrigeration, fish has become a significant export commodity with respect to food-short Zambia and famine-stricken Mozambique. Recognizing, however, that fish constitute an exhaustible resource, the Malawian government has been carrying out a systematic restocking and conservation program as well as the development of fish ponds.

There are other ways as well in which Malawi's water resources affect not only economic development in general, but agricultural growth in particular. The Shire and other rivers, which are fed by high mountain streams, contribute to the generation of hydroelectric power for commercial as well as limited residential use. This has reduced Malawi's need for imported oil. Of more direct benefit to agriculture, the piping of water from the year-long flowing streams of Mulanje and other high mountains has transformed life in rural Malawi (Liebenow, 1981). Over the past decade, more than a million villagers for the first time in their lives have immediate access to a clean, year-round water supply. This imaginative program, which utilizes gravity rather than expensive imported pumping and drilling equipment, has not only made rural areas more attractive, but control over water-borne diseases has significantly reduced the number of working days lost to malaria, bilharzia, dysentry, and other water-related illnesses. It has also liberated women from the arduous task of searching for and carrying huge containers of water long distances, during the dry season. These measures have enhanced the quality of rural life.

The effective use of surface resources may actually involve the removal of areas from production or changes in usage. One very significant action in this respect has been the curtailing of cultivation and cattle grazing on the badly eroded slopes of Malawi's many mountains (Edje, 1982, 11–13). While not on the same scale as Swaziland's efforts, Malawi's approach is far more comprehensive in consequences. The reforestation program has not only cut back on soil erosion and flooding at the lower levels, but the ancillary reforestation program associated with this move has altered the economic status of smallholders and others in a number of respects. Again, with reference to the two-pronged approach, it has led to the creation of a large-scale timber industry which has provided needed lumber for domestic construction, jobs for local workers and additional export earnings. Of equal significance, the many voluntary community-help schemes have led to the planting of blue gums and other trees by smallholders around homesteads, along the margins of roads and highways, and on abandoned slopes. Under a controlled cutting program this has provided a ready source of firewood and timber for traditional house construction. Animal forage, medicinal herbs, bark for cloth, salable fruits and nuts, and other economic benefits also accrue. The general improvement in the quality of life of rural residents which has come through reforestation has been a factor in reducing the rural-to-urban drift experienced in other African countries.

Agricultural diversification

Another plank in the platform of maintaining food self-sufficiency is the effort to diversify not only food crops but also the many fiber and other non-food crops which have increasingly provided the margin for survival to both the estate and the smallholder sectors. Diversification has several objectives. The primary concern, perhaps, is the avoidance of the fate which has befallen most post-colonial states, namely, the dependence of the national economy upon a relatively few export crops or minerals for economic survival. Under the colonial system, it made sense from the perspective of the imperial power to encourage territorial specialization and thereby reduce inter-colony competition. Within the British Empire, for example, Tanganyika grew sisal as its main export, Ghana concentrated on cocoa,

Nyasaland developed tobacco and tea estates, Nigeria produced groundnuts and rubber, Zambia mined copper, and Sierra Leone exploited its iron reserves. Once a colony became independent, however, it found itself vulnerable not only to competition from outside the former imperial system, but it painfully realized that the prices for its basic commodities were being set at the global level by the industrial consumer nations. Development plans which were based upon a past history of a rise in the price of its primary export had to be reduced or even scuttled, as the price for that commodity plummeted. The causes of the latter situation were varied, including new producers of that commodity, a recession or depression in the consumer nations, the development of synthetic and other substitutes, and other factors which reduced revenue earnings even as local production expanded. Diversification thus constitutes a hedge against the possibility of prices simultaneously falling for all exported commodities.

Crop diversification, particularly when tied to the creation of processing plants and import substitution industries, has wider implications, of course, than hedging against global price fluctuations. It provides new sources of rural income, new jobs for the urban unemployed, and reduced dependence upon external sources for the vital commodities which adversely affect the balance of payments.

Equally important, crop diversification provides protection against plant diseases as well as against drought and other vagaries of weather, since each crop has its own specific susceptibilities. Diversification tied with intercropping and alley cropping makes a more effective use of land (Edje, 1982, 8–10). Coffee stands in Malawi, for example, are frequently interplanted with varieties of aleurite trees, the fruit of which produces tung oil, an ingredient in fast-drying paints. Crop rotation is particularly important in Malawi since the main staples of the diet (maize, sorghum and millet) and one of the main cash crops (tobacco) are rapid exhausters of soil nutrients. Thus, the success in encouraging the cultivation of groundnuts and other nitrogen-fixing plants in rotation with maize and tobacco is sound agricultural practice. In many cases the awarding of government loans to credit groups has required the planting of groundnuts for rotation purposes, as a condition of the loan.

Diversification of agriculture in Malawi has been perceived as an effort not only to enhance the national economy but to improve the economic status of the individual smallholder and estate owner as well. Diversification has been accomplished in a variety of ways. One method has been the introduction of entirely new crops which are suited to the altitude, climate and other conditions in Malawi. The macadamia tree, for example, which is native to Australia and was hybridized in Hawaii, has proved to be a highly profitable new crop, since the confectionary nut has great value in small quantity and is in much demand in Europe and America. Most of the nuts are processed in Malawi itself, thereby extracting maximum value from the product by increasing local employment opportunities and reducing transport costs.

Diversification in Malawi has also been accomplished by returning to crops which had earlier been experimented with or had actually achieved a measure of commercial success at an earlier period. This is the case with rubber, coffee and cotton, which were leading estate crops at the turn of the last century until global competition, pests, and administrative mismanagement virtually eliminated them (McCracken, 1982, 20). Finally, diversification may take place by encouraging new smallholders to engage in production of a commodity which has already been successfully tried in the country. This is the case with respect to traditional staple crops which have recently lagged in production and consumption. Cassava, for example, regarded as a 'starvation' crop during the colonial era, is an excellent crop which survives despite shortages in rainfall. Because so much stigma was associated with the punitive rules established during the colonial era, production fell off once independence was achieved. Recently, the government has been encouraging expanded production for both export and domestic markets, as well as the use of cassava in producing ethanol as a fuel additive. Discussion in Malawi has recently focused on another traditional crop, sorghum, which has been hybridized in the Sudan and has proved to be more drought-resistant and tolerant of a spectrum of soils. Another traditional food source, the Nyala antelope, is now being bred for its meat, initially using animals from the overstocked Lengwe National Park.

Perhaps the outstanding example of expansion of a current crop is sugar, which is also relevant to Malawi's approach to import substitution. Prior to 1970 Malawi was a net importer of sugar, with the 36 049 short tons of local production hardly meeting carbohydrate needs of Malawians. As a consequence of reclaiming swamp and floodplain lands, production by 1981 had reached 145 000 short tons, ranking Malawi tenth among 29 African producers. It suddenly became a significant exporter of sugar. Even when a crisis arose following the withdrawal of the very favorable US import quota allocation, and the subsequent disruption of the shorter rail link to Nacala by Mozambican rebels, the government found ways to convert disadvantage into advantage. The European Economic Community, for example, purchased sugar from Malawi for distribution to refugees in countries such as Somalia. More importantly, some of the previously exported sugar was converted at Dwangwa into molasses and then into ethanol (Carroll, 1984, 558). The latter, when mixed with gasoline, has substantially reduced Malawi's need for imported petroleum.

Food preservation

It may appear to be a common sense observation that the goal of attaining self-sufficiency in food is enhanced when the advances in production actually lead to an increase in the amount of food available for consumption. Unfortunately, planners often overlook the fact that an estimated one-third to one-half of the food grown in Africa is consumed by rodents, birds, insects and other pests or is destroyed by rot, wilt or mildew. Thus, a key element in Malawi's strategy has been to guarantee improved facilities for food storage and preservation. Farmers have been encouraged, for example, to resume the practice of constructing traditional wooden or bamboo circular storage sheds, which are raised off the ground, rather than relying on the more expensive imported metal storage bins. The significant difference in food storage, however, has come with the construction of modern concrete silos in Lilongwe and other depots around the country. The 48 silos in Lilongwe, completed in 1982, can hold 180 000 tons of maize and other grains. Hence, during times of bumper yields the surplus grain can be stored, and then rolled over and sold back at roughly the same purchase price during lean harvest years. The storage program is central to Malawi's being able not only to feed itself but to fill some of the food needs of its neighbors in Zambia, Mozambique, Zimbabwe and elsewhere. Exports of maize, which had been stored and rolled over, averaged 38 000 short tons a year during the last decade. This stood in sharp contrast to the experience in Tanzania where the absence of storage facilities forced the country in 1979 to export its bumper crop, only to be compelled in the face of next year's drought to buy from abroad roughly the same amount of grain as previously exported but at highly inflated prices. While grain storage is the key to the food preservation program, Malawi has engaged in diverse methods for preserving fish, fresh fruit and other foodstuffs for domestic and export purposes. This includes canning, sun-drying and refrigeration.

Focusing on the needs of the smallholder

Since the central player in the goal of food self-sufficiency is the smallholder cultivator, special attention has been paid by the government to the broad complex of needs of the more than 80% of the farmers who cultivate 1–2 ha plots. Robert Bates and others have noted the tendency for leaders in other countries in Africa to extract an increasingly greater share from the earnings of the agricultural sector as a whole to pay both for instant industrialization and the subsidization of the food and other needs of the more politically volatile urban masses (Bates, 1981, 30–44; Eicher, 1982, 156). The potential for this tendency to have manifested itself in Malawi was great, considering the absence of mineral resources and the failure of the colonial government to establish agricultural processing industries. Malawi, nevertheless, has pursued a different course. While in the colonial era, coercion was used to bring about innovation, since independence more positive measures have been employed (Fetter, 1982, 83).

Improvements in the quality of rural life have been more than symbolic, and restraints upon the unorganized burgeoning of urban residence have been effective. Urban wages have not risen out of proportion to rural farm income. Malawi stands out among its central and southern African neighbors in terms of letting the upward adjustment of prices being paid to farmers serve as an incentive to expand production and provide something like a fair return on investment in capital and labor (Carroll, 1984, 558). Just what constitutes a 'fair' price is, of course, difficult to determine, given the relative absence of competitive market economies in neighboring states.

It is the opinion of many observers that the upward adjustment of crop prices by the government – even in the face of contrary advice from the International Monetary Fund – has produced the desired result of more food being offered for sale at the buying posts established by ADMARC (Archarya, 1981, 16; Lele, 1981). Rainfall alone cannot account for the increase in food production – 91 900 metric tons sold in 1980 compared to roughly 245 000 metric tons purchased by ADMARC in both 1982 and 1983. Further evidence of policy successes are revealed in the fact that in 1980 Malawi has had one of the lowest ratios of rural residents (nine to one) on the continent, and it was able to persuade people to remain in the rural areas far more successfully than Tanzania and other states had done.

Ancillary to price adjustment as a positive inducement is the provision of improved transport and marketing facilities. The expansion of the road and rail network in Malawi is impressive, not merely in contrast with the low level of development during the colonial era, but in qualitative terms as well. While not neglecting the politically important task of linking all three provinces of the 920 km-long country through the construction of a macadamized highway, Banda has also paid attention to the economic need for feeder roads to reach the small rural hamlets where the food and other agricultural wealth of the country is being produced. The linkage of road development and agricultural success has been a constant theme in Banda's speeches over the past two decades, with as much as 25% of some annual budgets being allocated to road construction. Rural roads have been the product of a combined effort, with the national government providing some of the basic initial construction costs and local development efforts maintaining the largely unpaved but nevertheless all-weather feeder roads.

The feeder roads make possible the establishment of some 800 temporary buying posts which the parastatal ADMARC has maintained during the buying seasons for particular crops. The maize, groundnuts, tobacco, cotton, cassava and paddy rice are stored in 72 depots located around the country. The ADMARC system has meant that farmers in the remotest areas of Malawi have close and convenient access to the national purchasing operations.

ADMARC is the chosen instrument for filling other roles vital to improved smallholder production. It provides financial and technical assistance (most of it with respect to tobacco), and it is responsible for the storage, fumigation and transporting of a wide range of export crops. For the smallholder in particular, ADMARC distributes, free of charge, improved or specialized seed, such as cotton and tobacco. The parastatal also sells, at subsidized prices, fertilizer, pesticides and small farm implements (such as insecticide sprayers, oxcarts, plows and hoes). ADMARC is not without critics. Many Malawians and expatriates charge that prices paid for food crops could be even higher; that ADMARC should not enjoy a monopoly with respect to the purchase of certain commodities; and that it should avoid coming into direct competition with smallholders, as it does when it engages in direct production of tung oil, cashew and macadamia nuts, and various fruits and vegetables.

Outside the framework of ADMARC, the Malawian government has boldly addressed another problem which is crucial to smallholder survival, namely, access to credit for innovation as well as meeting the annual need for seed, fertilizer, oxen and farm implements. Customary land tenure practices prevent land serving as collateral. Malawi, therefore, has linked the availability of credit to more effective agricultural extension and community development programs. Instead of relying on the coercive tactics that the colonial administrators employed in achieving agricultural innovation, recent efforts in Malawi have been directed to linking availability of credit to the voluntary acceptance of crop rotation, experimentation with new crops and improvements in agricultural techniques. The status of the main spokesmen for innovation, the extension agents, has been enhanced by credit being funneled through those agents. The agents, however, are spared the onerous task of securing repayment of loans through a creative device in community development: although loans are given in certain cases to individual smallholders, the interest rate that is charged to smallholder credit groups is 5 percentage points less than that charged to individuals. The formation of these credit groups (limited to a hundred members each) as well as the recruitment of new members and elections of officers is left entirely to community initiative. The return of the loan is the responsibility of the group as a whole and peer pressure has achieved dramatic results in terms of securing roughly a 95% recovery rate from these groups, as opposed to a far lower recovery from individual smallholders. In addition to the financial aspects, this strategy frees the extension agent to engage in more positive educative programs.

Leadership factor

Many of the factors which contribute to Malawi's efforts in achieving self-sufficiency in food production can be attributed to the leadership factor. Often bucking the tide of international attitudes with respect to South Africa, for example, Malawi secured the funds needed to build the new railroad to Nacala and to relocate the capital at Lilongwe, both of which have had significant impact on the development of agriculture. It also took political courage and demonstrated political independence when Banda in 1974 abruptly terminated the recruitment of Malawians to work in South Africa. As is true in many southern African countries, labor recruitment has been a significant factor in providing employment opportunities as well as securing government revenues. Although the recruitment of labor for South Africa was subsequently resumed, it has taken place at levels considerably reduced from the pre-1974 figures.

Domestically as well, Banda has taken positive steps to ride against the tide. The involvement of traditional chiefs and elders taps a source of legitimacy in advancing development schemes which other African leaders have sometimes unwisely denied themselves. Although coercion is certainly employed within the Malawi political system, the tools of persuasion, exhortation, innovation by example, and economic incentives are far more in evidence in stimulating agricultural innovation than is true even in Tanzania. The restraints imposed upon rural-to-urban drift have also manifested political courage, as has Banda's paternalistic curbing of consumer appetites on the part of government, the political elite and the rural masses. With only a few obvious exceptions, the lifestyle of the Malawian elite approximates the lifestyle of middle-class professionals elsewhere in Africa. Malawian officials seldom engage in the international junketing which characterizes the elite in so many other new African states. Imported manufactured goods are limited, and imports are not calculated to undermine an economy which of necessity is based upon agriculture. The housing of rural smallholders and urban workers is modest and follows traditional lines and uses traditional materials where possible. There are few areas in either the cities or the countryside that could be described as 'slums'. There is a penchant for neatness and cleanliness that is refreshing in a Third World country. One obvious index of reduced consumer expectations is revealed in the fact that Malawi is one of the few African countries that has not developed a television system. Privately owned cars are relatively few in number. Despite the preceding, one does not have the constant feeling that Malawians feel impoverished. There is great hope. Shops in both the cities and the rural areas are well stocked with the basic necessities of life at affordable prices.

One final comment on reduced expectations is the ability of the leadership to search for what is popularly called appropriate technology in coping with the problems of development. Reference has already been made to the use of gravity rather than expensive imported pumping equipment in bringing clean water to the rural poor. A further example is the encouragement of draft oxen for cultivation. The number of oxen went from 35 000 in 1964 to 91 000 a decade later. Unlike imported tractors, which are costly and require imported spare parts and expensive petrol, oxen are better adapted to cultivate on rocky slopes, can be utilized for rural transport, and provide an additional source of animal fertilizer. Such labor-intensive approaches to innovation are being encouraged by the Malawian extension service in other fields of agricultural development.

Problems and prospects

Malawian leaders reject the notion that their program of dealing with food self-sufficiency constitutes a 'miracle of development'. The country is still heavily dependent upon outside assistance and, despite the trade surplus in 1984, the balance of payments situation is precarious considering the general poverty of the country. Many Malawians, particularly underpaid estate workers and women who serve as heads of household in the absence of their migrant worker spouses, are living at or below the poverty line. Their life expectancy is still short compared to that in Western societies.

There are many political problems associated with agriculture, moreover, that have not been addressed. A new program on birth spacing only begins to attack population growth rates that are estimated at 3.2%. As noted previously, the government has not yet addressed the traditional attitudes toward cattle, which results in overgrazing without much change in the protein intake of Malawians.

Another potentially explosive political problem is the issue of land ownership. While giving support to the smallholder, the government of Malawi is reluctant to address the issue of land tenure which affects future agricultural growth. Traditional land allocation, as noted previously, serves social justice objectives but it leaves Malawian farmers with small, fragmented plots, in which usufructuary claims are to a considerable extent inherited through the female line. These factors deter more effective land utilization as well as provide disincentives to innovation. The other side of the coin is that the expansion of the estate sector, either through outright freehold grants or long-term leaseholds, creates the specter of either a class which is permanently landless or tenants on land which was traditionally theirs. Many of these new estate owners, moreover, are members of the political elite.

Successes in the achievement of food self-sufficiency, however, overshadow these and other issues that have yet to be resolved in Malawi. The basic question for Africa is whether the advances secured in one country can easily be translated to other parts of the continent. Indeed, there are some economists who question whether the attainment of self-sufficiency in food production must be a uniformly accepted goal for African states. Elliot Berg, for example, has argued that in Liberia the inefficiency of rice production is such that Liberia ought to concentrate on forestry, mining and other agricultural activities, while importing the basic food staple from countries that enjoy a comparative advantage in that respect. While this advice is dubious even for Liberia, it is certianly not an answer for countries that have severely limited economic options outside of food production. Feeding one's own people must remain a top political priority for most African leadership groups.

Transfer of successful models of agricultural innovation is, of course, limited by the accidents of history, the peculiarities of location and environment, cultural values and such non-replicable factors as dynamic leadership and a pronounced work ethic. Even where these factors can be scientifically controlled or discounted, the history of both colonial and post-independence Africa shows that experiments in African agriculture are extremely site- and time-specific. The experience of the Green Revolution has demonstrated this. What may work in one country will not necessarily work in a neighboring country or even in another district of the first country.

Success or failure, moreover, cannot be measured in terms of one or a few years of experimentation. Some innovations – especially with tree crops – only succeed over an extended period of time. There must be a determination to persevere through radical variations in annual rainfall, fluctuations in global pricing, and a conservatism of farmers who are surviving on the margin with respect to embracing high-risk innovations.

In many respects the preceding commentary is the real lesson of the Malawi experience. Malawian leaders reject a rigid philosophically integrated blueprint for economic development, particularly as it relates to agriculture. Both modern and traditional institutions and values have legitimacy and have reinforced a wide range of economic innovations. Hence, the program in expanded land utilization and crop diversification as well as the emphasis on preservation and storage of crops have been accepted on pragmatic rather than ideological grounds. Inputs to smallholders in terms of seed and fertilizer as well as the extension of group credit to farmers are judged primarily on their ability to sustain food security and advance agricultural development and, only secondarily, on the contribution of these measures to the goals of social justice and the strengthening of the sense of political community. The avoidance of a program of instant industrialization, moreover, does not constitute a moralistic or ideological rejection of modernization and urbanization but rather a recognition of the fact that, given the circumstances of Malawi, the best promise of improvement in lifestyle must come through development of the country's agricultural potential. Similarly, the emphasis on appropriate technology does not constitute a glorification of the assumption that the conventional wisdom of several centuries can be effective in coping with seemingly modern problems. Rather, the use of draft animals instead of tractors, the reliance upon gravity flow rather than expensive pumping equipment in providing clean water, and the reforestation program as an alternative to importation of fuel and building materials provide answers to a very critical problem faced by Third and Fourth World countries, namely the adverse balance of payments. Much of the innovation in agriculture in Malawi discussed in this chapter might fall under the category of 'common sense'. The latter, however, is a commodity that is sometimes in short supply when it comes to the foreign agricultural adviser or to the ambitious African politician who is eager to project his country immediately into the modern era. The quality of 'common sense', however, must be complemented by a quality which also is frequently in short supply: political courage. The matching of these two factors is perhaps the ultimate lesson of the Malawian experience in agricultural innovation.

References

Archarya, S. (1981). Development perspectives and priorities in sub-Saharan Africa. *Finance and Development*, **18** (March 1981), 16–19.

Bates, R. H. (1981). *Markets and States in Tropical Africa: The Political Basis of Agricultural Policies*. Berkeley: University of California Press.

Carroll, J. (1984). Malawi: economy. In *Africa South of the Sahara, 1984–85*, 14th edn, pp. 557–60. London: Europa Publications.

Chanock, M. (1977). Agricultural change and continuity in Malawi. In *The Roots of Poverty in Central and Southern Africa*, ed. R. Palmer & N. Parsons, Berkeley: University of California Press.

Edje, O. T. (1982). Agroforestry: an integrated landuse system for increasing agricultural productivity. Chancellor College Symposium on *Development in Malawi in the 1980s: Progress and Prospects*. Zomba: University of Malawi (mimeo).

Eicher, C. K. (1982). Facing up to Africa's food crisis. *Foreign Affairs*, **61** (Fall, 1982), 151–74.

Fetter, B. (1982). Malawi: everybody's hinterland. *African Studies Review*, **25** (June–Sept., 1982), 79–116.

International Bank for Reconstruction and Development (1984). *Toward Sustained Development in Sub-Saharan Africa: A Joint Program of Action*. Washington, DC: IBRD.

Kandawire, J. A. K. (1979). *Thangata: Forced Labour or Reciprocal Assistance?* Blantyre, Malawi: University of Malawi Research and Publication Committee.

Kydd, J. & Christiansen, R. (1982). Structural change in Malawi since independence: consequences of a development strategy based on large scale agriculture. Chancellor College Symposium on *Development in Malawi in the 1980s: Progress and Prospects*. Zomba: University of Malawi (mimeo).

Lele, U. (1981). Rural Africa: modernization, equity, and long-term development. *Science*, **21**, 547–53.

Liebenow, J. G. (1981). Malawi: clean water for the rural poor. *American Universities Field Staff Reports*, 40/1981, Africa. Hanover, New Hampshire: AUFS.

Liebenow, J. G. (1982*a*). SADCC: challenging the South African connection. *Universities Field Staff International Reports*, 13/1982, Africa. Hanover, New Hampshire: UFSI.

Liebenow, J. G. (1982*b*). Malawi's search for food self-sufficiency. *Universities Field Staff International Reports*, 30–32/1982, Africa. Hanover, New Hampshire: UFSI.

McCracken, J. (1982). Experts and expertise in colonial Malawi. *The Malawi Review*, **1** (1), 19–25.

Malawi, National Statistical Office (1977). *Compendium of Agricultural Statistics, 1977*. Zomba, Malawi: Government Printer.

Morton, K. (1975). *Aid and Dependence: British Aid to Malawi*. London: Croom Helm.

Nyasaland Protectorate (1911–1963). *Annual Report of the Department of Agriculture*. Zomba, Nyasaland: Government Printer.

Pike, J. G. (1968). *Malawi: A Political and Economic History*. New York: Frederick A. Praeger.

Pike, J. G. & Rimmington, G. T. (1965). *Malawi: A Geographical Study*. London: Oxford University Press.

Vail, L. (1982). The historical roots of Malawi's economic underdevelopment. Chancellor College Symposium on *Development in Malawi in the 1980s: Progress and Prospects*. Zomba: University of Malawi (mimeo).

17

Famine relief policy in India: six lessons for Africa

MICHELLE B. McALPIN*

Tufts University

For the second time in 15 years the world has responded to the cry that famine is killing the already abysmally poor people of sub-Saharan Africa. Distress has extended beyond Ethiopia to much of the Sahel and to parts of eastern and southern Africa. With the return of the rains in Ethiopia and the south, these pictures of distress fade from our minds and we turn our attention to the next crisis – a cyclone in Bangladesh. But the *climate* of Africa that brought about the current drought years has not changed – only the weather.[1] Now, while the weather may be better for a time, it is imperative that the policies and institutions be built that will permit the individual countries of Africa and the international donor community to contain the effects of the next drought.[2] Steps must be taken now to assure that there will be no more refugee camps like Korem (Ethiopia). What can be done? What are the crucial steps that need to be taken?

Africa's current food crisis has two components. First, there appears to have been a decline in per capita production of food in many countries. Second, much of Africa lives with uncertain rainfall and that rainfall has been especially sparse over the last several years. Many people with long experience are writing volumes on the mechanisms that might be used to deal with the first of these problems (Mellor, 1985; Cleaver, 1985). This chapter focuses on the second. Any efforts that may be made to deal with the recurrent threat of drought will certainly show more results, if the longer-term productivity of African agriculture can be increased. In addition, successful efforts to deal with drought will minimize the disruption of ongoing developmental work.

*   The author is Professor of Economics. For dicussion and comments on earlier drafts, she wishes to thank William K. Carruth, John O. Field, I. S. Gulati, Ann Helwege, Jean Mayer, Morris David Morris and Al Sollod.

The Indian experience

What experience can be brought to bear on the problems many African nations face today? For the last hundred years the governments of the dry and drought-prone areas of India have been evolving a series of policies for managing the social impact of droughts. These policies are now strikingly successful in three ways, as the recent experience of Maharashtra illustrates.[3] First, they have been able to prevent significant elevations of mortality, even during severe and prolonged droughts. The state of Maharashtra (with an area of 308 000 km^2 and a 1971 population of 50.4 million) experienced a three-year drought in 1970–73 during which the outputs of foodgrains (cereals and pulses) were 19% below the average of 1968–69 and 1969–70 in the first year, 27% in the second year, and 53% in the third year (see Table 17.1). These deficits were added to Maharashtra's normal need to import 2.6 million tons of foodgrains, about one-third of its consumption requirements.[4] Yet the vital data for this state show no significant rise in the death rate and only a small fall in the birth rate (India, 1973, 1976, 1978).[5]

Fig. 17.1. India

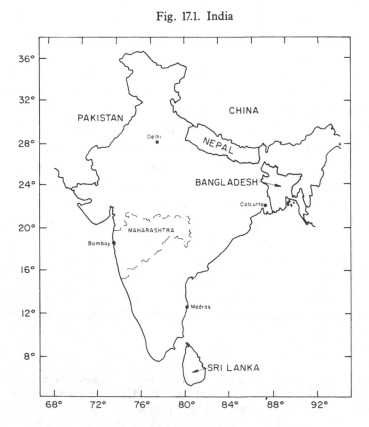

Table 17.1 *Food grain production in Maharashtra, 1968–79*

	Output of foodgrains (× 10³ tons)	Index of foodgrain output	Percent change (over previous year)	Foodgrain yield (kg/ha)	Index of yields	Percent change (over previous year)
Baseline (average 1968–69 and 1969–70)	6622.6	100	—	504.5	100	—
Drought years						
1970–71	5413.7	81.7	−18.3	420	83.3	−16.7
1971–72	4856.4	73.3	−10.3	392	77.7	−6.7
1972–73	3104.2	46.9	−36.1	272	53.9	−30.6
Post-drought years						
1973–74	7045.1	106.4	127.0	509	100.9	87.1
1974–75	7899.9	119.3	12.1	585	116.0	14.9
1975–76	9036.2	136.4	14.4	653	129.4	11.6
1976–77	9615.0	145.2	6.4	686	136.0	5.1
1977–78	10358.3	156.4	7.7	743	147.3	8.3
1978–79	9942.7	150.1	−4.0	716	141.9	−3.6

Maharashtra, India, Directorate of Agriculture, *Season and Crop Report*, appropriate years.

Second, government policies have been able to prevent major disruptions in the ongoing agricultural processes. Ways have been found to limit the depletion and insure the rapid rebuilding of working capital in agriculture. For instance, in 1973–74, the first post-drought year in Maharashtra, production of foodgrains rebounded to 106% of the average production of the 2 years immediately preceding the drought.

Third, the crisis-management policies of the state of Maharashtra and the Government of India have been able to leave the region of the drought with more infrastructure and more fixed capital in agriculture. That is, the famine-relief efforts have been designed to contribute to the improvement of agriculture in the state. In 1973–74, the first year after the drought, irrigated area in the state reached its highest level, with gross irrigated area equal to 9% of gross cropped area. In the next 4 years, foodgrain output increased by 12, 14, 6 and 8% over the preceding year. These increases were largely the result of increases in yields of 15, 12, 5 and 8% in those same years (See Table 17.1). Increased yields were made possible, because work done during the drought – bunding fields, constructing percolation tanks, well construction, clearing existing tanks of silt, terracing land and building roads – had improved irrigation facilities and access to markets.[6] These improvements had been made with the labor of people who were being given employment as part of the relief effort. Employment for all who wanted it was the major relief tool used by the Government of Maharashtra. By May 1973, in the third year of the drought but before the beginning of the good monsoon that ended it, the average daily employment on all forms of relief works was 4.95 million, i.e. 14% of the 1971 rural population of Maharashtra (Subramanian, 1975, 552–3).

Lessons for Africa

Can the experience of Maharashtra help in the formulation of policy for Africa? Unlike the countries of Europe which left food crises and famines behind by becoming high-income industrialized societies, India has been conquering famine while still a low-income and largely rural and agricultural society. Table 17.2 shows the GNP per capita, the Physical Quality of Life Index (PQLI), and selected other relevant figures for India, for the state of Maharashtra, and for a selection of African countries that have recently been adversely affected by drought. As can be seen, the most severely affected African countries (those in the panel beginning with Chad) include some that have per capita GNP less than one-third of India's and others that have per capita GNP 80% greater than India's. However, they all have

Table 17.2 Selected indicators for India and Africa

	GNP per capita 1982 (US$)	PQLI (1981)	Percent urban	Labor force in Agriculture (%)	Area (×10³ km²)	Population	Population Density (per km²)
India	260	46	24	71	3288	717.0	218
Maharashtra		49	35		308	62.8	204
Africa							
Most seriously affected							
Chad	80	27	19	85	1284	4.6	4
Ethiopia	140	31	15	80	1222	32.9	27
Mali	180	26	19	73	1240	7.1	6
Burkina Faso	210	20	11	82	274	6.5	24
Niger	310	27	14	91	1267	5.9	5
Sudan	440	38	23	78	2506	20.2	8
Mauritania	470	30	26	69	1031	1.6	2
Moderately affected							
Senegal	490	26	34	77	196	6.0	31
Uganda	230	49	9	83	236	13.5	57
Tanzania	280	61	13	83	945	19.8	21
Somalia	290	38	32	82	638	4.5	7
Kenya	390	56	15	78	583	18.1	31
Zimbabwe	850	64	24	60	391	7.5	19

PQLI for Maharashtra: Morris & McAlpin (1982). All other PQLI's, M. D. Morris & J. Morduch, unpublished manuscript, Alan Shawn Feinstein World Hunger Program, Brown University. All other data for Maharashtra, Tata Services Limited. All other data, World Bank (1984).

PQLIs that are well below India's. These lower PQLIs reflect both higher levels of infant mortality and lower levels of literacy than prevail in India.

In terms of the percentage of the population which is urban and the percentage of the labor force which is engaged in agriculture, the countries taken together are not strikingly different from India. The major difference is that all of these countries are very large ones with very small populations, resulting in extremely low population densities compared with India. The combined population of the seven worst affected countries is only equal to about 11% of India's population, while their combined area is nearly three times as great. Compared just to Maharashtra, these countries in 1981 had one-quarter more population in 29 times as much area. These low densities do not imply that these countries are underpopulated. Rather, they have within them very little land that is capable of supporting many people per km^2. It would be useful to ascertain the population density per km^2 when those areas which are nearly uninhabited are removed from the figures for total area. Nevertheless, the gross similarities of populations that are poor, rural and heavily dependent on agriculture should be enough to make some of the general lessons learned in over 100 years of development of drought management in India useful for Africa.

I suggest that there are at least six major principles of famine relief policy, drawn mainly from the Maharashtran experience. These are intended to be the *principles* which should underlay the development of famine policy. As such, I will state them here rather tersely. Further, while I am convinced that these principles will serve Africa well, their implementation in African countries will have to be based on local conditions and on the knowledge of programs that have worked best during the drought that is now ending. I do plan to address some of these issues of implementation in future work. These six principles are:

1) intervene early in the drought episode;
2) early intervention requires local knowledge;
3) relief should be in the form of employment that generates money incomes (employment should be created in ways and on projects that facilitate the development of the affected area);
4) working capital for agricultural and pastoral populations must be rebuilt at the end of the drought;
5) the government of the affected region must be in charge with all other actors' efforts supporting and supplementing governmental efforts;
6) development of successful policy takes time – possibly a generation or more.

I will examine and elaborate on each of these principles.

1) Intervene early in the drought episode

The distress that is produced by drought is not a sudden crisis like an earthquake. When the rains are late or scanty, when fields go unplanted, when grain prices begin to rise, when pastoralists begin to depart from their normal routine, when the number of wanderers suddenly increases, the signs are there that action needs to be taken if disaster is to be averted. While the precise forms of early intervention will vary from place to place, the need for intervention which can keep people in their homes, with their productive assets (however meager), with their families, with their independence and under as little nutritional stress as possible seems obvious.

Various Indian provincial famine codes stress the need to take action 'when the rains fail and anxiety is felt' or 'on the first indication of a likely crop failure owing to deficient or irregular rainfall or any other causes'. In addition to visiting affected areas to 'put heart into the people and to make them aware that Government is fully alive to the situation', the relevant officials are required to make sure that the most complete and accurate reports of the condition of crops are available and generally to make sure that all preparations are ready, should it become necessary for government to intervene. Such preparations include preliminary lists of village-level public works that could provide employment, lists of persons who will need relief other than employment, advertisement of sources of employment outside the district, and the opening of 'test-works' to ascertain if the population is yet sufficiently distressed to accept such employment (Bihar and Orissa, 1930, 5–7; Maharashtra, 1962, 11–13).

It is the failure to intervene early that leads to scenes like those that the BBC filmed in 1984 in the camp at Korem. The people inhabiting that camp had exhausted their own resources and then had begun to migrate in search of food. The combined stresses of traveling, of social dislocation, and of exposure to more disease vectors, when added to severe malnutrition, brought daily death rates in that camp to frightening levels. The solution to avoid such elevations in the death rates does not lie in better-run camps; it lies in steps to provide people with relief before desperation forces them from their own homes.[7]

2) Early intervention requires local knowledge

While sophisticated methods of remote sensing can indicate where vegetative growth is abnormal and where problems may arise, only people on the ground (literally) can spot the trends that indicate that a need for relief is developing. A knowledgeable local administrator may be able to see the local signs of distress, but to be able to mobilize resources for early intervention, higher levels of government are likely to have to be convinced. It is generally meaningless to say that grain prices are rising, if there is no historical record to show what prices have been for the last 10 years. If herds are returning early from seasonal pastures, *early* can only be measured if documentation on what is *usual* exists. It cannot be determined whether more people (or different people) are migrating, unless it is known how many (and which) people usually migrate. Local knowledge is needed to be able to perceive the build-up of distress and to be able to convince those without such knowledge that the distress is real. Such local knowledge cannot be acquired in a year when the paucity of the early rains suggests trouble. It must be built up every year by local officials collecting information on what is 'normal'.

Different provinces of India have used different means to create an historical record by which current conditions could be compared with 'normal'. In Bombay Presidency the Famine Relief Code required the creation of

> a statistical account with map for each district, showing the areas
> therein, which are liable to famine, the causes which make them
> so liable, their accessibility or otherwise for relief purposes, the
> economic condition of the population of such areas, and a rough
> forecast of the maximum numbers of persons within each likely
> to require employment on relief works in a year of famine from
> extreme drought or other causes. (*Bombay State, 1950, p.i*)

The 1950 edition of this work provides, among its many other tables, average annual wholesale prices of four major grains from 1924 through 1947, for each district.[8] The Bihar and Orissa Famine Code (1930, 2–3) required the Director of Agriculture 'To prepare and submit to the local Government for publication in the *Bihar and Orissa Gazette* . . . every five years, a schedule of normal rates of prices of . . . foodgrains . . . ' In addition, the Director was to prepare regular statements on the 'agricultural circumstances of each tract of the province' and to make suggestions for 'improvement of the system for famine warning'.

Local knowledge is also needed to decide how the mounting distress is best relieved – best in the sense that relief is both acceptable to the affected population and useful to the community in the longer run. For example, in the nineteenth century in Bombay Presidency (much of which is included in the modern state of Maharashtra) separate systems of relief were needed to reach agricultural laborers and agriculturists, tribal people living in the hills and artisans. Reaching the tribal populations was especially difficult and emphasis was laid on the importance of using civil servants 'selected for the knowledge of, and sympathy with, the tribes' as well as on using 'their own village headmen or . . . their natural leaders' to administer relief (Bombay Presidency, 1927, 37).

3) Relief should be in the form of employment that generates money incomes (employment should be created in ways and on projects that facilitate the development of the affected region)

Drought in India has been met by offering employment during virtually every scarcity since the 1870s. However, the virtues of this form of relief were also hotly debated during the formative years of the famine codes. The basic question was: would it not be cheaper just to feed people minimum maintenance diets rather than organize works of questionable longer-term value to the society?[9] Whatever the concern for economy, the answer of the Government of India and its various provincial parts has been 'cheaper, perhaps, but not better'. As one Indian civil servant wrote:

> The chief and immediate object of relief was to mitigate distress and to prevent loss of human life and useful cattle. With this end in view Government opened a large number of works scattered over the affected tract to give employment to the able-bodied persons and distributed gratuitous relief to those who were physically incapable of any kind of work . . . The second, and an equally important objective, was to minimize the charity aspect of relief by associating it with some kind of useful work so as to prevent a dole mentality and consequent demoralisation. (*Bombay State, 1958, 15*)

In the Maharashtra drought of 1970–73, fewer than 200 000 individuals were given relief for which no work was exchanged (Subramanian, 1975, 167).

It needs to be stressed that using employment to provide relief can only work if two conditions are met. First, relief must begin early. It is inhumane in the extreme to wait to begin relief until some section of the population has exhausted all of its reserves, and has, in fact, begun to suffer from acute want of food and then to offer only employment. If governments at various levels, by failing to intervene early, have permitted a full-scale famine to develop, then the only recourse is to direct all efforts to minimizing the human cost by feeding programs and medical intervention.[10]

Second, it is essential that food supplies should flow into distribution channels in the affected region. If the food distribution network is generally very heavily controlled and regulated in normal years, it would be foolish to leave it suddenly unfettered in a year of major food shortages. The Indians have long used the distribution of some ration of grain at subsidized prices via 'fair price shops', as a means of limiting speculative excesses and of assisting the poorer sections of the population. Maharashtra had over 24 000 such shops in 1969 and over 29 000 by the end of the 1970–73 drought. Care was taken to assure that there was a fair price shop near every site where relief employment was ongoing (Subramanian, 1975, 256–7). Fair price shops were in turn supplied by deliveries from the parastatal organization in charge of procurement and sale of grain between surplus and deficit regions.[11]

If these two conditions can be met – early intervention before people are emaciated and exhausted and the supply of foodgrains through a distribution network that serves those in need – relief offered via employment has several advantages. First, if wisely chosen and well-organized, it produces something – perhaps an improved road, perhaps some miles of railroad bed, perhaps a deepened and repaired well for drinking water, perhaps a building for a clinic or an elementary school, perhaps a small irrigation facility. All of these forms of social overhead capital and agricultural capital are in very short supply in Africa, so the selection of useful projects should not present insurmountable obstacles. Second, employment can generate new skills, preserve old ones and maintain the self-respect and independence of the affected population in ways that no program of free food can.

Third, the payment of money wages (rather than food wages) permits the private sector to absorb some of the newly effective demand.[12] While vigilance is always needed to assure that more reliance than is warranted is not placed on private grain traders, strengthening normal channels for marketing grain in rural areas can assist in mitigating future shortfalls of production by encouraging the normal market process to move grain from where it is cheap to where it is costly. Supervision of private grain trade may be necessary and certainly surveillance of grain markets to observe available supplies and prevailing prices (and changes therein) is needed.

Finally, payments in cash permit recipients to choose their own preferred allocation of income between food and all other consumption, as well as among possible food items. While there may be some systematic tendency among workers to spend less on food than seems wise to outsiders, this might be partially overcome by the use of supplementary feeding programs for pregnant and lactating women and for small children, as well as by provision of a midday meal at worksites.[13]

Employment may be created in several ways. If village communities are strong and well-governed, they can be given loans with which to carry out labor-intensive projects that they clearly see to be of value to the village. If individual landowners usually employ significant amounts of labor, the landowners might be given loans to make capital improvements to their farms – like levelling and bunding fields, digging, deepening, or repairing wells. Considerations of equity might dictate that loans to communities should be given on more generous terms than loans to individuals already above the average for wealth and income in their locale. A third alternative is the direct provision of employment by government on public works projects – building of major roads, irrigation works and the like.

There are three disadvantages to this third alternative: first, supervision of such works requires many administrators who may be in rather short supply; second, it may be more difficult for the government, working from the top down, to design useful works that will be completed even after the drought ends; third, if the works are large, they may force many of the employed to travel from their homes and live on the work site. This is undesirable, both because it makes it more difficult for them to go back to their usual occupations and because of the increased risk of disease in any major congregation of poor, nutritionally stressed people in temporary housing. Government-organized employment relief may be necessary (and desirable) for artisans. If people in the drought-stricken area need blankets and if weavers need relief, it just makes sense to put the weavers to work making blankets.

Unfortunately, the experience of India gives few suggestions for the relief of pastoralists by employment. To the extent that pastoralists have the capacity to transport foodgrains and a tradition of carrying goods to trade, they might need less relief, if the agricultural population had more money to buy grain and if the grain trade were partly left in private hands.

4) Working capital for agricultural and pastoral populations must be rebuilt at the end of a drought

While early intervention in the drought episode may help to stem the depletion of all working capital, it is still essential to direct significant attention and funds to rebuilding of working capital, so that the affected population has the capacity to return to its former occupations with reasonable chance of success. In the 1970–73 drought in Maharashtra, loans to agriculturists to enable them to rebuild their working capital were still being made under Acts dating from 1883 and 1884 (Subramanian, 1975, 124–6). The practice of 'soft' loans from government for these purposes originated, however, well before th British arrived in India. Indeed, these sorts of loans were considered an obligation of traditional Indian rulers.

The rebuilding of working capital is not just a matter of loans, nor can it be left until the return of the rains is imminent. Stocks of seed should ideally be purchased and kept in government stores on a preventative basis. Seed grains, more than almost any other input, are locally specific – imported seed usually will not be as suitable as local seed. If significant numbers of agricultural implements have been destroyed, early planning may permit their local manufacture rather than their importation – thereby saving foreign exchange, employing craftspeople who might otherwise need relief and stimulating the domestic economy.

A conscious policy is also necessary to preserve reasonable populations of all major types of livestock. However, shortages of fodder are likely to preclude preservation of all or even most livestock. Attention should be devoted to preserving the animals most valuable in rebuilding herds and to assuring that owners find the best possible markets for animals that cannot be carried over. Increased production of fodder is, of course, extremely desirable, to the extent that it permits the system to recover more rapidly from a period of drought by permitting survival of more valuable animals.

It is important to realize that the good health and nutrition of the population are also inputs into the production process. Energy must be directed to forestalling declines in the health and nutritional status of the population and, where such declines occur, to providing inputs to assure that people have the necessary strength and good health to return to their usual occupations at the end of the drought.

5) *The government of the affected region must be in charge with all other actors' efforts supporting and supplementing governmental efforts*

In the development of policies for coping with drought in India, the issue of the roles of the government versus international agencies did not arise for the simple reason that a century ago such agencies were largely absent. However, the issue of competition between government relief efforts and those of various private groups did arise. Government officials were concerned that, if private groups dispensed free food on any significant scale, people would be unwilling to labor at relief works and would instead flock to feeding sites. To deal with this possibility the government strongly encouraged those persons or organizations that wished to make charitable contributions to do so through one of the several 'famine relief funds'. Such funds were normally established by the head of each district, as well as by the head of each state. The funds so collected were used to supplement governmental efforts in a variety of ways, including the provision of free food – generally as uncooked grain – to those unable to go to the relief works, the provision of clothing, the distribution of fodder and grants of seed grain to those who were deemed too poor to be likely to be able to repay loans.

In the period since independence, the same pattern has been continued, although the funds and activities of international donors have been added into the system. For instance, in the Bihar famine of 1966–67, CARE made its contribution by feeding school children through a regular school lunch program (Singh, 1975, 109–10). In the Maharashtra drought, CARE contributed corn, soybean oil and milk powder for the production of a nutritionally rich dietary supplement distributed to all persons on the relief works. The mixture was produced by a private charitable trust, under the direction of the scion of a Bombay textile firm, with the state government contributing some additional raw materials and about one-fourth of the cost of production (Subramanian, 1975, 261–2).[14]

There are a number of reasons, beyond the example of India and beyond the preservation of national sovereignty, why having the control of government clearly present and effective is important to both the short-term efficiency of relief efforts and to the longer-term development effort. First, the nature of the policy I have been suggesting – early intervention based on local knowledge, use of employment to provide relief, employment to be generated wherever possible on works that contribute to the development of the affected region – requires knowledge which is very likely to be had only by the regular civil service (if by anyone). In any case, the accumulation of this knowledge must be supervised and maintained by the regular civil service. Expatriates, whatever their access to other resources, are unlikely to be able to command this knowledge, without establishing a quasi-colonial relationship with host countries.

Second, the government has the strongest incentives to use relief expenditures for development. National governments, however venal particular ones may be perceived to be, still have, as institutions, the most to lose, if relief aid cannot be used to further development. It is they who will have to go begging to the international agencies in the next drought, if they cannot successfully use resources mobilized for relief in this drought. On the other hand, international agencies, however virtuous particular ones may be, derive much of their visibility and ability to command resources from their intervention in crisis situations.

Third, local governments have a comparative advantage in early intervention, while international agency skills generally become useful only as the crisis worsens. Local knowledge is necessary to spot early signs of distress and to implement programs to prevent that distress from becoming famine. But once a true famine has developed – people wandering in search of food, all of their resources gone – local knowledge is much less needed. The ability to create employment on useful public works is irrelevant when people are already at the edge of starvation. At that stage what is needed is the ability to organize and manage famine camps (skills in which there is still room for improvement) and to deal with logistical and political problems for assuring the delivery of needed supplies.

Fourth, it takes more authority to require people to work for relief than it does to hand out free food. Governments are more likely to be able to exercise this authority than are expatriate agencies. That is, demands that work for development be exchanged for money to buy food are more likely to be seen as legitimate when they come from the government than when they are made by charitable institutions. The government also has the advantage that it will still be in place (revolutions and the like aside for the moment) when the emergency is over. It will still have to govern when the massive aid effort to save lives has passed. As such, it has greater incentives not to dispense relief at levels and in ways which profoundly alter the population's expectations of how they will, in the future, obtain their sustenance.[15] It has a much greater incentive than an international agency to avoid creating a population permanently dependent on food handouts.[16]

Finally, in the long run, the ability to deal successfully with drought (i.e. to prevent elevations of mortality, to minimize disruption of the ongoing agricultural and pastoral activities, to add to the capital stock in agriculture and to the social overhead capital) requires that governments be reasonably strong administrative states. The development of a strong administration (i.e able to collect information on which decisions may be made, able to implement those decisions with reasonable efficiency) requires experience. It is of little value to the nation concerned if a significant portion of the experience in any major crisis accrues to expatriates who will be somewhere else when the next crisis strikes.

6) Development of an effective famine policy takes time – perhaps a generation or more

The components of a successful famine-relief policy are neither particularly quick nor easy to develop. First, there is the need for good statistical series that permit governmental decision makers to diagnose an impending crisis. Because the particular factors that are the best predictors may differ quite dramatically from one place to another, it is likely to require some experimentation to select the most effective sets of indicators. As was pointed out above, however, indicators are only useful to the extent that they permit the comparison of the current year/season with a 'normal' year/season. In addition to the problems of determining what series are most useful, ways must also be found to create and maintain a network of agents that can feed the necessary information into the system. Particularly in countries where population is more sparse and rates of literacy are extraordinarily low, the development of an adequate reporting system is unlikely to be a small task.

Second, even when the reporting system is working up to capacity, the effective range of early interventions must still be found. For instance, the Government of India in the nineteenth century went through a long process of finding ways to generate employment or other means of relief that were acceptable to the population, contributed to capital formation and did not overwhelm the budget. In Bombay Presidency it was found that the large numbers of tribal people would generally starve rather than leave their native districts for employment at relief works. Over the course of 60 years, policies were developed that involved government payments for collection of forest products – a form of employment that both kept tribal people in their home areas (which in any case they refused to leave) and in occupations which were familiar to them. In the interim some success was achieved with small irrigation projects and road-building projects in tribal areas, where the presence of only tribal people made it possible to have a different standard for the day's task than at works where agriculturists formed most of the workforce (Bombay Presidency, 1927, 37–9). The relief of craftspeople who normally sold their wares to agriculturists also presented different problems which took time and experimentation to solve.

In the countries of Africa that face the threat of drought, the specific forms of relief employment as well as the means of organizing that employment will have to be worked out for each affected population – pastoralists, agropastoralists and craftspeople. Even as these governments gain administrative strength, the creation of effective policy will still involve trial and error.

One activity which can help to shorten the time that it takes to develop effective policy is a systematic process of report and analysis. If we do not know what worked and did not work this time, what was tried and what was not (and why), we are condemned to begin the process over when the next drought comes. Ideally, this process should be one which is 'error embracing'; that is, when some particular process or policy appears not to be working, people involved should be free to suggest alternatives. In addition, the felt needs of the population receiving assistance should be ascertained and their suggestions given full consideration in the development of a drought-relief strategy.[17]

Report and analysis – the creation of an institutional memory – is one key to making sustained progress in the development of effective famine policy. Again, this is a matter which is best undertaken by the government of the affected region. It, after all, will be on the front line in the next drought. It has to be the agency that collects the information needed for early diagnosis of distress, since it is the most likely source of early intervention which can prevent drought and scarcity from becoming famine.

Report and analysis will be of little use, however, unless the findings of such a process are fed back into the development of government manuals that specify the steps to be taken at each stage of a crisis from the beginning of the season (when only data collection may be appropriate) through the beginning of the next year (when the primary activities may be the termination of relief employment, reconstruction of working capital and preparation of reports). The mere fact of the preparation of a comprehensive set of procedures for dealing with drought, scarcity and famine should not be seen as likely to mire the relief process in bureaucratic red tape. One of the more interesting elements of Indian famine policy, since early in its development, has been the tendency to suspend normal bureaucratic requirements in periods when alleviation of distress required various officials to be able to act rapidly.[18] At the same time, after a period of scarcity, each official responsible for relief was required to give a full account of those measures undertaken and their success or failure. These reports typically were summarized in a final report which regularly contained suggestions for revising the famine codes and relief operations.

While the emergence of successful famine-relief policy is likely to be a fairly lengthy process, there are some ways in which progress towards such a policy can be measured. These include (*a*) development and use of a regular system of reporting the relevant variables, (*b*) preparation of plans for response to droughts, and (*c*) successful management of the smaller, local crises that sometimes occur between the more serious droughts that affect most of the country.

The role of international aid

What can the international donor community do (*a*) to aid in the development of better famine policy in Africa and (*b*) to play a constructive role within that policy?

To the extent possible, international donors can encourage the African governments with which they deal to work now, as the current emergency subsides, on the preparation of a famine-relief policy for the future. This encouragement could include (*a*) subsidies to permit key individuals to have time to write up their analysis of those aspects of the current relief efforts which were most and least successful and their recommendations for improvements, (*b*) provision of training and start-up funding for improvement of systems of data collection, and (*c*) assistance in developing 'famine-relief' or 'drought-relief' courses for permanent civil servants and other relevant people. The international donor community should also, of course, undertake review and analysis of its own efforts.

To the extent that many members of the international donor community are also involved in long-term development projects for agriculture, they need to encourage the rationalization of national food policies in Africa. They need first to try to make sure that their own aid does not act as a disincentive to the achievement of reasonable self-sufficiency in food production. Beyond that they need to assist countries in the process of providing remunerative prices to agriculturists, while easing the burdens of rising food prices on urban dwellers. This suggestion is placed within the framework of developing drought or famine relief policy, because of the intimate connections between the 'normal' productivity of the agricultural sector and the country's likely need for relief. To the extent that agriculturists become more prosperous, they are, within limits, better able to provide for themselves and less likely to need relief.

When a drought emergency develops, the international donor community needs to act with restraint. For instance, to the greatest extent possible, whatever aid it gives should be channeled through local institutions. Local institutions have two advantages – they already exist and they are part of the society and will remain. However inefficient they may be thought to be, as Alan Berg has written about his experience during the Bihar famine of 1966–67, 'energies should be devoted to grafting the necessary innovations onto them to make them work' (1971, 125). If schools are in session, they should be used to distribute supplemental food to children – and the teachers' salaries should be supplemented so that they can continue to eat in the face of rising food prices. If cooperatives usually sell grain to members, they should be used as the basis of a network of subsidized (not free) grain distribution. Roles should be found for religious institutions to play.

As much as possible, inputs should be supplied which will enable the national government to conduct successful employment relief. For instance, if road construction is to provide lasting benefits, the earthwork needs to be consolidated by road rollers. These can be supplied by an international donor. If tools could be manufactured locally, were steel available, the steel may be supplied by a donor agency. If textiles are needed and there exists in the country a tradition of weaving, cotton and wool fibers could be supplied by donor countries. It should be stressed that the provision of supplies to help people meet their own needs is only a workable policy when intervention has come early in the drought episode. The inhabitants of the Korem camp needed blankets; they were cold, they were weak and they had sold any weaving equipment they possessed. But earlier, when they were still in their homes, employment relief could have worked.

Another way in which the international donor community should be restrained is in the demands it places upon the time and energy of the local civil servants. Probably, the most effective act of self-restraint here is cooperation among all donors so that each and every agency does not demand equal time from harassed local officials. In addition, donors should make every effort to minimize the number and scope of conditions (including time-consuming paperwork) they impose in return for their aid. For instance, if the harvest in a particular country is adequate to feed the rural population but not the cities, it would seem rational to use food aid in the cities. This would keep urban demand from drawing food out of the countryside and forcing up rural prices. Donors often are unwilling to see food aid used in the cities, where incomes are commonly higher than in the rural areas. As a result, food aid must be used to fill the gaps in rural needs created by the flow to urban areas. Distribution of food to rural areas is more costly and difficult than getting the same resources into urban areas. One solution might be to store foodgrains in rural areas in normal times. However, flexibility on the part of donors (perhaps, brought about in part by efforts to make the real situation clearer to donor governments) would be very helpful.

Finally, donors should realize that supporting governments' efforts to develop a drought- or famine-relief policy which is based on early intervention is cost effective. The total volume of aid required during the drought may or may not be less under a program of early intervention, but the amounts required *after* the drought ends will certainly be less. It will be less because one of the great benefits of intervening early and using relief employment to build capital is that the system can regain its former level of productivity much more quickly than when a full-fledged famine – high mortality, great social and geographical dislocation, depletion of resources – has been permitted to occur.

Notes

1. The actual constancy of the climate of sub-Saharan Africa is at the moment a matter of some dispute among meteorologists. No one, however, seems prepared to argue that the Sahelian region (extending into Ethiopia and Sudan) is likely to experience either a significant increase in long-term average rainfall or a reduction in the year-to-year fluctuations in rainfall (Winstanley, 1985).
2. The need to develop policies and institutions for constructive and effective famine relief was addressed eleven years ago by Jean Mayer in an essay which raises some of the same issues as this chapter (see Mayer, 1974).
3. I have chosen to use the Maharashtran experience because (*a*) it is the most recent example of a major drought in India and (*b*) because the sources of data and analysis for that drought are unusually good. Maharashtra's experience is not unique. Alan Berg has written of the sucessful management of famine in Bihar in 1966–67 (Berg, 1971).
4. The estimate of the normal deficit is from Subramanian, (1975, 244–5). Subramanian was in charge of coordinating relief efforts for the entire period of the drought. His account is an extremely detailed and valuable one. He was very nearly the 'relief dictator' which Jean Mayer recommended in 'Coping with famine'.
5. For the kinds of increases in mortality and reduction in birth rates that occurred in much of this region in the famines of 1876–78 and 1899–1901, see McAlpin (1983).
6. 'Bunding' is the operation of creating an earthen dam or ridge to slow the runoff of rainwater from fields. 'Percolation tanks' are built on the upper reaches of water courses to hold water that would otherwise be lost downstream and thereby to recharge the nearby water table to supply more water to wells for both irrigation and drinking water.
7. There is increasingly widespread awareness of the need to provide aid to drought-affected regions while people are still in their homes (see Society of International Development, 1985).
8. The first edition of the atlas was published in 1888, with revisions in 1906 and 1925.
9. The exception to the massive organization of relief works was the Bengal Famine of 1943–44 (see Greenough, 1982, 127–9). For a brief discussion of the debates that surrounded the early famine codes see Brennan (1984).
10. See Mayer (1974, 106–8) for an outline of how the distribution of food and medical services might be arranged when true famine has already emerged.
11. For a description of the sources of foodgrains and their distribution during the drought in Maharashtra, see Subramanian (1975, 245–5).
12. For a full and sophisticated discussion of the relative importance of effective demand for food (pull factors) and supply of food (reponse factors) in generating famine conditions, see Sen (1981). He argues that we are mistaken, if we think that starvation results only from a decline in the overall availability of food in a society. Rather, he argues, we need to realize that a variety of events may deprive some section of the population of its (normally effective) claims to food. It may be that they can normally exchange 1 day's labor for 2 day's food. In an abnormal year, when food prices rise, they can only earn one-fourth as much food for a day's labor. If the situation continues, they will sell their assets, then begin to starve. This may be prevented by giving them a way of earning more income and, hence, generating a 'pull' of food into the affected area, assuming that the market will respond by supplying more food. This strategy can only work, of course, when there are supplies of food that can be drawn out of storage or into the region at prices such that local demand remains effective.
13. At a UNICEF pilot project in Ethiopia using cash relief, recipients frequently saved

part of the cash they received to purchase agricultural implements and livestock (Padmini, 1985). The various reports on this project have indicated the advantages of cash (as opposed to food) for relief. While UNICEF conceived the idea of using cash relief to enable distressed people to gain access to local food stores, it was the Ethiopian government which insisted that cash could not be given away to able-bodied people but rather that they must do some work in community development to earn cash.

14. No real history exists of integration of public and private relief efforts in India. Both Singh (1975) and Subramanian (1975) mention this integration and Subramanian provides a separate chapter (1975, 411–29) on non-governmental efforts. The role of the Maharashtra Central Famine Relief Committee 'as a link between the affected people on the one hand and the administration on the other', merits further study.

15. Clifford D. May has written

> The places where reality is most drastically
> altered for famine victims may be in Ethiopia's
> many relief camps and feeding centers. Not far
> from Maskel, for example, is a camp outfitted
> with large water-proof tents, street lights,
> generators, a hospital, free, hot food, and guards
> at the gate. It is a functioning town with far more
> amenities than its inhabitants have ever known
> before or are likely to know again. To leave such
> an environment, Mr Salole [Save the Children]
> observes, 'can be a traumatic transition'.
> (*May, 1985, p. E9*)

However splendid conditions may be in this camp, one wonders if its creation was wise.

16. Cuny (1983) provides an extensive discussion of the problems that can arise when a host of international donors descend on an underdeveloped country and are not kept firmly under the control of the host government. Among others, he cites the tendency for donors to impose different requirements for giving aid to the local populace, a process sure to generate resentment (at a minimum).

17. I have borrowed the term 'error embracing' from Korten (1980).

18. Subramanian describes the high-level streamlining of 'normal channels' that was used during the 1970–73 drought in Maharashtra in his chapter on 'Organisational management'. The various famine relief codes and handbooks stress the need for the district collector to act 'on his own responsibility' (Bombay Presidency, 1905, 7) and to proceed if waiting for official sanctions 'may, in his judgment . . . lead to privation which endangers life' (Bombay Presidency, 1927, 7).

References

Berg, A. (1971). *Famine Contained: Notes and Lessons from the Bihar Experience*. Washington: Brookings Institute.

Bihar and Orissa (1930). *The Bihar and Orissa Famine Code, 1930*. Patna: Superintendent of Government Printing.

Bombay Presidency (1905). *A Handbook on Famine Relief Administration in the Bombay Presidency*. Bombay: Government Central Press.

Bombay Presidency (1927). *Famine Relief Code, 1927*. Poona: Yervada Prison Press.

Bombay State (1950). Bureau of Economics and Statistics, Statistical Atlas of Bombay State. Bombay: The Examiner Press.

Bombay State (1958). *Final Report on the Scarcity of 1952–53 in Bombay State* by M. R. Yardi. Bombay: Director of Government Printing.

Brennan, L. (1984). The development of the Indian famine codes: personalities, politics, and policies. In *Famine as a Geographical Phenomenon*, eds Bruce Currey & Graeme Hugo, pp. 91–111. Dordrecht: D. Reidel Publishing.

Cleaver, K. (1985). The impact of price and exchange rate policies on agriculture in sub-Saharan Africa. *Staff Working Paper No. 728.* Washington: The World Bank.

Cuny, S. C. (1983). *Disasters and Development.* New York: Oxford University Press.

Greenough, P. (1982). *Prosperity and Misery in Modern Bengal: The Famine of 1943–1944.* New York: Oxford University Press.

India, Government of. (1973, 1976, 1978). Ministry of Home Affairs, Office of the Registrar General, Vital Registration Division, *Sample Registration Bulletin*, vol. VII, no. 2, April–June 1973; vol. X, nos 3 & 4, July & Oct. 1976; vol. XII, no. 2, Dec. 1978.

Korten, D. C. (1980). Community organization and rural development: a learning process approach. *Public Administration Review*, **40**, 480–511.

McAlpin, M. B. (1983). *Subject to Famine: Food Crises and Economic Change in Western India 1860–1920.* Princeton: Princeton University Press.

Maharashtra, Government of (1962). Revenue Department, *The Bombay Scarcity Manual (Draft).* Bombay: Government Central Press.

May, C. D. (1985). Article in *The New York Times*, 9 June 1985, p. E9.

Mayer, J. (1974). Coping with famine. *Foreign Affairs*, **53**, 99–120.

Mellor, J. W. (1985). *The Changing World Food Situation.* Washington, DC: International Food Policy Research Institute.

Morris, M. D. & McAlpin, M. B. (1982). *Measuring the Condition of India's Poor: The Physical Quality of Life Index.* New Delhi: Promilla & Co.

Padmini, R. (1985). *The Local Purchase of Food Commodities: 'Cash for Food' Project.* Addis Ababa: UNICEF.

Sen, Amartya (1981). *Poverty and Famines: An Essay on Entitlement and Deprivation.* Oxford: Oxford University Press.

Singh, K. S. (1975). *The Indian Famine, 1967.* New Delhi: People's Publishing House.

Society of International Development (1985). *Report of the North–South Food Roundtable on the Crisis in Africa.* Privately printed.

Subramanian, V. (1975). *Parched Earth: The Maharashtra Drought 1970–1973.* Bombay: Orient Longman Limited.

Tata Services Limited (1984). *Statistical Outline of India.* Bombay: Tata.

Winstanley, D. (1985). Africa in drought: a change of climate? *Weatherwise*, April, 74–81.

World Bank (1984). *World Development Report.* Washington, DC: Oxford University Press.

18

Famine and famine relief: viewing Africa in the 1980s from China in the 1920s

LILLIAN M. LI

Department of History, Swarthmore College

During 1984 and 1985, as the tragedy of the Ethiopian famine has been played out in Africa, another human drama has unfolded in the United States and Europe. Although anticipated by experts for years, and in progress for months, the famine in Ethiopia did not reach the American public's attention until October 1984, when NBC evening news aired a BBC special about Ethiopia. As they ate dinner, Americans could watch with horror the spectre of emaciated, fly-ridden bodies dying of starvation before their eyes. During the following winter and spring, millions of dollars poured into relief organizations such as Oxfam America and Catholic Relief Services, completely overwhelming their staffs. Rock stars, having already made a best-selling record, *We Are the World*, donated their talents to the ultimate transoceanic media event, 'Live Aid'– grossing millions more for African relief.

A year later, the crisis in Ethiopia has peaked. Although several million remain 'at risk', homeless and severely malnourished, summer rains in 1985 have brought the hope of a successful harvest in some areas. The flow of millions of dollars of international assistance has helped to limit the number of human fatalities. Yet, as Africans and African specialists know, the deep underlying causes of famine have not been addressed, and the deteriorating economic conditions in much of sub-Saharan Africa suggest that hunger and famine will continue to haunt Africa for the foreseeable future.

Just as Africa seems to be the 'basket case' of the world today, half a century ago, it was China that was called 'the land of famine'. From the late nineteenth century, massive famines hit China like relentless waves, taking millions of lives. The 1876–79 drought-related famine in north China may have cost 9–13 million lives. Floods in the 1890s cost additional thousands. Each decade of the twentieth century brought major catastrophes. Nature seemed cruel and unforgiving, as droughts and floods alternated to create what seemed by the 1920s to be a chronic condition of famine in one part of China or another.

The American public was well aware of 'the starving Chinese'. Pictures of ragged and wide-eyed Chinese children filled the American newspapers. Unlike today, however, the real medium of fund-raising was neither journalists nor rock stars, but missionaries. In an era when thousands of young Americans went out to Asia to serve Christ, churches were the backbone of the relief effort. Collections were taken, sermons preached, relief stamps sold. The China Famine Fund of 1921 churned out slogan after slogan to nag the American conscience. 'Famine relief is a sermon without words', the posters said, 'Pick a Pal in China', 'Give China a chance to live!' '15 million starving – Every minute counts'. Articles explained, 'How your dollar reaches a starving Chinese'. 'Self-Denial Week' was proclaimed. No contribution was too small. One could buy 'Life-saving Stamps'. 'Each mercy stamp purchased for 3 cents provides food for one day for a Chinese' (Presbyterian Historical Society, 82/20/11).

In many respects the problems faced by Africa today resemble those experienced by China in the first half of this century. First, recurrent African famines take place in a physical environment whose natural instability and vulnerability have been exacerbated by human behavior. In the Sahel, the effect of drought has been greatly magnified by the spread of the desert southward, which, in turn has probably been caused by overgrazing of livestock, deforestation, and other land-use practices. In north China, similarly, since at least the mid-nineteenth century, the natural tendency of the Yellow River to overflow its banks had been greatly increased by neglect of dike repairs, and also by silting generated by continual deforestation of the upland areas.

Second, famine in Africa occurs in the context of a population explosion, which is sometimes mistakenly taken to be the cause of the famine itself. Despite poverty and hunger – some would say because of them – Africa's population is growing faster than that of any other region of the world. Unlike Africa, China by the early twentieth century had already experienced centuries of high population density, but the rate of population growth seemed to many contemporary observers to have accelerated and to be creating Malthusian pressures on the land.

Third, the very low standard of living of large sectors of the population in Africa was also found in China in the 1920s and 1930s, and was frequently observed by foreigners. Chinese peasant life was characterized by malnutrition and poverty, high infant mortality, and low life expectancy.

Fourth, low productivity in agriculture is held largely responsible for

Africa's increasing inability to feed itself, but the reasons for this low productivity are disputed. Similarly, both Chinese and foreigners in the 1920s and 1930s agreed that Chinese agriculture could be more productive, but disagreed about the causes of agricultural stagnation.

Fifth, wide income inequalities in Africa are intensified by a growing urban–rural disparity in living standards and opportunities. In China before 1949, an ever-widening urban–rural gap seemed even more stark because most of the major cities were treaty ports where foreign privileges and the foreign presence were prominent.

Finally, Africa's serious economic problems are unfolding in a political context that is, in most African countries, quite unstable. In Ethiopia, of course, full-scale secessionist wars have greatly contributed to the severity of the famine. Likewise, China between 1911 and 1949 was in a state of political disorder, in which the major actors were militarists whose primary concern was their own survival.

Such apparent similarities – although on further examination they may be more apparent than real – strongly suggest that Africans may well wish to consider what lessons the Chinese experience with famine may contain for them. China has, after all, managed to avoid any major famine in the last 20 or more years. Although still a very poor country, China is proud of its self-sufficiency in food. With the recent economic reforms, there is every hope that the material life of the Chinese people will continue to improve. So far has China come from being 'the land of famine' that last spring the Chinese Red Cross received donations from thousands of ordinary Chinese people, including school children, to aid famine victims in Africa (*China Daily*, 23 May 1985).

International relief in China

International involvement in famine relief for China began in the 1870s, when a young Welsh Baptist missionary named Timothy Richard began to work in Shantung and Shansi provinces in north China, where successive years of drought had produced a devastating famine. Richard and his colleagues saw that 'the famine itself has given us unprecedented opportunities for the preaching of the Gospel'. Although Governor Tseng Kuo-ch'uan of Shansi had at first been reluctant to accept aid from Western missionaries, fearing that foreigners would only stir up trouble among the people, through persistence and sincerity, Richard finally received permission to dispense famine relief in some villages. He and his colleagues conducted house-to-house surveys to make sure that relief, in the form of a cash dole, was given to those who were truly needy. At the same time, he never lost an opportunity to pass out religious tracts. Despite 2 years of hard work, Richard later had to admit that his efforts had ultimately accomplished little either to win converts or to alleviate famine. As he gained more experience in China, Richard began to realize that famine relief was not enough. He became convinced that only economic development would allow China to prevent the recurrence of famines. To this end he later advocated the building of railroads, the development of mining and other industries, and other reforms that would make China stronger. Richard turned from direct proselytizing to trying to educate the Chinese elite to the importance of both political and economic reforms. In the 1890s he played a key role in shaping the ideas of Chinese reform leaders through his publication, the *Wan-kuo kung-pao* (*The Globe Magazine*), a magazine that offered translations of key works of Western philosophy and science.[1]

Most other foreign missionaries in the late nineteenth century, however, saw famine relief as a definite opportunity to spread the Gospel. The Rev. Arthur H. Smith, an American Congregationalist, reported in 1890 that the work of his mission in Shantung province had been greatly helped by 'the judicious use of famine relief in the past winters' (ABCFM, 21/112, 8 December 1890). His colleague C. A. Stanley said that he had spent much time in the distribution of famine relief during the flood of that summer, which had caused great suffering, but 'it is hoped that we may be able to follow up this opening with evangelistic effort later on' (ABCFM, 14/44, 30 April 1891). Smith appealed to his superiors at the American Board of Commissioners for Foreign Missions (ABCFM) in New York for more relief funds, observing shrewdly that 'there are many in the U.S. as in China, who will give to philanthropic objects, who will not give to missionary objects' (ABCFM, 21/111, 25 August 1890). Missionaries thus saw the distinct material benefits, as well as spiritual ones, in the giving of famine relief.

Missionary involvement in famine relief produced, however, much misunderstanding. Just as Chinese had sometimes thought Catholic missionaries were stealing babies (when, in fact, they were rescuing abandoned infants to place in an orphanage) so, too, famine relief provoked unfortunate incidents. At one Shantung mission in 1893, for example, on one Sunday groups of Chinese men from villages as far as 15 or 20 miles away came to attend church services. Their travel expenses had apparently been subsidized by people in their villages, who thought they would be able to obtain famine relief. When no such relief was given, there was great consternation. It emerged that when a young Chinese minister had recently taken the names of people in these villages professing an interest in Christianity, those enrolled mistakenly thought they were being registered for famine relief (ABCFM, 14/135, April 1894). Chinese converts whose motives were material rather than spiritual were referred to as 'Rice Christians'. Whatever their motives, all Chinese Christians, and the foreign missionaries who guided them, became the targets of waves of anti-foreign hostility that culminated in the Boxer Rebellion at the end of the century.

After 1900, as famines continued to strike in one part of China or another, the foreign relief effort grew from fragmented local efforts into more coordinated and organized campaigns. During the north China famine of 1920–21, caused by a drought that threatened to be as devastating as the 1876–79 crisis, foreign and Chinese relief organizations amalgamated to form the Peking United International Famine Relief Committee, with primary responsibility for west Chihli province, while corresponding organizations, such as the North China International Relief Society of Tientsin, the American Red Cross, the Honan Famine Relief Committee of Kaifeng, etc., had jurisdiction over other regions.

The 1920–21 north China famine provoked a tremendous outpouring of generosity in the United States. At the request of President Woodrow Wilson, the banker Thomas W. Lamont agreed to chair the China Famine Relief Fund. Through a vigorous nationwide fund-raising campaign, the Fund collected over 4 million dollars, which were transferred to the Peking United International Famine Relief Committee for its use (Nathan, 1965, 6). John Earl Baker, one American prominent in this relief effort, wrote in his memoirs that in China also there was a 'contagion of philanthropy' in both the Chinese and Western communities in Peking and Tientsin. Among the Westerners, the more people gave to relief, the more others joined in. 'It became the socially correct thing to donate bridge winnings to some relief fund, and one became sure of a moment in the spotlight by letting it be known that shortly one was "going down to the famine area"' (Baker, 1943, 71, 81–2).

Individual generosity was not limited to bridge winnings. Baker once received a donation of 40 cases of chewing gum, with instructions that their contents be distributed to the 'starving Chinese'. Baker was caught on the horns of a dilemma. He was on the one hand alarmed by the prospect of what hungry Chinese, totally unfamiliar with chewing gum, might do with it – and what it might do to them. On the other hand, he could not afford to offend the American donor of the 40 cases. Nor could he risk scandal by giving it away to the American volunteers in the field. Baker finally hit upon an inspired plan: he instructed the American volunteers – a rag-tag group of marines, engineers, infantrymen, etc. – to use every opportunity to "demonstrate" the use of chewing gum to the famine victims, thus assuring that Americans would speedily consume the 40 cases without offence to the well-meaning donor (Baker, 1943, 72–3).

Chewing gum aside, the 1920–21 relief effort was considered by the Western volunteers to have been a great success. All told, the estimated mortality in this famine was half a million, a relatively low toll compared to the estimated 9–13 million who died in the 1876–79 famine. Rail lines that could speed grain shipments to some of the famine areas were a critical new factor. Techniques of survey, inspection, food distribution, etc., that had been developed by missionaries in previous famines were employed to good effect in this famine. John Baker masterminded a large-scale work-relief program in western Shantung that built many miles of roads and was considered a model for future operations.

So successful was the fund-raising side of this campaign that it still had not peaked when the famine was declared over. The China Famine Relief Fund was embarrassed to find itself with 2 million dollars of unexpended funds. A substantial portion of this money was turned over to Nanking and Yenching Universities for agricultural research to support famine prevention. Another portion was used to underwrite the establishment in September 1921 of the China International Famine Relief Commission (CIFRC) by the Peking United group and other relief organizations as a permanent relief organization (Baker, 1943, 81, 157–66). The CIFRC functioned for almost two decades as the key private voluntary organization for relief operations. Its directors and constituents were both Chinese and foreigners, and it had branches and projects in most of the provinces.

The CIFRC saw itself not as an emergency relief organization but an organization dedicated to seeking a 'permanent improvement' of conditions in China. It sought to define famine broadly, as a condition 'where drought or flood has reduced any considerable portion of the respectable countryside to a diet of unwholesome substitutes'. Relief should be given in such conditions, even if there was no increase in the death rate. In addition, the CIFRC stressed that the principle of labor relief, rather than free relief, should be applied whenever possible. The public works that resulted from such labor should benefit the local community, but the community should in due time repay the CIFRC for the cost of the project. Finally, the CIFRC stressed that emergency relief was not sufficient but instead efforts should be directed toward the prevention of famines, particularly through river control projects (Nathan, 1965, 13–16).

Thus the ultimate objective of the CIFRC was to foster individual and community self-reliance. People should be required to work for their relief so that they would not become dependent on a dole and thus become permanently 'pauperized'. Borrowing the concept of 'pauperization' from the English Poor Laws of the nineteenth century, the CIFRC principles subscribed to a fundamental assumption about human nature that also prevails in American welfare legislation – that people will cheat and be lazy if given the opportunity, and that work and morality are linked.

In promoting community self-reliance, the CIFRC devoted its greatest efforts to public works projects, particularly the building of roads, bridges and dikes. John Baker had set a high standard of productivity in 1921, when he supervised the construction of 128 km of mountain road in Shansi province in 164 days by 20 000 laborers. This two-lane, paved road crossed five mountain ranges, ranging between 750 and 1500 m in height, and crossed twelve rivers (Baker, 1943, 99). By 1936 the CIFRC had built a total of 3200 km of new roads in fourteen provinces, repaired 2000 km of old road, sunk 5000 tube wells, dug 3 large irrigation canals, and built 1600 km of river embankment (China International Famine Relief Commission, 1936, 9–12).

In addition to the bricks-and-mortar approach to famine prevention, the CIFRC also promoted social reform. The establishment of rural cooperatives of various types – primarily credit cooperatives – became a major thrust of its activities in the 1930s, and by 1936, some 20 000 cooperatives had been sponsored. Between 1922 and 1936, the CIFRC disbursed a total of about 50 million Chinese dollars, of which more than half went to such rural reconstruction projects, and about 22 million was spent on free relief. More than half of the CIFRC funds came from the Chinese government or individual Chinese donors (China International Famine Relief Commission, 1936, 9–12).

Although famine relief work had become fully secularized, the overwhelming majority of Americans involved in it were missionaries, and there were close ties between church groups and famine relief organizations. Ninety-five out of 125 foreigners on CIFRC committees were missionaries (Edwards, 1932, 695). While there were still some hard-core evangelical types who criticized missionary participation in famine relief work on the grounds that it drew attention away from spreading the Gospel, most missionaries saw relief as charitable work, a 'ministry of loving deeds', that must be performed, even if no evangelical results were accomplished (Blom, 1932, 696–9). More critically, missionaries saw that fundamental structural reform of the Chinese economy and society would facilitate the long-term prospects for Christianity in China. To this end, they engaged in a wide range of secular activities in addition to famine relief, such as the building of hospitals, schools and universities. Finally, most missionaries tacitly understood that famine relief campaigns were extremely useful in attracting the American public's support for church work in China (Baker, 1943, 189). Their critics, in turn, accused them of creating 'missionary famines' simply to raise more funds for missionary work (Baker, 1943, 80).

Whether their motives were religious or secular, Americans approached the reform of China with characteristic energy and enthusiasm. Despite the grim conditions in China, Americans were invariably full of optimism. O. J. Todd, the Chief Engineer for the CIFRC from 1923 to 1935, perhaps best exemplified this 'can-do' attitude. Known as the 'River Tamer', he supervised numerous flood-control and road-building projects. He regarded the Chinese as hardworking and easy to teach; traditional Chinese methods, he felt, needed only the extra benefit that could be provided by Western technology and good leadership. So spectacular were his accomplishments, and so large his ego, that in the foreign community he was known as 'Todd Al'mighty'.[2]

Through the reports of missionaries or the writings of influential authors such as Pearl Buck, Americans at home developed a special sympathy for China. The Chinese peasant, as depicted by Pearl Buck in *The Good Earth*, was a simple creature, but essentially virtuous and hardworking – a worthy object of American patronage and charity. By perceiving Chinese as honest and hardworking, and seeing the potential in Chinese villages for self-sufficiency and even democracy, Americans were essentially re-creating China in their own image. Many Americans developed a deep and sentimental attachment to China, one that was difficult to sever or alter after 1949.

The international famine-relief effort in China flourished in the 1910s and 1920s, when civil disorder was most rampant. After 1928, the new Nanking government under Chiang Kai-shek sought to impose its political control over such foreign activities. The CIFRC, and other such Sino-foreign organizations, continued to function, but with the clear understanding that foreigners participated under Chinese supervision. Some Americans were uneasy about the close association between their charitable efforts and the new government. In 1929 the American Red Cross sent a major commission to China, which produced a report attacking the CIFRC for transcending its original objectives and becoming a permanent, all-purpose philanthropic organization. Famine relief, the Red Cross asserted, should be given only in disasters where the cause was unmistakably 'natural', i.e. a flood or drought, and not in cases where the cause was demonstrably 'political'. If China could count on foreign assistance under any circumstances, then a dangerous situation of dependency would develop, the Red Cross argued. 'Sympathizing deeply as we do with the efforts of patriotic Chinese to bring about these happy conditions, we nevertheless believe that China should be permitted to work out her own salvation and that to extend relief in the absence of conditions plainly due to an act of God – natural causes – but retards her ultimate recovery' (Nathan, 1965, 16–22; American Red Cross, 1929, 30).

The Red Cross also criticized the CIFRC's concentration on labor relief and public works. In real crises, it argued, labor relief did not help those in greatest need; only grain should be used in such situations. Moreover, labor relief tended to benefit the wrong people and the wrong localities. Rich landowners tended to reap an unintended benefit from irrigation projects funded by foreigners. The locality where the project was built was not necessarily the same as the area needing relief the most. In any case, the report asserted, the Chinese government should assume full responsibility for the type of public works sponsored by the CIFRC, and it should not rely on foreign assistance.

The CIFRC's public response to this report was: first, that it was a Sino-foreign organization representing Chinese interests as well as foreign. Second, the criterion for the giving of relief should always be need, and not politics. Third, the basic causes of recent disasters were fundamentally 'natural'; politics had merely exacerbated the situation. Privately, however, the CIFRC staff regarded the Red Cross report as an attack on the Nationalist government.

Infuriated by the report, William Johnson, an American missionary active with the CIFRC in Kiangsi province, drafted a sharp rebuttal entitled 'Politics and the Red Cross', in which he openly denounced the report as a politically motivated attack on the Kuomintang government.[3] Johnson wrote that 'the American Red Cross has lost its soul' (Johnson papers, 18 January 1930, 27/11). Although he tried to get his article published in the United States, more moderate members of the relief community suppressed the report for fear that a public airing of these issues would interfere with fund-raising (Johnson papers, 7 March 1930, 14/235). Indeed in the previous year, 1929, there had been considerable reluctance to launch a major fund-raising campaign for famine in northwest China because the American public's interest was at a low ebb. Some were opposed to giving relief when 'brigandage' was so rife in the famine area. No matter which side of the political fence they were on, the foreign relief organizations all realized that the American public would more readily give money for disaster relief than for long-term development projects.

The 1931 flood of the Yangtze River provided the Nationalist government with an opportunity to assert its control over disaster relief. Probably one of the largest floods in world history, the Yangtze disaster affected all of central China and parts of the north and east, covering a territory of 87 000 km² and resulting in damage of billions of dollars of property. The National Flood Relief Commission that was set up employed many foreigners. Indeed, its director-general was Sir John Hope Simpson, who had long experience with relief administration in India and Greece. John Earl Baker, Dwight W. Edwards and several other Americans active in the CIFRC were also recruited to service, but these foreigners were considered to be advisors to the Chinese government, and the Chairman of the Commission was none other than T. V. Soong, Minister of Finance and brother-in-law of Chiang Kai-shek.

The relief effort was greatly aided by the purchase of 450 000 tons of wheat and flour from the United States, purchased on long-term credit from the Federal Farm Board. Although the costs of shipment and other relief work were substantial, the Commission received only about 1.25 million dollars from foreign donations, and raised the rest – a sum of 20 million dollars – through private Chinese contributions and a 10% customs surcharge (China, National Flood Relief Commission, 1932, 7–13; Stroebe, 1932, 676).

This relief effort was remarkably successful on the whole. No serious food shortage resulted, and the price of grain was kept low. Repairs to 7000 km of dikes were completed by 30 June 1932. The Nationalist government, and its foreign supporters, regarded these accomplishments as another sign of its political legitimacy (Stroebe, 1932, 678; Baker, 1943, 371). Like the Imperial rulers of the past, the Nationalists celebrated their success in river control through the publication of a commemorative volume (China, National Flood Relief Commission, 1932).

This success was, however, short-lived. After 1932, Japanese invasion, Communist insurgency, and then civil war, totally preoccupied the Nationalist government. Warfare and occupation made conditions even more desperate for millions of Chinese. In their classic of wartime reporting, *Thunder Out of China*, Theodore White and Annalee Jacoby (1946) described graphically how millions starved in Honan province in 1943 while tons of grain in neighboring provinces were blocked by opposing warlord factions.

International relief in Africa

Most of the problems and controversies experienced by Americans in China in the 1920s have also been encountered in Africa in the 1980s. Granted, there are two fundamental differences. One is that Americans do not play the dominant role among foreigners in Africa nor do they have the sentimental attachment to Africa that they had toward China. For better or worse, Africa is little on the American consciousness. The second is that religious motives and church groups do not figure so strongly in development work in Africa. Catholic Relief Services and World Vision are among the largest of the American private voluntary organizations working in Africa, but they operate in a context that is largely secular.

The most overwhelming problem for foreign governments and agencies who wish to give aid in Africa is that of *political recognition* and *political authority*. In Ethiopia, a hotly contested and protracted internal war has rendered the giving of relief not only difficult, but also dangerous. Some agencies such as Lutheran World Relief and Oxfam America, holding to the principle that need should be the principal criterion for the receipt of aid, have sent relief both to the areas controlled by Ethiopian government and to territory controlled by the Eritrean and Tigray Liberation movements. They have transported food into Eritrea and Tigray at night through neighboring Sudan. Such maneuvers have repercussions both in the United States and in Ethiopia. Oxfam America, for example, was under intense pressure from some of its former employees, who have formed an alternative group, Grassroots, to support the Liberation movements alone and not the Ethiopian government. Catholic Relief and World Vision, for their part, are being privately criticized by other agencies for working too closely with the Ethiopian military.

The war in Ethiopia takes place in the larger context of the superpower struggle for influence in Africa. Although the socialist Mengistu government had appealed for international aid as early as 1983, when it foresaw the impending crisis, the United States Agency for International Development (USAID) did not send significant quantities of food directly until forced by the public pressure after the BBC–NBC news broadcast in the fall of 1984. Allocations by USAID have always been closely tied to political considerations. Over the last few years, the largest recipients of aid in sub-Saharan Africa have been the Sudan, Liberia, Somalia and Kenya – each critical to the United States' strategic interests in the Middle East and the Indian Ocean (Lancaster, 1985, 183). Food aid must also be seen in relationship to the very substantial amounts of military assistance being given to Africa. The Soviet Union, for example, provided over three billion dollars of military equipment to Ethiopia between 1977 and 1984, while providing only a small amount of food aid in the 1984 crisis – the equivalent of about 3% of the amount given by the United States (Schwab, 1985, 223; Shepherd, 1985b, 7).

Some Americans familiar with Ethiopia regard its government as primarily a military junta, whose socialism should not be taken too seriously. It is a 'nasty, brutal government', and 'the basis of all its policies is coercion', says one Ethiopian specialist. On the other hand, the same specialist as well as some relief workers have given high marks to the Ethiopian Relief and Rehabilitation Commission, the agency that coordinates the relief effort, saying that its officials are highly experienced professionals. High marks are also given to its counterparts, the Eritrean Relief Association and the Relief Society of Tigray. Other relief workers have been much more critical of the Ethiopian government but dare not speak their minds. One French relief organization, Médecins Sans Frontières (Doctors Without Borders), that has openly criticized the government's resettlement policy has now been expelled from the country (*The New York Times*, 4 December 1985). Only in comparison to the Haile Selassie government, which suppressed news of the 1974 famine, can the current Ethiopian government be said to have reacted promptly and responsibly to the current famine.

The *organizational difficulties* that the CIFRC experienced in China have been substantially manifested in the current African crisis, and indeed may be considered an unavoidable part of the international relief business. Like the missionaries in China, the private voluntary organizations (PVOs) in Africa realize that the American public would rather give money for emergencies than for long-term economic development. Donations that are necessarily spent on administrative costs or held over to the following year for development spending, rather than spent on food directly fed to starving children, are a potential source of embarrassment to PVOs. Bob Geldof, who generated millions of dollars through the 'Live Aid' concert, now has to go on television to explain why his committee cannot spend the money as fast as it has been coming in. Catholic Relief Services, the largest American PVO in Ethiopia, had by June 1985 spent only US$9 million out of the US$52 million it had received (*Newsweek*, 26 August 1985, p. 68).

Sensitive to public criticism, PVOs have a tendency to suppress unfavorable news for fear that the public will stop contributing. News of food that gets diverted to the military, or trucks that cannot be driven for lack of spare parts, are the current-day counterparts to the chewing gum cases that might not reach famine victims.

Rivalry among PVOs can also be intense. In Ethiopia, foreign and Ethiopian PVOs are organized under an umbrella organization similar to the CIFRC in China: the Christian Relief and Development Association. To date, any unflattering stories about competition have not been publicized, and are likely to remain suppressed until the end of the crisis. Relief workers seem to think, however, that the unseemly competition that existed among PVOs in Cambodia – as described in William Shawcross's indictment, *The Quality of Mercy* (1984) – has been avoided in Ethiopia. Even those who think Shawcross's account extremely biased admit that the atmosphere along the Thai border, where some PVOs chose to work, was 'poisonous', and that there was 'too much money for the problem'. In Ethiopia, by contrast, PVO and government efforts have been generally well-coordinated.

Finally, the *criticisms of food aid* voiced by the American Red Cross in China in 1929 seem to have anticipated the much larger controversy in the 1980s about the value of food aid. The giving of food aid, except during crises, is criticized because it is said to create dependency by the host government on cheap foreign solutions, and to create disincentives in the agricultural sector by lowering the producer prices for food. In addition, food-for-work, an important aspect of many international food programs, has been criticized by Tony Jackson in *Against the Grain* (1982), among others, for the same reasons the American Red Cross gave in 1929: it benefits the wrong people and the wrong localities.

Interestingly, modern concepts of *economic development* and *development aid* also seem to have been anticipated by Western ideas in China earlier this century. In Africa, the focus of development efforts has shifted from large-scale engineering or industrial projects that were initiated in the 1960s to small-scale, local projects. The current wisdom among development experts is that such projects are likely to produce local initiative and self-sufficiency, which in turn will generate greater economic development. Whether the projects have been large or small, however, the fact remains that after 20 years and billions of dollars of foreign assistance, the African economy is in more trouble now than ever before. During the 1970s, sub-Saharan Africa received US$22.5 billion in development aid from the West, but per capita food production declined by an average of 1.2% per year. According to Jack Shepherd, production of major food crops falls by about 2% per year, while the volume of food imports has increased 9.5% per year. By 1981, sub-Saharan Africa received 3.7 million tons of food aid each year, but this year it will have received 9.6 million tons of food (Shepherd, 1985*a*, 43).

The causes of Africa's persistent underdevelopment, the subject of several other papers in this volume, are a matter of great controversy. Some experts tend to stress the unfavorable international context in which African countries have had to operate their economies, pointing to such factors as the declining world market for Africa's cash crops, unfavorable terms of trade, and high oil prices. Others tend to fault the behavior and policies of African governments, especially with respect to agricultural pricing and currency overvaluation that tend to favor the urban consumers at the cost of the rural producers. What all these analyses share is the assumption that in the foreseeable future Africa will continue to need substantial amounts of international assistance. Even those who have been critical of past uses of foreign aid continue to advocate its maintenance at the levels of the past (Shepherd, 1985a, 46).

'Lessons' of the Chinese experience

While it would be foolish for an outsider to pass judgment on the relative merits of these African issues, the Chinese conquest of famine over the past decades does, I believe, contain some lessons for Africa today. The appropriate lessons, I shall argue, are not the obvious ones.

There are several key aspects of the Chinese developmental experience in the period since 1949 that should be considered. First, with the establishment of the People's Republic of China (PRC), virtually all forms of Western assistance and trade that had been so prominent during the earlier Republican period were curtailed and, after about 1959, all forms of technical and financial assistance from the Soviet Union were also terminated, leaving China to pursue an independent path, free of foreign interference. Second, through rapid steps, the organization of agriculture became collectivized into large-scale communes. In this it was the mobilization of labor rather than new technology that was emphasized. Third, the distribution of grain was strictly controlled by the state through a system of rationing in the urban areas and a minimum guarantee in the countryside. Fourth, both food production and food distribution were managed by a highly centralized and powerful state apparatus that placed high priority on eliminating famine.

While it is the first three aspects that comprise the distinctive characteristics of the 'Maoist' model of development, in my view it is the fourth characteristic – state policy – that may have been the most critical to the Chinese experience and that may be the most relevant to the African crisis. Almost complete economic self-sufficiency, as the ultimate expression of Chinese nationalism, may have been indispensable in establishing the legitimacy of the new national government, but it can hardly be said to have contributed directly to the elimination of hunger and famine. Collectivization may also have had greater political benefits than economic. Although grain output in China increased 75% from 1952 to 1977, agricultural growth barely kept pace with population growth (Tang & Stone, 1980, 13), and per capita grain output in 1980 was probably no greater than in the 1930s (Li, 1982, 701). Although it is too soon to evaluate the commune system definitively, the spectacular increases in output since the beginning of de-collectivization in 1978, strongly suggest that the communes may have inhibited growth by stifling individual initiative and motivation.

The system of food rationing, backed by a state reserve system, was probably the most important factor in the elimination of famine in China. Although the average per capita caloric availability of food in 1980 was probably no better than that in the 1930s, the critical difference between the two periods was that strict controls under the PRC assured the most equitable distribution of extremely meager resources. In a very real sense, then, the Maoist model gave higher priority to the *social* goal of equitable distribution than to the purely *economic* goal of growth. This degree of control over the distribution of food resources has probably never been achieved by any other government in world history, and it could not have been achieved in China without a highly powerful state system. Our growing understanding of China's state-granary and grain-price reporting system in the eighteenth and nineteenth centuries, moreover, permits us to understand that the food-distribution policies under the PRC represent an intensification of state policies from previous eras of Chinese history rather than a completely new direction (Li, 1982, 702).

The Maoist model has, of course, had a broad appeal to radical movements all over the world. Policies of isolation, at least from Western trade and aid, have been adopted in Cuba, Burma and other socialist countries, while land reform at least, if not collectivization, has been on the agenda in countries as distant and different as Ethiopia and Nicaragua. It is becoming painfully clear, however – at least to some observers – that such policies have often failed to raise the level of productivity. Even more painful should be the recognition that the Maoist model has now been repudiated by China, the very country that created it.

By contrast, the policy of strict rationing, which did work remarkably well in China to spread meager resources, is unlikely to be attempted on such an ambitious scale by any other country because it would be politically unpopular and, therefore, impossible to implement. The critical factor is not the type of ideology, political system or social policy, but a state policy that places the very highest priority on eliminating hunger and famine. It is state policy, together with the political capability to enforce it, that I believe are the transferable lessons of the Chinese experience.

Such a view is likely to meet serious objections. In Africa, many enlightened people regard the state and bureaucracy as the cause of the problem, not its solution. They see the clumsy manipulations of agricultural marketing boards and the corruption of their politicians as the very source of food distribution problems and, consequently, advocate a free market to eliminate the bottlenecks and disincentives that have occurred. To this, one can only respond that political control seems unavoidable in a situation crying for rapid solution. The relevant choice is between good government and bad, not between having political controls and not having them.

Second, there are those on the Chinese side who will surely object that in China, too, overcentralization of state power has had disastrous, indeed tragic, consequences. The Chinese government has now acknowledged that during the 'three lean years' of the Great Leap Forward, 1959–61, a massive famine did occur in China. Some American demographers now calculate that as many as 30 million may have died of hunger and malnutrition during those years – making the Great Leap Forward the largest famine ever recorded in world history (Ashton *et al.*, 1984). Although bad weather certainly was a factor, this famine was primarily the result of overwhelming pressure put on communes to say they had fulfilled the unreasonable quotas of the Great Leap, when in fact they had not. It was, in short, truly a man-made famine (Bernstein, 1984).

State power can be a terrible force for evil, but whether it must necessarily be so, and whether the Great Leap famine was an inevitable consequence of overcentralization, or an aberration, is not yet clear. Here again, it seems that the choice must be between enlightened state policy and unenlightened policy, and not between policy or no policy.

Finally, some may object that the highly politicized model of famine prevention and control developed in China may be totally inappropriate for Africa and other areas of the world. China, after all, has had a unified state and culture for thousands of years, and bureaucratic centralization has not been difficult to achieve there. But African states lack the tradition of national unity and the political culture of bureaucratic rule.

Certainly, Africa has a great disadvantage in this respect, but it is not the Maoist model, or even a Chinese model, that I am advocating, but simply state policy that will place the highest priority on eliminating hunger and famine. Such policy must necessarily be appropriate for its culture. India may serve as an example of another populous and poor country that has eliminated famine through appropriate state policy, but a policy distinctly different from the Chinese model.

In the 1960s and earlier, it was India, not Africa, that was considered the most dangerously food-deficit area of the world, the 'basket case' of its time, and among the largest recipients of grain from the United States under Public Law 480. Today, India is self-sufficient in food and even an exporter of rice. Although this happy turn of events has often been attributed to the recent successes of the Green Revolution, India has, in fact, succeeded in avoiding famine for a far longer period of time, virtually since its independence. The Bengal Famine of 1943 constituted such a psychological trauma, as well as a human tragedy, for the Indian people that, in the view of many observers, no Indian government since then could afford to permit famine to recur. To this end, India possesses a Public Distribution System for food, a key element of which is a system of fair price shops in urban areas (Chopra, 1981, chapters 1 & 27). In addition, as Michelle McAlpin's chapter (this volume) illustrates, India has effective famine-warning and famine-relief systems. Despite devastating drought and severe crop shortages in Bihar in 1966–67, and again in Maharashtra in 1970–73, for example, no actual famine took place, if famine is measured by excess mortality.

Despite this commendable record in famine prevention, India is still tormented by widespread hunger, malnutrition and poverty. According to one estimate, perhaps one-third or more of India's population is malnourished (receiving fewer than 2100 calories a day; Sanderson & Roy, 1979, 107). As Amartya Sen has pointed out, there is a profound irony in the fact that India's life expectancy is much lower than China's (Sen, 1984, 501).[4] Measured by all standards of human welfare, life in China for the very poor is far more secure than life for India's poor. Yet, it is India that has completely avoided famines over at least the last 30 years, while China produced the Great Leap famine. In India, Sen asserts, a famine such as the Great Leap's could not have occurred because the more open political system would not have allowed it. Yet, from the African perspective today, what is most important is what the Indian and Chinese experiences share: a high priority assigned to the prevention of famine, and a state apparatus able to implement food control.

Africa's path?

My emphasis on policy and politics has several implications for issues raised by others in this volume. First, it suggests that the emphasis given to economic development may be misplaced. Current economic development projects that stress local initiative and self-sufficiency assume a bottom-up-type of development process, whereby economic development will be achieved gradually and political development will follow. The expectation that political democracy will necessarily emerge from economic development is based on Western liberal assumptions that may well prove to be disappointing. My stronger objection is that this model is too slow to address the immediate threat to millions of Africans of hunger, disease and starvation. To meet the African food crisis, strong and enlightened political leadership must take precedence over gradual economic planning.

Second, economic development models often bypass the very poor and ignore their immediate problems, a point stressed by Randall Baker (this volume). Like him, I believe that the urgent questions posed by hunger and famine must be addressed as issues separate from, and prior to, long-term economic development projects. Both the examples of China and India show – albeit in strikingly different ways – that even very poor countries can do what Baker has suggested: move national food security, especially for the very poorest, up to the highest priority and solve successfully that problem even before agricultural production 'takes off'. While the economic development of poor countries will eventually solve the problem of widespread hunger and malnutrition, the elimination of famine need not wait for that higher stage of development.

Third, the priority assigned to policy and political development places the question of international assistance, the original focus of this chapter, in its proper context. Foreign aid need not be summarily rejected by African nations as a precondition to their true political independence, but it can be used effectively if closely controlled by a responsible host government. China's use of foreign assistance after 1928, and its selective use of World Bank and other international financing at present, are two examples of use of foreign aid conditional on domestic Chinese political control.

In conclusion, an international perspective, and particularly a Chinese one, suggests that there is both good news for Africa and bad news. The good news is that famine in Africa *will* eventually end. In modern times each region of the world has, in turn, broken out of its famine cycle, and Africa will not be an exception. The experiences of China and India in particular should bring hope to Africa. The bad news, however, is that it may be much more difficult for Africa than for China or India, primarily because it lacks a tradition of political unity and bureaucratic experience. In addition, the militarization of politics and the superpower competition for influence in Africa greatly handicap the efforts of governments to implement a 'food first' policy.[5] What is important for Africans is that their governments' political fortunes should be linked to their ability to put a stop to famine, not just for the urban middle class, but for the rural poor as well.

Acknowledgments

I wish to thank the following people who generously shared their insights about various topics discussed in this chapter: Joel Charny, James Field, Shirley Holmes, Raymond Hopkins, John Kerr, James McCann, Michael Scott, Robert Snow, Subramanian Swamy, Deborah Toler and Homer Williams. None, however, should be held responsible for the views expressed. Swarthmore College and the National Endowment for the Humanities provided support during the period when I worked on this chapter.

Notes

1. This material is summarized from Bohr, 1972.
2. Conversation with John K. Fairbank, Cambridge, Mass., March 1985.
3. Johnson was a fitting defender of the Nationalists, since he knew Chiang Kai-shek and his wife personally, and was later to work closely with them on rural projects designed to challenge the Communist initiative in Kiangsi (Thomson, 1969, 58–65). Still later, after 1949, he continued to speak for the Nationalist government by playing a major role in the China Lobby.
4. Sen states that life expectancy in India is 52 years, while life expectancy in China is 66–9 years. According to the *1984 World Population Data Sheet* (Population Reference Bureau, Washington, DC), life expectancy in India was 50 years, and in China 65 years.
5. This term is borrowed from the title *Food First* by Lappé & Collins (1977).

References

ABCFM (American Board of Commissioners for Foreign Missions) (1860–1950). Archives at Houghton Library, Harvard University. ABC 16.3.12, North China Mission, 88 vols.

American National Red Cross (1929). *The Report of the American Red Cross Commission to China*. Washington, DC: American Red Cross.

Ashton, B., Hill, K., Piazza, A. & Zeitz, R. (1984). Famine in China, 1958–61. *Population and Development Review*, 10, 613–45.

Baker, J. E. (1943). *Fighting China's Famines*. Unpublished manuscript. New York: Burke Library, Union Theological Seminary.

Bernstein, T. P. (1984). Stalinism, famine and Chinese peasants: grain procurements during the Great Leap Forward. *Theory and Society*, 13(3), 339–77.

Blom, C. F. (1932). The values of famine relief work. *The Chinese Recorder*, 63(11), 696–9.

Bohr, P. R. (1972). *Famine in China and the Missionary: Timothy Richard as Relief Administrator & Advocate of National Reform, 1876–1884*. Cambridge, Mass.: East Asian Research Center, Harvard University.

China International Famine Relief Commission (1936). *The CIFRC Fifteenth Anniversary Book, 1921–1936*. Peiping: CIFRC.

China, National Flood Relief Commission (1932). *The Work of the National Flood Relief Commission of the National Government of China, August 1931–June 1932*. Shanghai: National Government of China.

Chopra, R. N. (1981). *Evolution of Food Policy in India*. New Delhi: Macmillan India Limited.

Edwards, D. W. (1932). The missionary and famine relief. *The Chinese Recorder*, 63(11), 689–96.

Jackson, T. (1982). *Against the Grain*. Oxford: OXFAM.

Johnson, W. R. Papers, deposited at Day Missions Library, Yale Divinity School, China Records Project, Record Group 6.

Lancaster, C. (1985). Africa's development challenges. *Current History*, April, 145–9.

Lappé, F. M. & Collins, J. (1977). *Food First: Beyond the Myth of Scarcity*. Boston: Houghton Mifflin.

Li, L. M. (1982). Introduction: Food, famine, and the Chinese state. *Journal of Asian Studies*, **XLI**, 687–707.

Nathan, A. J. (1965). *A History of the China International Famine Relief Commission*. Cambridge, Mass.: East Asian Research Center, Harvard University.

Presbyterian Historical Society. Philadelphia. Record Group 82. China Mission, 1890–1955. Box 20, Folders 11–12. China Famine Fund.

Sanderson, F. H. & Roy, S. (1979). *Food Trends and Prospects in India*. Washington, DC: The Brookings Institute.

Schwab, P. (1985). Political change and famine in Ethiopia. *Current History* (May), 221–3.

Sen, A. (1984). Development: which way now? In his *Resources, Values, and Development*, 485–508. Cambridge, Mass.: Harvard University Press.

Shawcross, W. (1984). *The Quality of Mercy: Cambodia, Holocaust and Modern Conscience*. New York: Simon & Schuster.

Shepherd, J. (1985a). When foreign aid fails. *The Atlantic Monthly*, April, 41–6.

Shepherd, J. (1985b). Ethiopia: the use of food as an instrument of US foreign policy. *Issue*, **14**, 4–9.

Stroebe, G. G. (1932). The great central China flood of 1931. *The Chinese Recorder*, 63(11), 669–80.

Tang, A. M. & Stone, B. (1980). *Food Production in the People's Republic of China*. Washington, DC: International Food Policy Research Institute.

Thomson, J. C., Jr (1969). *When China Faced West: American Reformers in Nationalist China, 1928–1937*. Cambridge, Mass.: Harvard University Press.

White, T. H. & Jacoby, A. (1946). *Thunder Out of China*. New York: William Sloan.

19

Denying famine a future: concluding remarks

RHYS PAYNE, LYNETTE RUMMEL* and MICHAEL GLANTZ†

*African Studies Center, University of California, Los Angeles, presently at National Center for Atmospheric Research and †Environmental and Societal Impacts Group, National Center for Atmospheric Research‡, Boulder, CO 80307

Of course, drought itself is not the fundamental problem in sub-Saharan Africa. After all, drought prevails in many parts of the world and, in affluent societies, need be no more than a nuisance. The real problem in Africa is poverty – the lack of development – the seeds of which lie in Africa's colonial past, and in unwise policy choices made in the early days of independence by national governments and external donors.
Bradford Morse (Foreword, this volume)

The lack of sufficient and timely rainfall across the continent during the last decade and a half has thrown into stark relief for our generation (as similar situations have for past generations) the fact that many African societies are finding it difficult to cope with the demands of their natural environment. As a result, drought has most recently assumed a heightened degree of prominence and concern in contemporary global consciousness, as we are coming to accept the fact that the occurrence of a drought somewhere in Africa in any given year is not unusual and that runs of drought years in certain regions are also to be expected.

Until recently, the African famines have been popularly portrayed as a direct result of drought. Just what is the relationship between famine and drought? Do the recent African famines result primarily from climate variability or from social dysfunction? To what extent are Africa's agrarian problems a result of natural factors and to what extent do they result from human activities and decisions?

As some of the contributions to this volume suggest, African droughts, with their devastating societal and ecological impacts, appear to be the result of relatively long-term climatic trends, and interannual and decadal variations about those trends. These atmospheric fluctuations, occurring on various time scales, may result from such natural processes as the Milankovitch mechanism (related to the season during which the northern

‡ The National Center for Atmospheric Research is sponsored by the National Science Foundation.

hemisphere is closest to the sun), natural changes in land surface reflectivity (albedo), and natural stochastic variations in global atmospheric processes affecting African climates.

As other contributions to this volume suggest, the *impacts* of African droughts and the occurrence of African famines are less the result of climate fluctuations than the result of human interventions, such as poor land-use practices, high population growth rates, inappropriate economic and political policies pursued by national governments, by foreign governmental and nongovernmental donors, or the international economic system.

Looking at either the geophysical phenomenon or the effects of human intervention yields only partial explanations. Obviously, both the geophysical and societal contributions to each famine being investigated must be identified if there is any hope of denying African famines a future.

Debates about the causes and impacts of Africa's agrarian crises, droughts and famines exist not only between the physical and social sciences, but within the physical and social sciences as well. As noted in the scientific literature on drought, the recent 17-year drought in the West African Sahel has been the longest and perhaps the most severe this century. Ethiopia, too, has been plagued by prolonged meteorological drought conditions since the early 1970s. Questions arise, however, about how the current climatic regimes in Africa differ from past rainfall regimes. While there have been other such extended droughts in West Africa in past centuries (Maley, 1981; S. E. Nicholson, unpublished) as well as lesser droughts this century (Sirculon, 1976; Kates, 1981), observers are not yet certain what this means for future African climates.

One problem in resolving the controversy over whether the recent African drought is symptomatic of large-scale global atmospheric changes resulting ultimately in a shift to different (possibly less favorable) climate regimes in the future, relates to meteorological data. Unfortunately, for many parts of Africa existing climate information is either of relatively short duration, is incomplete, or is of questionable reliability. Compounding the absence of good quality (reliable) meteorological information is the fact that governments, faced with hydra-headed crises hampering development prospects, have generally placed a low priority on maintaining, let alone expanding, national meteorological networks. As a result, it becomes increasingly difficult for the meteorological community to provide accurate or timely warnings about potential drought situations.

Even if the current drought situation does prove to represent a shift to a climate regime less favorable than the one that existed before 1968, meteorological explanations alone cannot account for Africa's present famines. In fact, only a small number of areas have been designated as areas of famine: Ethiopia, Angola, Mozambique, Chad and, most recently, Sudan and Uganda. Thus, only in a few of the more than 25 to 30 countries designated at the height of the drought by the UN FAO as being in need of food imports in the mid-1980s did famine situations occur.

Furthermore, the geographic scope of the African agrarian crisis, which is marked by a sustained gap between increases in agricultural productivity and rates of population growth, is even broader than that delimiting the meteorologically drought-affected areas. Yet, the broader agrarian crisis has not captured the attention of the media's photographic essays and news releases, as chronic malnutrition is not as dramatic as emergency refugee camps and relief centers.

The most recent meteorological droughts have made it clear that African governments, with but few exceptions, are unable to assure their citizens adequate nutrition. Since African societies remain largely agrarian, the impact of drought is primarily felt through a direct reduction in agricultural productivity. Importing foodstuffs, therefore, has become a vital element of nutritional as well as political survival. Scarce foreign exchange, however, is needed to pay for such food imports, causing a diversion of funds from development to 'stop-gap', relief-oriented food imports. Controversy about the impact of emergency food aid on long-term recovery and development in sub-Saharan Africa continues unabated (Hopkins, 1986).

Rapid population growth rates mean that greater numbers of people must support themselves on a degraded, dwindling, resource base. As a result of this and other demographic, political and economic pressures cited in the general African development literature, an increasing number of people have been forced to cultivate areas considered by most standards marginal with respect to topography, soil quality and rainfall, making human activities there even more vulnerable to drought. This factor might help to explain why the societal impacts of drought in this century seems to have become increasingly severe.

The impacts of weather on society can either be mitigated or exacerbated by societal mechanisms (Timmerman, 1981). Thus, the severity of the impacts of a particular drought is, in most instances, determined by society's preparation for it. Whatever else climate anomalies do, their impacts serve to highlight existing vulnerabilities at all levels of social organization.

Adaptability to climate variability, especially drought, occurs at various levels of social organization. For each one of these levels, however, there is debate about the policies required to reverse the decline of per capita food production in Africa.

At the local level, several researchers suggest that the household represents the key production unit of agricultural production in contemporary Africa, and that the way households are organized and how, when and why they interact with others affects their ability to cope with food shortages. It also determines which households fare well during prolonged drought situations and which do not. Many researchers now suggest that a primary cause of the agrarian crisis has not been a lack of skill or ingenuity on the part of African peasants (farmers or herders). On the contrary, the resilience of traditional modes of production (and livestock management) can be seen as a testament to their relative success. While prescriptions differ among researchers, there appears to be general agreement that a key constraint on agricultural production is the lack of production incentives for many African rural producers. The potential of African smallholders has been constrained by the broader social context in which they operate and, therefore, the removal of such constraints represents a hope for an end to Africa's agrarian crisis.

At the national level, state policies have generally been acknowledged to have had an unfavorable impact on agricultural productivity. These policies have weakened the ability of society and of the household to escape extended breakdowns in production, such as occurred during the recent prolonged droughts. African governments have, for example, 'taxed' food crops and encouraged cash crop production. This has been an extremely controversial issue, as witnessed by the Berg Report (World Bank, 1981) and the widespread reaction to it (African Studies Review, 1984). Inappropriate pricing policies, inefficient (often corrupt) parastatals (marketing boards), artificial foreign exchange rates, and the diversion of inputs (and credits) toward cash crop exports sectors form a set of pervasive and persistent disincentives for increasing food production.

The urban bias of government policies and a preference for industrial strategies of development is another national-level problem. There is much debate about the validity of urban-biased strategies of development and the relative priorities given to the industrial and agricultural sectors. Whatever the individual motives for policies biased against the agricultural sector, agricultural production sectors in different African countries must cope with similar constraints. Such policies have well-known adverse consequences; yet, they are consistently pursued.

Many researchers argue that the international economic system has reduced the ability of African governments as well as that of African peasants to cope with fluctuations and changes in their physical and societal environments. Declining terms of trade for primary agricultural products, protectionism on the part of industrialized countries, commodity price fluctuations in the international market, and enormous debt-servicing burdens, among other factors, create a hostile environment that restricts the ability of African governments to take control of their future. While many aspects of the international environment, such as the declining terms of trade for Africa's agricultural exports, are resistant to change, some may be amenable to reform. Debt-burden demands of the World Bank and the International Monetary Fund as well as protectionism practiced by developed countries could be reformed to favor African exports in the broader interest of development.

The physical environment, too, has been affected by different local-, national- and international-level factors. For example, there has been much scientific speculation about the impact of localized overgrazing and deforestation on local and regional-scale atmospheric processes (e.g. the albedo effect). Such hypotheses, if substantiated, could lead to policies restricting local-level human activities that increase surface albedo. It can also be argued that desertification is a national-level problem, as desertified areas have been shown to follow national political borders. The relationship between drought and desertification processes has yet to be closely explored. Climatic fluctuations over a particular region are often the result of global atmospheric processes. A warming of the lower atmosphere as a result of, for example, increased carbon dioxide loading of the atmosphere by industrial activities in the developed countries, will affect atmospheric processes over the African continent. Some countries will benefit by such changes, while others will lose.

One of the attractions of focusing on one specific set or level of variables is that particular aspects of a more complex problem can be isolated and its policy implications better understood. Yet, it is increasingly recognized that reform at any one level, however necessary, will not by itself be sufficient to engender development. Comprehensive reform (at all levels) is clearly needed to provide more effective mitigation strategies to buffer societies from the vagaries of climate and other changes in the physical and societal environments.

The recent African drought exposed problems that Bradford Morse referred to as a 'crisis in development'. The African food production crisis today is the result of the interaction of complex societal and geophysical factors that minimizes the success of selective reforms. Development is not a process that 'takes off' following the addition of a few subjectively defined key inputs or the correction of a single policy perceived to be errant; faddish development policies have had little success, no matter how well intentioned.

The diversity of levels, methods, and foci of analyses in the African development literature shows that there are many ways to view the African 'crisis in development'. While different approaches may be employed to identify processes that influence the degree of severity of a drought's impacts, most researchers still stress the importance of the historical context. Typifying a prevailing sentiment, Baker (this volume) notes, ' . . . it is the interplay of external, local and historical factors which accounts for the acuteness of the African dilemma, rather than any one element in isolation'.

Social organizations are the result of historical processes. What in Africa's history left the continent so vulnerable to the development policies of national governments as well as to the vagaries of climate? What are the historical roots of the various elements of Africa's agrarian crisis?

In pre-colonial Africa, technological constraints caused social organizations to develop an adaptability to their natural environments. These societies were distinguished by their successful adaptation to climatic variations. While sedentary civilizations predominated in areas well-endowed for stable agricultural production (either naturally or as a result of human intervention), pastoral societies developed according to the dictates of seasonal climatic fluctuations (e.g. Baier, 1980).

Pre-colonial African societies are best understood as agrarian, because their dominant patterns of social organization were based on control over key factors of agricultural production. Localized systems of control over the essential agrarian resources allowed for flexibility and adaptation toward the natural environment. Today, however, many African societies appear to have lost their ability to adjust to the variability of climate. What, then, has interceded between agrarian societies and their environments to produce severe drought impacts and famines of such magnitude?

A most discernible change in the agrarian basis of African societies occurred during colonialism, which fundamentally altered the relationship of rural Africans to one another and to their environments. In colonial areas, control over essential elements of agricultural production was centralized by force and coercion. This centralization was of immediate and dramatic consequence for society's ability to respond to drought, because complex social relations based on local environmental requisites were radically changed. The centralizing thrust of colonialism alienated societies from their environments, upsetting traditional, localized patterns of coping with uncertainties.

The centralization of fragmented agrarian societies came as a result of an integration into a broader, albeit disjointed, world system. Traditional practices were superseded by an administrative system designed to facilitate the expropriation and exploitation of Africa's best agrarian (and livestock) resources for foreign interests. Agrarian social structures had been permanently altered, as control over the critical factors of agricultural production became the province of a central state bureaucracy created and maintained by the interests of an external world system. The centralizing forces did not emerge from the interaction of indigenous societies with their environments, but were the result of impulses external to Africa.

Despite the fact that centralization was often carried out through local African intermediaries, long-established indigenous political systems of rule and accountability were profoundly altered. Intermediary African elites became responsible to their colonial sponsors, not to their own people. The *raison d'être* of the new administrative structures were at odds with the needs of the local inhabitants.

The newly imposed centralized systems failed to bring about widespread development and, instead, generated a gap between agricultural production (both cash and food crops) and the basic food needs of most of the population. Although a small number of Africans benefited from European patronage, most found their means of coping with their physical and social environments endangered by the new arrangements. Thus, colonialism can be said to have led to a severe disarticulation of Africa's agrarian transformation.

The concept of disarticulation was employed by Alain de Janvry in his study of agrarian crises in Latin America. Disarticulation highlights a disjuncture between the driving forces of a system and the needs of its component parts. de Janvry shows how the integration of Latin America into a world economic system caused domestic relations between production and consumption to become skewed in favor of extractive export sectors to the detriment of domestic basic needs. de Janvry (1981, 175) writes:

> In the periphery, accumulation under dependent disarticulation creates the objective need for cheap labor and hence cheap food. And the cheap labor-cheap food logic creates the food and hunger crises. On the production side, the pressure for cheap food is met more systematically by expedient extractivist policies (unfavorable terms of trade, taxation, forced deliveries) than by the induction of developmental sequences (technological change, infrastructure investment). The result is uneven development and global stagnation . . . On the consumption side, malnutrition and hunger result from cheap labor (unemployment, low wages, and low productivity in the informal sector) enforced by repressive regimes. In peasant agriculture the mass of rural population is

relegated to poverty, which induces both demographic explosion and ecological destruction as individually rational components for survival strategies.

Clearly, there are important differences between the evolution and structures of the Latin American and African agricultural sectors. Yet, certain aspects of Africa's experience suggest that the concept of disarticulation may be of some comparative value. In Africa, as in Latin America, the dominant central impulses of the world economic system have impinged upon the ability of peripheral societies to provide adequate food supplies for their populations.

Despite the precipitous demise of colonialism, the structural changes created by the colonial metropoles remain essentially intact. Newly appointed African leaders were obliged to rely on European conceptions of frontiers and administrative unity, as well as on inherited political structures that had disrupted the delicate balance that had existed between human activities and the environment. The foundations on which the post-independence systems were built apparently continue to plague independent African governments. As a result, African governments have been unable to make good on the promises of independence, as the basic needs of the population remain threatened by the continued deterioration of their traditional means for survival.

The disarticulated transition of agrarian society has left Africa exceptionally vulnerable to climatic variations. Since independence, African governments have had to cope with a prolonged, intense drought which, for several countries in West Africa, has spanned about two-thirds of their years of independence. The drought has underscored the inability of many African societies to adapt to the physical constraints of their environment.

By stressing the historical significance of the current African crisis, we are recognizing the global dimensions of the challenges of development and of Africa's 'crisis in development'. Agrarian societies in sub-Saharan Africa have been integrated into a world system. Since the colonial era, Africa's peasants, farmers and herders found their economic opportunities increasingly dependent on, and affected by, the decisions made by national governments that are, in turn, part of a global community.

The recent African drought has exposed a global crisis, casting into stark relief the lack of accountability at an international level. Africa's famines as well as its agrarian crisis may be a foreboding of the consequences that might lay in store for other parts of a global environmental system whose social components deny responsibility for the use of their environment.

If the African drought is to have a positive impact, perhaps it will be as a catalyst to action. The recent televised suffering of Ethiopian victims expose the scale of the consequences that are bound to follow, unless development can be realized. The current global response to Africa's agrarian crises, droughts, and famines should be seen as an historic opportunity for sorely needed progress and a break from the past.

References

African Studies Review (1984). The World Bank's Symposium on Accelerated Development in Sub-Saharan Africa, 27(4), entire issue.

Baier, S. (1980). *An Economic History of Central Niger*. Oxford: Oxford University Press.

de Janvry, A. (1981). *The Agrarian Question and Reformism in Latin America*. Baltimore: The Johns Hopkins University Press.

Hopkins, R. F. (1986). Food Aid: Solution, palliative, or danger for Africa's food crisis. In *African Agrarian Crisis: Roots of Famine*, ed. S. Commins, M. F. Lofchie & R. Payne, pp. 196–209. Boulder, CO: Lynne Reinner Pub.

Kates, R. W. (1981). Drought in the Sahel. *Mazingira*, 5(2), 72–83.

Maley, J. (1981). *Etudes palynologiques dans le bassin du Tchad et paleoclimatologie de l'Afrique nord-tropicale de 30 000 ans à l'époque actuelle*. Paris: Office de la Recherche Scientifique et Technique Outre-Mer (ORSTROM).

Sirculon, J. (1976). Les donnes hydropluviometriques de la sécheresse recente en Afrique intertropicale: comparison avec les secheresses 1913 et 1940. *Série Hydrologie*, 13(2), 75–174 (Paris, ORSTROM).

Timmerman, P. (1981). Vulnerability, resilience and the collapse of society. *Environmental Monograph* No. 1. Toronto: Institute for Environmental Studies, University of Toronto.

World Bank (1981). *Accelerated Development in Sub-Saharan Africa*. Washington, DC: World Bank.

INDEX

Printed in the United States
By Bookmasters